Springer Proceedings in Materials 86

Springer Proceedings in Materials publishes the latest research in Materials Science and Engineering presented at high standard academic conferences and scientific meetings. It provides a platform for researchers, professionals and students to present their scientific findings and stay up-to-date with the development in Materials Science and Engineering. The scope is multidisciplinary and ranges from fundamental to applied research, including, but not limited to:

- Structural Materials
- Metallic Materials
- Magnetic, Optical and Electronic Materials
- Ceramics, Glass, Composites, Natural Materials
- Biomaterials
- Nanotechnology
- Characterization and Evaluation of Materials
- Energy Materials
- Materials Processing

To submit a proposal or request further information, please contact one of the following Springer Publishing Editors according to your affiliation location:

European countries: **Mayra Castro** (mayra.castro@springer.com)
India, South Asia and Middle East: **Swati Meherishi** (swati.meherishi@springer.com)
South Korea: **Smith Chae** (smith.chae@springer.com)
Southeast Asia, Australia and New Zealand: **Ramesh Nath Premnath** (ramesh.premnath@springer.com)
The Americas: **Michael Luby** (michael.luby@springer.com)
China and all the other countries or regions, as well as topics in materials chemistry: **Maggie Guo** (maggie.guo@cn.springernature.com)

This book series is indexed in **SCOPUS** and **EI Compendex** database.

Giacomo Goli · Francesco Negro ·
Ignazia Cuccui · Luigi Todaro · Dennis Jones ·
Holger Militz · Paola Cetera
Editors

Proceedings of the 11th European Conference on Wood Modification

Springer

Editors
Giacomo Goli
Department of Agriculture, Food, Environment and Forestry
University of Florence
Florence, Italy

Francesco Negro
Department of Agricultural, Forest and Food Sciences
University of Turin
Grugliasco, Italy

Ignazia Cuccui
National Research Council
Institute of Bioeconomy
San Michele all'Adige, Italy

Luigi Todaro
Department of Agricultural, Forestry, Food and Environmental Sciences
University of Basilicata
Potenza, Italy

Dennis Jones
Department of Engineering Sciences and Mathematics
Luleå University of Technology
Skellefteå, Sweden

Holger Militz
Wood Biology and Wood Products
Georg-August-Universität Göttingen
Göttingen, Niedersachsen, Germany

Paola Cetera
Department of Agricultural Sciences
University of Sassari
Sassari, Italy

ISSN 2662-3161 ISSN 2662-317X (electronic)
Springer Proceedings in Materials
ISBN 978-3-031-99417-3 ISBN 978-3-031-99418-0 (eBook)
https://doi.org/10.1007/978-3-031-99418-0

This Springer imprint is published by the registered company Springer Nature Switzerland AG
The registered company address is: Gewerbestrasse 11, 6330 Cham, Switzerland

Preface

The European Conference on Wood Modification provides a periodic forum for individuals from research organizations and companies as well as stakeholders, architects and project managers to learn about the latest developments in this rapidly evolving field, to discuss with renowned experts and to build networks in an informal and friendly atmosphere.

The conference covers all forms of wood modification, including, but not limited to, thermal, oleo thermal, thermo-hydro-mechanical, chemical and impregnation modifications, as well as emerging technologies such as microwave, plasma and enzymatic modification. The conference also deals with testing methods and standards, environmental issues, quality control and various aspects of commercialization and markets. It is now 20 years since the first European Conference on Wood Modification took place in Ghent in 2003. Since then, there have been many advances in the area of wood modification and the market for commercial modified wood products has expanded. In recent times, the conference has also started to explore how modified wood products might contribute to the development of a more resource efficient future, providing new challenges and opportunities to mitigate climate change.

The 11th European Conference on Wood Modification was co-organized by the Department of Agriculture, Food, Environment and Forestry (DAGRI) of the University of Florence, the Institute for Bio-Economy (IBE) of the Italian National Research Council (CNR) and the School of Agriculture, Forest, Food and Environmental Sciences (SAFE) of the University of Basilicata. The conference was also supported by the Wood Value Chain division of the Italian Society of Silviculture and Forest Ecology (SISEF) and by IUFRO.

The conference brought together more than 150 participants from more than 25 countries, demonstrating the vitality of the sector. A total of 96 contributions were presented during the conference, divided into 49 oral presentations and 46 posters. Thanks to a strong response from the industrial sector, an "Industrial" session was held to show that wood modification is far from being at the research stage. This gave researchers a view of the commercial reality, as well as provided industrialists a view of future opportunities. A special session on "new trends" was also introduced. The vitality of the sector was also demonstrated by the strong response from industrial sponsors, whom both the Organising and Scientific Committees would like to thank for their help in supporting the conference. It is the first time that the proceedings of the conference are published in indexed format. This will give authors the opportunity to have their work recognised in terms of bibliometric indexes, as well as giving the sector a better understanding of the impact of the conference on the scientific world, as the knowledge produced will be tracked.

Giacomo Goli
Chair of the Organising Committee

Holger Militz
Chair of the Scientific Committee

Organization

Scientific committee

Allegretti, Ottaviano	Institute of Bioeconomy, National Research Council, Italy
Gerardin, Philippe	Laboratory for Studies and Research on Wood Materials, University of Lorraine, France
Goli, Giacomo	Department of Agriculture, Food, Environment and Forestry, University of Firenze, Italy
Hill, Callum	JCH Industrial Ecology Limited, North Wales
Jones, Dennis	Wood Science and Engineering Department, Luleå University of Technology, Sweden
Militz, Holger	Department of Wood Biology and Wood Products, Georg-August-Universität Göttingen, Germany
Milne e Carmo, Julia	Carmo Wood, Portugal
Van Acker, Joris	Laboratory of Wood Technology, Ghent University, Belgium
Van der Zee, Marina	SHR, The Netherlands

Organising Committee

Brunetti, Michele	Institute of Bioeconomy, National Research Council, Italy
Cetera, Paola	Department of Agricultural Sciences, University of Sassari, Italy
Conti, Elena	CATAS, Italy
Cuccui, Ignazia	Institute of Bioeconomy, National Research Council, Italy
De Francesco, Fabio	Institute of Bioeconomy, National Research Council, Italy
Goli, Giacomo	Department of Agriculture, Food, Environment and Forestry, University of Firenze, Italy
Lo Giudice, Valentina	Department of Agricultural, Forestry, Food and Environmental Science, University of Basilicata, Italy
Mazzanti, Paola	Department of Agriculture, Food, Environment and Forestry, University of Firenze, Italy

Negro, Francesco	Department of Agricultural, Forest and Food Science, University of Torino, Italy
Palanti, Sabrina	Institute of Bioeconomy, National Research Council, Italy
Todaro, Luigi	Department of Agricultural, Forestry, Food and Environmental Science, University of Basilicata, Italy
Tondi, Gianluca	Department of Land, Environment, Agriculture and Forestry, University of Padova, Italy

Reviewers

Alao, Percy Festus	Department of Material and Environmental Technology, Tallinn University of Technology, Estonia
Allegretti, Ottaviano	Institute of Bioeconomy, National Research Council, Italy
Antov, Petar	Faculty of Forest Industry, University of Forestry, Bulgaria
Bak, Miklós	Faculty of Wood Engineering and Creative Industry, University of Sopron, Hungary
Báder, Mátyás	Faculty of Wood Engineering and Creative Industries, University of Sopron, Hungary
Can, Ahmet	Wood Science and Engineering Department, Bartin University, Türkiye
Cetera, Paola	Department of Agricultural Sciences, University of Sassari, Italy
Chin, Yi Hien	Institut Pascal, University Clermont Auvergne, France
Cirule, Dace	Laboratory of Wood Biodegradation and Protection, Latvian State Institute of Wood Chemistry, Latvia
Cota, Holta	Department of Wood Industry, Agricultural University of Tirana, Albania
Cuccui, Ignazia	Institute of Bioeconomy, National Research Council, Italy
Daher, Hussein	Navier Laboratory, Gustave Eiffel University, France
Del Menezzi, Claudio	Department of Forestry Engineering, University of Brasilia, Brazil
Fodor, Fanni	Institute of Wood Technology and Technical Sciences, University of Sopron, Hungary

Gerardin, Philippe	Laboratory for Studies and Research on Wood Materials, University of Lorraine, France
Gindl-Altmutter, Wolfgang	Department of Materials Science and Process Engineering, BOKU University, Austria
Goli, Giacomo	Department of Agriculture, Food, Environment and Forestry, University of Firenze, Italy
Hill, Callum	JCH Industrial Ecology Limited, North Wales
Jones, Dennis	Wood Science and Engineering Department, Luleå University of Technology, Sweden
Jäger, Elena	Institute for Materials and Wood Technology, Architecture, Wood and Civil Engineering, Bern University of Applied Sciences, Switzerland.
Kupfernagel, Carlo	BioComposites Centre, Bangor University, UK
Marcon, Bertrand	Laboratory of Materials and Processes, Arts et Métiers Institute of Technology, France
Mascarenhas, Fernando	Department of Civil Engineering, University of Coimbra, Portugal
Militz, Holger	Department of Wood Biology and Wood Products, Georg-August-Universität Göttingen, Germany
Negro, Francesco	Department of Agricultural, Forest and Food Science, University of Torino, Italy
Nemeth, Robert	Institute of Wood Technology and Technical Sciences, University of Sopron, Hungary
Olaniran, Samuel	Department of Wood Biology and Wood Products, Georg-August-Universität Göttingen, Germany
Palanti, Sabrina	Institute of Bioeconomy, National Research Council, Italy
Pambou Nziengui, Claude Feldman	Ecole Polytechnique de Masuku, Masuku University of Science and Technology, Gabon
Scheiding, Wolfram	Institut für Holztechnologie Dresden, Germany
Schnabel, Thomas	Department of Green Engineering and Circular Design, Salzburg University of Applied Sciences, Austria
Sedliačik, Ján	Department of Furniture and Wood Products, Technical University in Zvolen, Slovakia
Sugimoto, Hiroyuki	Department of Forest Resources, Ehime University, Japan
Swaboda, Christiane	Institut für Holztechnologie Dresden, Germany
Terziev, Nasko	Swedish University of Agricultural Sciences, Department of Forest Biomaterials and Technology, Uppsala, Sweden

Todaro, Luigi	Department of Agricultural, Forestry, Food and Environmental Science, University of Basilicata, Italy
Tondi, Gianluca	Department of Land, Environment, Agriculture and Forestry, University of Padova, Italy
Valkonen, Mikko	Department of Bioproducts and Biosystems, Aalto University School of Chemical Engineering, Finland
Van Acker, Joris	Laboratory of Wood Technology, Ghent University, Belgium
Van der Zee, Marina	SHR, The Netherlands
Wu, Muting	Department of Wood Biology and Wood Products, Georg-August-Universität Göttingen, Germany
Wuijtens, Inge	WOOD.BE, Belgium
Zauer, Mario	TU Dresden University of Technology, Germany
Zwanger, Christina	Material Development and Processing, University of Applied Forest Sciences Rottenburg, Germany

Contents

Modification with Chemicals

Industrial

Certification of Thermally Modified Timber - The Experience and View of an Industrialist

Bror Moldrup(✉)

COO, IWT-Moldrup, Vejle, Denmark
bm@moldrup.com

Abstract. Quality control and certification of thermally modified timber (TMT) has undergone a number of changes after the product was introduced on the industrial market in the 1990'ies. But the standards and certification still leave a lot of questions, misunderstandings and faults to be resolved before it will provide a sound and true basis for product and quality assessment. The paper deals with these issues based on theoretical and practical experience by an equipment/process supplier (Moldrup-SSP) as well as plant operator. Even the latest versions of EN113-2 from 2020 and EN350 from 2016 have major flaws and fail to address the basic question of how to provide quality assessment and certification for durability based on the use areas of TMT. Treated product control as mostly excuted today may not offer a good picture of the quality of the finished product when bought by the final user and may result in disappointed expectations that will have an unnecessary negative impact on the product in the long term.

Keywords: EN113-2:2020 · EN350:2016 · TMT certification · TMT product control · TMT and use classes

1 Introduction

Thermal modification of wood (TMT) took off on an industrial scale in the middle of the 1990 in countries like Finland, Holland and France. Thermal modification improves a number of wood properties such as dimensional stability, durability in the form of greater resistance to fungi, rot and some insects, better thermal insulation, a uniform pleasant brown colour throughout the wood without adding any chemicals to the process for a wide range of wood species including refractory species that are difficult to treat by traditional biocide impregnation.

IWT-Moldrup has since 1957 been active in industrial wood impregnation using biocides such as CCA, Copper Organics, creosote and other organic biocides dissolved in solvents and oil, and supplying plant equipment and processes for this since 1978. Only in 1999 IWT-Moldrup went into thermal modification with its own thermal modification process with autogenerated steam under pressure called Moldrup-TMT (earlier Moldrup SSP) - a socalled closed, wet process, and more recently, IWT Moldrup has also introduced chemical modification with their unique concept called biobiowood. The

G. Goli et al. (Eds.): ECWM 2024, SPM 86, pp. 3–9, 2025.
https://doi.org/10.1007/978-3-031-99418-0_1

company is now run by the second and third generation from offices in Singapore and Denmark.

The market for TMT has seen steady growth for the past 30 years in many countries in Europe, whereas the development was more slow in Oceania, United States and South America, and the total production of thermally modified timber will soon reach 750,000 m^3 annually world-wide and has important economic impact in the wood industry.

Unfortunately, thermal modification is sometimes called heat treatment, but this donomination can often be confused with heat treatment of wood for temporary protection killing insects for ISP15 packing material and similar, but is a totally different form of temporary treatment. The industry and the scientific world ought to come to an agreement with a single easily defined denomination, such as thermal modification, or TMT.

The market growth suffered a set-back after initial enthusiasm, as TMT was often in the beginning believed to and oversold as offering protection in ground contact use (UC 4) and against termites.

In a number of markets, the selling point for TMT is often the improved durability, although the buyers may actually value the other properties such as dimensional stability and the pleasant brown colour just as high as the improved durability. In other more recent markets, there is a greater understanding of the real value of TMT in dimensional stability compared to a relatively smaller improved durability. The wanted durability is then provided by a second layer of protection which can be a paint, or a transparent stain or treatment with biocides for more efficient protection against rot, fungi and termites.

This paper will mainly deal with challenges in understanding, defining and quality assurance/certification of improved durability.

IWT-Moldrup has a unique background in both biocide impregnation and vacuum drying of wood, and knowledge in both areas is important for a better understanding of the challenges in defining and understanding the improved durability of TMT.

Unfortunately, testing for improved durability of thermally modified wood is almost exclusively based on standards relating to biocide impregnation of wood, although the mode of protection is quite different in biocide impregnated wood compared to thermally modified wood. For an industrialist it is the impression that industry and many parts of the scientific world are trying to adapt standards for biocide impregnation to be used on thermal modification of wood almost to the point of not being entirely faithful to methods and intention of the testing standards.

1.1 TMT, EN350 and Use Areas?

In Europe, EN350 has been used to try to define and assure the durability of TMT in durability classes 1–5. This is a questionable practise. In other parts of the world, there has been more reluctance in defining the durability alone based on the understanding and the principles of EN350. EN350 is used together with testing according to EN113, which is only laboratory testing and not field testing.

In IWT-Moldrup's experience, it would be better and perhaps more honest to the final consumer/specifier to employ a system based on Use Classes in EN 335, as the use area would be more important for choosing TMT, since the deterioration of TMT (fungal and mechanically), and probably also some of the chemically modified timbers, is closely related to the presence of high moisture in the wood, whereas deterioration of biocide impregnated wood has another mode of reaction where protection - as proven in laboratory in EN113 - is provided even under high, constant moisture levels such as in ground contact and sea water.

1.2 The Shortcomings of EN350 for TMT

EN350 used to be only applicable for defining the expected durability of heartwood and not the sapwood in wood. It is recognized for many species that the durability of the heartwood of considerably better than for the sapwood in the same piece of timber.

EN350 was changed in 2016 and is now also being used for defining the durability of sapwood of TMT. It seems to be overly optimistic to assume that the resistance to fungi and rot is identical in TMT heartwood and TMT sapwood for many species such as many pines, including notably scots pine, and it would be good if there is/would have been solid scientific evidence for this very broad assumption of equal durability in TMT heart- and sapwood, but we have failed to find this.

In EN350, this apparent shortcoming has been addressed in the notes Annex D (informative) "Wood anatomy and heartwood extractives can affect its permeability and biological durability". Furthermore, in Annex F it is said "Most notably in use class 3, the permeability of the material, the design of the component/structure to shed water and avoid water trapping, and in some cases the maintenance of a component or structure can all greatly improve the performance of wood in use". "The durability tests enable a classification of performance against "benchmark" materials that have been tested in the same way". But the "benchmark" is often not wood treated with biocode that are generally in Durability Class 1 in EN350, but natural durable species like Western Red Cedar in Durability Class 2 acc. to EN350. But we have never - not even one time - noticed this being taken into consideration in the industry when marketing and selling TMT, and it might be too much to expect each individuel specifier or buyer to study, address and take into consideration the Annexes of EN350, when the thermal modification industry does not do it by itself.

The industry should - but this may be a bit too optimistic - address the issue together with the scientific world as the consumers are exposed to a marketing which is probably not covering a vital issue necessary for them to make a correct decision.

1.3 The Development of EN113-1 - Sample Moistures - Heartwood and Sapwood Discussion

As mentioned, EN113–1 was developed for laboraty testing of biocides for impregnation of wood. Originally EN113 had requirements for the testing to provide an environment where the test samples are exposed to at least 25% wood moisture content (EN 113:1996 8.6.3). How it has been possible to obtain 25% wood moisture contect as required by the standard on properly treated TMT remains a mystery to us, although numerous reports have been published for testing of TMT according to the old EN113. That the moisture content may in reality not have been reached according to the requirements of the standard in many reports may be reflected in the fact that this requirement was removed in the

CEN/TS 15083-1:2005 that for some years was used for testing of durability for TMT. In the EN113 from 2020, the requirement is maintained for biocide impregnation (EN113-1) but has been removed for modification testing (EN113-2) - it is not unreasonable to assume that it was removed also because the requirement is more than difficult to meet for TMT.

The arguments for the decision to remove the essential requirement for modified timber and not for biocide impregnated wood has not been explained clearly in the new version of EN113 and weakens weight of the standard to be used as a comprehensive testing result for TMT.

EN113-2:2020 also does not distinguish between heartwood and sapwood in testing.

It appears to be possible to obtain a testing report from EN113-2:2020 with a good result for durability using samples which are 100% heartwood and then use this testing result for a production with wood which is a mixture of heartwood and sapwood and even almost full sapwood. It is not unreasonable to assume that it may lead to unexpected early failures of TMT wood in use, consisting of primarily sapwood and result in a great disappointment for the consumer.

1.4 EN113 - Selection of Testing Fungi and Experience from Oceania

EN 113-2:2020 has defined 2 mandatory fungi for testing for softwoods and 2 mandatory fungi for testing for hardwoods. A number of other fungi have been mentioned as options, but typically testing is only performed with the 2 mandatory fungi.

In our experience, coniophora puteana is rather benign to TMT, whereas for softwood rhodonia placenta is seen as the great "terminator" in testing. For hardwoods it is again the benign coniophora puteana together with trametes versicolor. But the trametes versicolor is also rather benign for TMT. If another test fungi would have been used e.g. rhodonia placenta, it is our impression from some tests that the results would be less favorable for hardwoods.

The choice of rhodonia placenta probably also originates from testing of biocides used in a quite humid environment - typically in ground contact. If rhodonia placenta at all is relevant for TMT in use class 3.1 acc. to EN335-1 (occasionally wetted/limited wetting conditions) is worth a discussion, although it would probably be relevant for TMT in use class 3.2 (frequently wetted/prolonged wetting conditions).

Testing reports according to EN113 are used for evaluation of TMT in many countries around the world despite the shortcomings of the testing standard for TMT both in the old (EN113:1996) and the new version (EN113-2:2020). Unfortunately, we also often see that only the results for corniphora puteana are being presented to the customers without a clear mentioning that it is not a complete EN113 testing since it is without one or more additional mandatory fungi, and such a practise of selecting the most favorable fungi for testing and proving durability is not proper marketing.

In Oceania, EN113 testing results are also being used for marketing, although the types of brown fungi often present in Australia and New Zealand are different. In Australia and New Zealand, testing according to local standards would require at least 3–5 fungi from a selection relevant to the two countries. If the testing is carried out strictly according to AWPC (Australian Wood Preservation Committe), in the reports we have been, TMT performs very poorly with any type of process and at any practical temperature (also at 230 C) especially against F. liiacino-gilva which supposedly is a common brown rot in buildings, at least in Australia. Although termites in greater amounts are not found everywhere in Australia, all TMT in Australia usually has a second layer of protection in form of a biocide treatment against termites which at the same time can be a protection against the brown rots that would otherwise deteriorate the TMT. In this case, going to high temperatures in the TMT process is probably an overkill with an important negative side-effect in form of unnecessary increased brittleness on many species since the durability in real life is assured by the the second layer of protection and not the thermal modification.

In the US the AWPA (American Wood Potection Association) has struggled for years to agree on testing procedures for TMT under their regime. In light to the above, this is probably not surprising.

1.5 Is Present Certification Just a Questionable Marketing Tool - or a True Quality Assurance?

Several certification schemes are available to plant operators, specifiers and consumers.

The most well known are Thermowood Association from Finland, EPH Quality Mark TMT from Dresen, Germany, and NTR. KOMO from Holland also has a certification scheme, although it is probably not used much outside Holland, and there may be other certification systems that we have not encountered.

The question is, however, if the certification schemes provide a genuine quality assurance. It is not our impression. In many countries in Europe, the customers are not asking for certification for TMT, and they may in light of the shortcomings of standards be right in doing so, instead of believing in a quality control that seems to have serious shortcomings. In other countries outside Europe with less developed markets, durablity certification seems to be playing a bigger role at the present moment, but the question is if the users are not deceiving themselves by believing in the certification systems.

The most widely used certification system by far is the Thermowood Association. The association publishes a handwood to give an insight into how the certification is executed. It appears that the certification is based on EN350 and EN113, although it is not clear on which testing reports the certification system is based - perhaps including outdated EN113 results which do not completely meet the requirements in the standard. The inspection under the certification does not seem to be based on the durability of actual treated wood ("Finished products are tested once a year by an approved external testing laboratory. The following properties are inspected: moisture level, surface and interior cracks, colour and process parameters") but only on checking the plant equipment and personnell for their capability and treatment schedules developed from testing reports for wood that may be even very old. The handbook does not mention that the certification can be based on testing results for almost pure heartwood whereas the production in reality is a more even mix of sapwood and heartwood. Moreover there seems to be an unfortunate confusion of commercial interest between the private third party control company and the Association.

The German IHG from Dresden system works on some of the same principles, but at least with individual testing of samples of thermally modified timber from each single production unit before accreditation, and a limited quality testing of the thermally modified timber from production after 3 years.

An interesting new addition to the certification possibilities is NTR. The NTR scheme has been developed by the Nordic Wood Preservation Council. The certification is based both on laboratory results and field testing with e.g. EN330 L-joint or CEN/TS 12037 lap-joint, which is a major improvement over the other alternative certification systems. But also this certification systems suffers from the assumption that the natural durability of TMT is the same for sapwood and heartwood (NTR Document 2 part 4:2017 page 8) offering the possibility of using results from pure heartwood testing samples for mixed heart and softwood production. The NTR system also does not specify a system for third party control of the thermally modified timber from actual production. There is an added requirement for the Moisture exclusion efficiency (MEE - above 40%) of the wood to be used for quality control. But testing reports seem to indicate that the defined MEE is not correct or cannot be used, since wood that fulfills the MEE suffers from deterioration in laboratory tests. Finally there seems again to be an unfortunate mixture of commercial interest of one of the private third party control companies and the technical group taking all important decisions for the certification system combined with a completely untransparent decision making process, and a refusal to answer any questions of concern.

One paper has analyzed thermally modified wood from industrial production as opposed to laboratory samples used by the Thermowood Association. From the industrial production the durability class according to EN350 was disappointing compared to laboratory samples, and Scots pine actual production wood only reached durability class 3 (moderately durable - mass loss of 16%) with the Oligoporus placenta test fungi ("materials originating from industrial scale production fell short of the quality expected on the basis of laboratory scale treaments").

2 Conclusions

2.1 Working towards Believable and True Quality Control?

Without doubt, the scientific world has a genuine interest in trying to develop solid tools for quality control and certification. Numerous are the papers on various ideas and systems for checking the quality of thermally treated timber from production. Less numerous are the papers on problems in the basic foundation of testing thermally (and chemically) modified wood. So problems exist with both the validity of the laboratory testing of samples and third party control of wood from actual production.

IWT-Moldrup suggests to leave the EN350 durability class base and work on a use class certification where any use higher than 3.1 (occasionally wetted) is suspected to require a second layer of protection in form of e.g. a paint or a biocide treatment.

The present situation with testing using EN113–2 in today's version and defining durability according to the durability classes in EN350 is not sustainable and will lead to consumers being disappointed in their expectations that are created by some of the present industry marketing of certification.

Unfortunately the industry seems to take little interest in developing proper and solid quality assurance for their product in order to protect its future, although the revenue on thermal wood modfiication could support a much more active role to secure future income for the industry.

References

CEN/TS 12037: Wood preservatives - Field test method for determining the relative protective effectiveness of a wood preservative exposed out of ground contact - Horizontal lap-joint method (n.d.)
CEN/TS 15083-1: Durability of wood and wood-based products - Determination of the natural durability of solid wood against wood-destroying fungi, test methods - Part 1: basidiomycetes (n.d.)
EN350:2016: Durability of wood and wood-based products - Testing and Classification of the durability to biological agents of wood and wood-based products (n.d.)
EN113:1996: Test method for determining the protective effectiveness against wood destroying basidiomycetes - determination of the toxic value (n.d.)
EN113-1:2020: Durability of wood and wood-based products - Test method against wood destroying basidiomycetes - Part 1: assessment of biocidal efficacy of wood preservatives (n.d.)
EN113-2:2020: Durability of wood and wood-based products - Test method against wood destroying basidiomycetes - Part 2: assessment of inherent or enhanced durability (n.d.)
EN330:2014: Wood preservatives - Determination of the relative protective effectiveness of a wood preservative for use under a coating and exposed out-of-ground contact - Field test: L-joint method (n.d.)
EPH: Quality Mark TMT Certification Programme (2021)
Jacobs, K.: Durability testing of thermally modified timber: methods and results (2023)
NTR Document 2 part 4:2017: Conditions for approval of industrially protected wood in the Nordic Countries. Part 4: modified wood (n.d.)
ThermoWood® Handbook: International ThermoWood Association (2021)
Welzbacher, C.R., Rapp, A.O.: Comparison of thermally modified wood originating from four industrial scale processes - durability. IRG. **2002** (2021)

STYL+WOOD® System for Thermal Modification of Wood

Ignazia Cuccui[1](✉), Sonia Marchiori[2], Nasko Terziev[3], and Ottaviano Allegretti[1]

[1] CNR-IBE National Research Council - Istituto per la Bioeconomia, Via Biasi 75, 38098 San Michele all'Adige TN, Italy
{ignazia.cuccui,ottaviano.allegretti}@cnr.it
[2] BIGonDRY srl, Via G. Falcone, 30, 31037 Castione TV, Italy
sonia.marchiori@bigondry.com
[3] Department of Forest Biomaterials and Technology, Swedish University of Agricultural Sciences, Uppsala, Sweden
nasko.terziev@slu.se

Abstract. The paper reports the characteristic of hardwood and softwood wood species thermally modified at three intensity with the STYL+WOOD system, a process patented by BIGonDRY s.r.l, developed and optimized in the framework of a research project with the scientific support of the BioEconomy Institute of the National Research Council of Italy.

Among the results it stands out the improvement of durability from 5 to class 1 or 2 (EN 113) achieved for treated ash, poplar. Based on this results a standard production and quality control protocol was developed and adopted by the factory on voluntary base to guarantee the quality and performance results as indicated by the product sheets allowing the customers and engineers the choice and design of the best product for a given end-use.

Keywords: wet close system · durability · physical-mechanical properties

1 Introduction

BIGonDRY s.r.l, is a SME enterprise located in Veneto, north-east of Italy, operating for more than 30 years in the field of kiln drying manufacturing. The Company, has gained great reputation in the thorny sector of hardwood wooden floors and decking developing personalized solutions in processing of various wooden species and end-use requirements.

In the last decade, part of this background has gradually moved towards an integrated approach between drying and thermal modification, developing a proprietary system and process for the thermal modification of wood named the STYL+WOOD® system. In this system, the oxygen concentration is controlled by means of a slight overpressure given by the natural production of gases from the wood (water vapour and gases from pyrolysis) and, if needed, by injection of a small quantity of saturated steam. The process

G. Goli et al. (Eds.): ECWM 2024, SPM 86, pp. 10–18, 2025.
https://doi.org/10.1007/978-3-031-99418-0_2

is controlled by software, which considers inputs such oxygen concentration, pressure, and temperature difference between the core of the wood and the surrounding medium [1]. According to Hill [2] it is classified as a wet close system, even if it is actually a semi-wet system since the moisture in the reactor during the process is lower than in systems with saturated or superheated steam.

With the technical and scientific support of CNR-IBE, in the framework of an industrial project, STYL+WOOD® process has been optimized with regards to the wood species and final end-uses and the STYL+WOOD® product has been extensively characterized. The project involved an extensive sampling of process inputs and outputs and standard laboratory tests for the characterization of the process (consumption, emissions) and product (physical-chemical, mechanical, durability). The results have been condensed in product sheets for 9 wood species treated by a variety of processes and modification intensity.

2 Methods

2.1 Materials

Sawn boards (30 × 150 × 1000 mm) of following wood species were tested (species, acronym, density kg/m^3): spruce (AR) 425, Douglas fir (DG) 511, maritime pine (PM) 663, scots pine (PS) 533, oak (RV) 796, beech (FG) 746, ash (FR) 747, poplar (PP) 380, tulipier (TP) 607. For each species, an average of 25 to 30 boards was thermally treated in mixed stack in 5 different treatment batches. For each treated boards a reference sample was kept untreated.

2.2 Process

A total of 5 STYL+WOOD thermal treatments were performed. The modification batches named A, B, C, D and E have different maximum temperature reached T_{max} (160, 185, 200 and 215 °C). The modification batch named E has the same T_{max} value of D (215 °C) but a faster T rate during heating (Fig. 1).

2.3 Characterization Tests

After the treatments, small samples were machined from each board for the standard characterization tests. The number and size of the samples vary according to the technical reference standards.

Mass Loss. Mass loss *(ML)* was determined according to the: ML = $(M_0 - M_{HT}) / M_0$ when, M_0 is the weight of the boards before treatment (but after drying phase when the moisture MC ~ 0%) and M_{HT} is the weight of the boards immediately after it and expressed in %. 20 boards were measured for each species and treatment.

Density, EMC and ASE. Density, Equilibrium moisture content (*W*), linear *(βR, βT)* and volumetric swellings, were measured on oriented samples whit dimensions of 20 × 20 × 40 mm according to the standards [3]. The samples were conditioned in a climatic

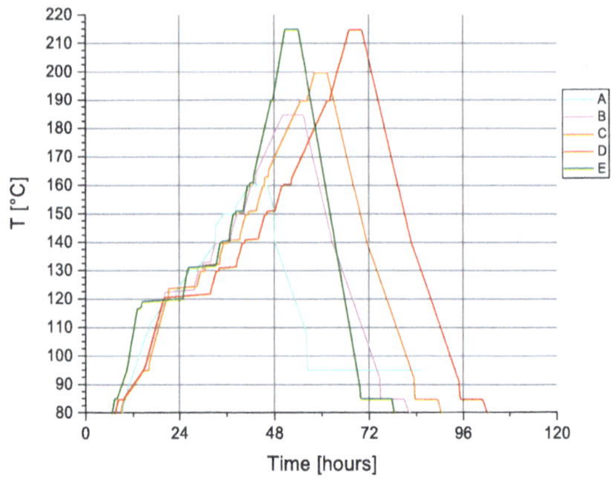

Fig. 1. Temperature profiles of the thermal treatments

chamber at a temperature of T = 20 °C and relative humidity (RH) of 50%, 65% and 99%. Samples swelling was measured and calculated in the two anatomical directions from the state of 0% MC to the nominal MC of 12%. A total of 1130 samples were measured cut from a number of boards for each species (treated and untreated), ranging from 3 to 6.

Color. A 400–700 nm visible light spectrophotometer (Micro Flash 200D Data Color Int) was used for color measurements on planed boards without defects before and after the treatments; 4 boards for species and treatment were selected, 20 measures were collected for each board. The colorimetric coordinates used are relative to the CIE L * a * b * three-dimensional color space in which L * specifies the lightness, a * and b * are the positive/negative color coordinates [4].

Mechanical Properties. 4 to 7 boards were selected for each species and treatment. Small, free of defects specimens, were cut from both treated and untreated boards for mechanical characterization of the material. The samples were kept in a controlled environment (20 °C temperature and 65% RH) until they reach constant weight. Compression (σ_c) shear (τ) and, bending strength (σ_b), bending modulus of elasticity (E) and hardness (HB) were measured according to the technical standards [3, 5]. A total of 630 specimens were used to determine each property. The data were analysed by the calculation of the descriptive statistics (means and variation coefficients) and by statistical tests of mean comparison.

Durability. Durability assessment of the treated and control samples against fungal attack by basidiomycetes was carried out according to the European standard EN 113 [6]. The samples were leached prior to the test according to the standard EN 84 [7]. Basidiomycetes used were three brown rot (*Coniophora puteana* BAM Ebw. 15, *Gloeophyllum trabeum* BAM Ebw. 109, *Rodhonia placenta* FPRL 280) and a white rot (*Trametes versicolor* CTB 863A) fungi. Criterion used for the classification of wood durability is given in the EN 350–1 standard [8]. The evaluation of laboratory decay test is based on recorded mass loss after a period of exposure of 16 weeks.

3 Results and Discussion

In Fig. 2 are presented results of characterization test. Data for untreated (NT), and thermally treated samples at 160 (A), 185 (B), 200 (C), 215 (D) and 215 fast (E) °C are reported to compare temperature of treatment. No mechanical characterization test was carried out on the samples treated at 160 °C (A).

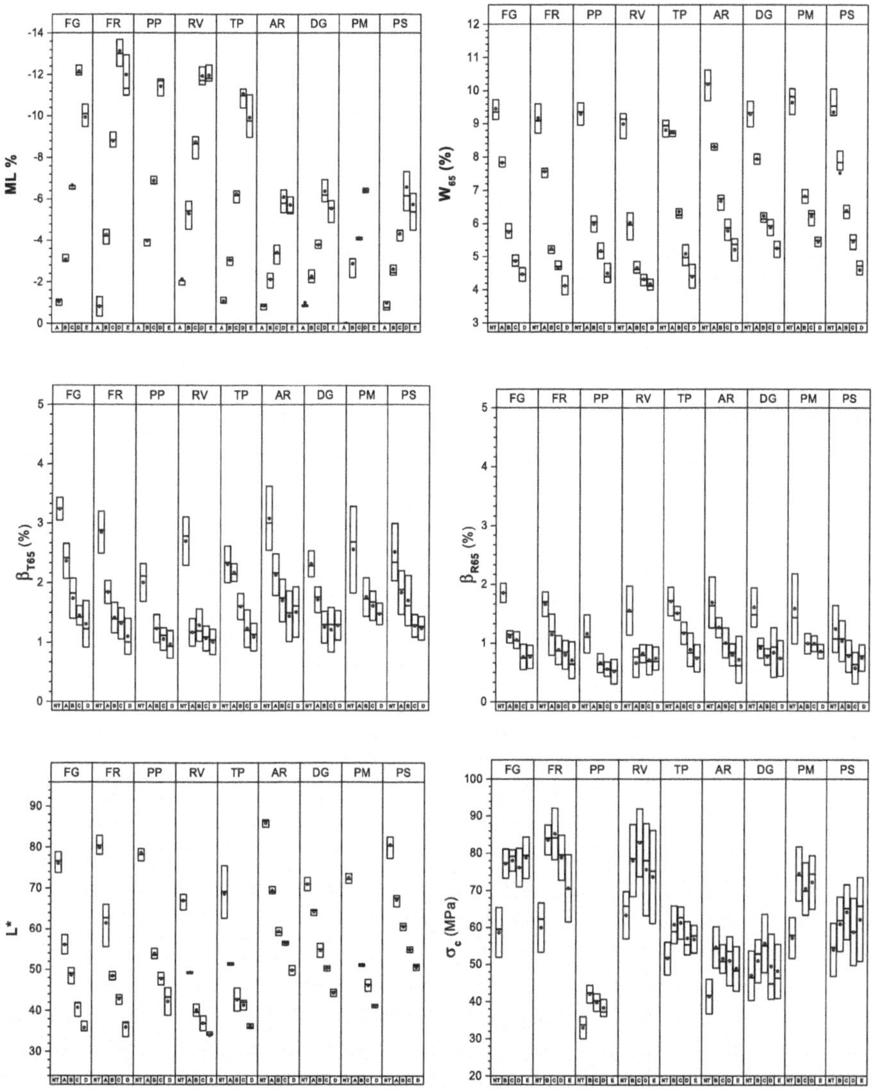

Fig. 2. Boxplot of measured properties. The bold line indicates the median value, the box the interquartile range (from first to third quartile), and the circles indicates the average. Each column shows the results obtained for a wood species (acronym above) at different treatment temperatures (below). Sections 2.1, 2.2 and 2.3 for details.

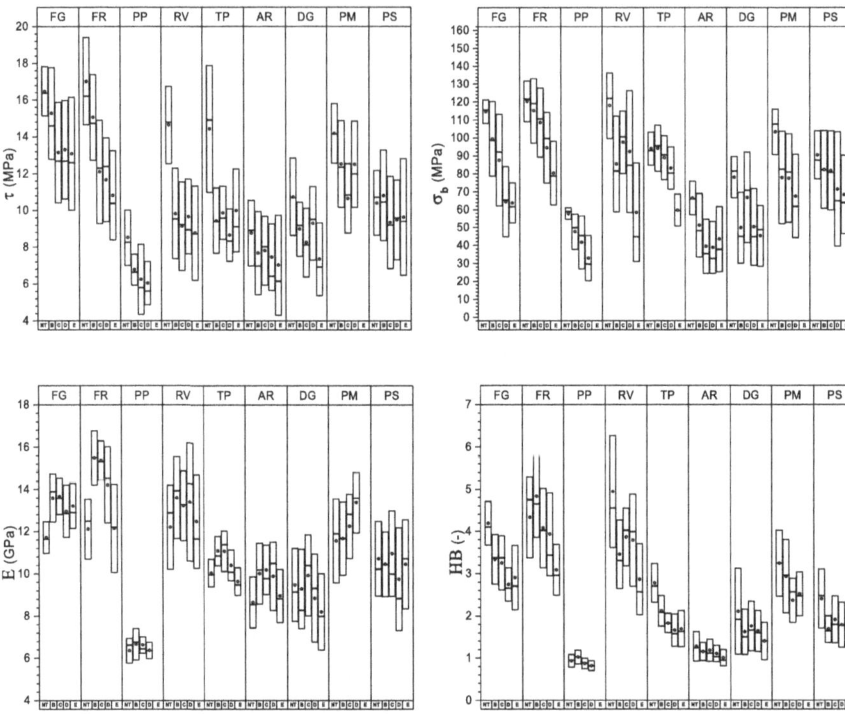

Fig. 2. *(continued)*

More emphasis is given to the results of durability tests.

Regarding the physical properties, the data are in line with the literature [2]. As expected, ML is related to the modification intensity. Hygroscopicity, dimensional stability and colour are always correlated with the ML [9].

Regarding the mechanical properties, the compression strength of Douglas fir (DG) and Scots pine (PS) control does not differ significantly between all treatments. For the other species, there is a 25–30% increase in compression strength (only tulipier (TP) increases by only 14% compared to the control). In case of ash (FR) the effect of the treatments differs from each other. Shear strength of spruce (AR) and Scots pine (PS) was not significantly different when comparing the control and treated samples. Differences between control and treated by 20–30% are shown. There are few differences between treatments. For FG and FR treatment B differs (is equal to control).

The bending strength of tulipier (TP) and Scots pine (PS) control samples does not differ significantly from that of the treated samples despite the treatment intensity. For the other species there is a 30% decrease in bending strength between control and treated samples. In some cases (FG, FR, PP, RV) treatments differ from each other. Regarding the modulus of elasticity, the control samples does not differ significantly from the treated ones by all treatments. Higher modulus of elasticity (15%) for treated specimens of FG and FR was observed. The hardness of the control samples was 25–30% larger when compared to the hardness of the treated samples of FG, PM, PS, RV, and TP. For the other species no difference was found (Fig. 2).

Table 1. Average mass loss, correction values (%) and durability class (red digit) of thermally modified wood, leached according to standard EN 84. The blue fields indicate treatment temperature that should be applied to improve the durability of wood. Red digit after species is the natural durability of wood according to EN350.

		D (215 °C)		*C* (200 °C)		*B* (185 °C)	
Correction values for beech		***1.4***		***1.6***		***1.6***	
	Rh. placenta	0.1	1	7.1	2	15.9	3
BEECH *(5)*	*Con. puteana*	2.2	1	5.2	2	7.7	2
	Tram. versicolor	7.2	2	10	3	15.4	3
	Gl. trabeum	1.0	1	2.7	1	5.1	2
Correction values for ash		***1.1***		***1.4***		***1.5***	
	Rh. placenta	-0.3	1	0.1	1	9.6	3
ASH *(5)*	*Con. puteana*	0.5	1	1.3	1	3.9	2
	Tram. versicolor	1.1	1	5.2	2	9.7	3
	Gl. trabeum	-0.1	1	0.1	1	1.5	1
Correction values for poplar		***2.1***		***2.3***		***2.5***	
	Rh. placenta	1.0	1	15	3	33.4	4
POPLAR *(5)*	*Con. puteana*	0.2	1	0.5	1	7.1	3
	Tram. versicolor	2.0	1	10.3	3	19.4	4
	Gl. trabeum	1.6	1	3	1	9.2	2
Correction values for tulipier		***0.9***		***1.2***		***1.3***	
	Rh. placenta	*1.0*	1	3.7	1	13.8	3
TULIPIER *(5)*	*Con. puteana*	*1.1*	1	1.8	1	14.5	4
	Tram. versicolor	*5.4*	2	7.3	2	18	4
	Gl. trabeum	*0.8*	1	1.4	1	2.8	1
Correction value for spruce		***0.9***		***1.0***		***1.0***	
	Rh. placenta	10.8	2	17.7	3	26.1	4
SPRUCE *(4)*	*Con. puteana*	3	1	7.4	2	10.5	3
	Tram. versicolor	3.4	2	5.1	3	8.2	3
	Gl. trabeum	1.9	1	6.2	2	10.5	3
Correction value for Douglas fir		***1.1***		***1.0***		***1.0***	
	Rh. placenta	6.1	2	11.6	3	17.8	3
DOUGLAS FIR *(4-5)*	*Con. puteana*	1	1	4.5	2	8.7	3
	Tram. versicolor	2	1	1.9	1	5.7	3
	Gl. trabeum	0.5	1	1.3	1	4.2	2
Correction value for Scots pine		***1.1***		***1.0***		***1.0***	
	Rh. placenta	7.2	2	14.9	3	17.2	3
SCOTS PINE *(5)*	*Con. puteana*	1	1	4.7	2	8.6	3
	Tram. versicolor	1	1	2.8	2	3.9	2
	Gl. trabeum	0.9	1	4.0	2	5.4	2

A summary of the thermo-modification effect on the wood mass loss caused by selected fungi and durability classes (EN 350) is shown in Table 1. Prior to the test, all treated samples were leached in deionised water according to standard EN 84. Correction values indicate the changes of sample's mass *due to leaching*. The correction values of the treated samples of ash and poplar decreased with increasing treatment temperature; for beech, Douglas fir, Scots pine and spruce, no relationship was found between treatment temperatures and correction values. Some insignificant decrease of the correction values was found with increased treatment temperatures in the case of tulipier. The mass loss of the untreated samples fulfils the validation criteria of the EN 113-2 for the test fungi and the tests are valid. Mass loss and respectively durability class achieved by thermal modification vary with regard to wood species, treatment temperature and acting fungus. The application of the temperature of 215 °C inhibited totally the growth of the test fungi on the thermally modified (TM) ash wood and durability class 1 was achieved. The mass loss was insignificant and less than 2%. The most aggressive fungi on the tested TM wood were the white rot fungus *T. versicolor* and the brown rot fungus *Rh. placenta*. Both fungi caused 9.7 and 9.6% for the respective fungus when the wood was treated at 185 °C and thus, the durability was only slightly improved to durability class 3. The actions of the two brown rot fungi *Gl. trabeum* and *Con. puteana* are similar and significantly inhibited in TM material at 215 °C but also treated wood at 200 °C (durability class 1, see Table 1). Growth of the brown rot fungi caused low mass loss at temperature of 185 °C; durability classes 1 (*Gl. trabeum*) and 2 (*Con. puteana*) were achieved.

The most aggressive fungus on the tested TM beech wood was the white rot *T. versicolor*. This is in line with data available about the thermal treatment of hardwood species as well as previous tests. *T. versicolor* caused an average mass loss in the range of 7.2 to 15.4% with regard to the treatment temperature. The treatment at 215 °C classify the wood in durability class 2. The actions of the two brown rot fungi *Rh. placenta* and *Con. puteana* are similar and significantly inhibited in TM material at 215 °C (durability class 1). Growth of the brown rot fungi caused mass loss of 15.9 and 7.7% at temperature of 185 °C for the respective fungus. Durability classes 3 (*Rh. placenta*) and 2 (*Con. puteana*) were achieved at 185 °C. Thermal modification at temperatures of 215 and 200 °C was the most efficient against the fungus *Gl. trabeum*; the mass loss was in the range 1.0–2.7%. The durability of beech wood was improved to class 1 and further to class 2 after treatment at 185 °C.

The most aggressive fungus on the tested TM Douglas fir wood was the brown rot *Rh. placenta*. This is in line with data available in the literature and all previous tests. *Rh. placenta* caused an average mass loss in the range of 6.1 to 17.8% with regard to the treatment temperature. The treatment at 215 °C can classify the wood in durability class 2 as maximum. The actions of the two brown rot fungi *Con. puteana* and *Gl. trabeum* as well as the white rot fungus *Tram. versicolor* are similar and are all strongly inhibited in TM material at 215 °C (durability class 1, see Table 1). The growth of *Tram. versicolor*, *Con. puteana* and *Gl. trabeum* caused a mass loss of 0.5–2.2% at the treatment temperature of 215 °C. Thermal modification at temperatures of 200 and 185 °C ensured decreased mass loss of the samples to 1.3–8.7% and thus, improved the durability of Douglas fir wood to class 2 and 3. In the case of poplar, the growth of the brown rot fungus *Con. puteana* caused smallest mass loss in the range 0.2–7.1% depending on the treatment temperature. Thermal modification at temperature of 200 °C decreased the mass loss of the samples to 0.5% and thus, improved the durability of the test species to class 1, i.e. very durable when exposed to *Con. puteana*. The action of the brown rot fungus *Gl. trabeum* is similar to that of *Con. puteana* by showing that a treatment temperature of 200 °C also ensured low mass loss of 3.0% for poplar wood. Accordingly, the durability was shifted to class 1 (very durable). The white rot fungus *Tram. versicolor* demonstrated a gradual decrease of its ability to degrade wood with increased treatment temperature. Thermal treatment at 200 °C ensures durability class 3 while the treatment at 215 °C improved further the durability to class 1. The most aggressive fungus on the tested TM wood was the brown rot *Rh. placenta*. This is in line with data available in the literature and previous tests. *Rh. placenta* caused an average mass loss of 1.0, 15.0 and 33.4% for treatment temperatures of 215, 200 and 185 °C and thus, only the highest temperature of 215 °C can classify the wood in durability class 1.The application of the temperature of 215 °C did not show significant differences in the mass loss of the samples and the achieved durability class. The most aggressive fungus on the tested TM Scots pine wood was the brown rot *Rh. placenta. Rh. placenta* caused an average mass loss in the range of 7.2–17.2% with regard to the treatment temperature. The treatment at 215 °C can classify the wood in durability class 2 as maximum. The actions of the two brown rot fungi *Con. puteana* and *Gl. trabeum* as well as the white rot fungus *Tram. versicolor* are similar and are all strongly inhibited in TM material at 215 °C (durability class 1, see Table 1). Growth of the three white and brown rot fungi caused small mass loss found in the range 0.9–1.0% at the treatment temperature of 215 °C. Thermal modification at temperatures of 200 and 185 °C ensured decreased mass loss of the samples to 2.8–8.6% and thus, improved the durability of Scots pine wood to class 2 with an exception of the fungus *Con. puteana* at 185 °C showing durability class 3.

The most aggressive fungus on the tested TM Norway spruce wood was the brown rot *Rh. placenta.* This is in line with data available in the literature and previous tests. *Rh. placenta* caused an average mass loss in the range of 10.8 to 26.1% with regard to the treatment temperature. The treatment at 215 °C classify the wood in durability class 2. The actions of the two brown rot fungi *Con. puteana* and *Gl. trabeum* as well as the white rot fungus *Tram. versicolor* are similar and all are significantly inhibited in TM material at 215 °C (durability class 1 and 2, see Table 1). Growth of the three white and brown rot fungi caused small mass loss found in the range 1.9–3.4% at the treatment temperatures of 215 °C. Thermal modification at temperatures of 200 and 185 °C was less efficient leading to decreased mass loss of the samples to 5.1–26.1% and thus, the durability of Norway spruce wood was slightly improved to class 2 and 3 with an exception of the fungus *Rh. placenta* at 185 °C showing durability class 4.

Regarding tulipier wood, the application of the temperature of 200 and 215 °C showed some similar mass loss of the samples as well as in the achieved durability classes when the samples were exposed to *Rh. placenta, Con. puteana* and *Gl. trabeum.* The mass loss varied in the range of 0.8–3.7%; durability class 1 was achieved. Growth of the above three brown rot fungi caused mass loss of 2.8% (*Gl. trabeum*), 13.8% (*Rh. placenta*) and 14.5% (*Con. puteana*) at treatment temperature of 185 °C which led to durability class 3 (*Rh. placenta* and *Con. puteana*) but class 1 for *Gl. trabeum.* The most aggressive fungus on the tested TM wood was the white rot *T. versicolor* causing an average mass loss in the range of 5.4–18.0% with regard to the treatment temperature. The treatment at 215 and that of 200 °C classify the wood in durability class 2. The durability of tulipier wood was hardly improved to class 4 after treatment at 185 °C.

4 Conclusions

With regard to the mechanical properties, the tested wood species react differently and according to the treatment temperatures, (ash seem the most affected). Significant differences between the control and treated samples were noticed, e.g., compression strength and the modulus of elasticity increased with 25–30% and 15% respectively due to the thermal treatment. The other mechanical properties, shear strength, bending and hardness decreased with 20–30%, 30% and 15% respectively.

The study clearly shows the effect of the thermal modification on the durability by increase from a low durability class of 4 or 5 to durability class 1 or 2. Only 2 species namely, ash and poplar reach durability class 1, which requires a thermal treatment temperature of 215 °C. The other wood species have increased durability to class 2 and must be treated at the same temperature, i.e. 215 °C. Thus, the required treatment temperature to ensure the higher durability class is 215 °C with an exception for the wood species tulipier that shows the maximum at 200 °C (see the blue field). *Tr. versicolor* is the most aggressive fungus on heartwoods and it determines the durability class while *Rd. placenta* governs the durability class of the softwood being the most aggressive case.

Acknowledgements. – Financial supported by Veneto region in the framework of POR FESR STYL$^+$WOOD project 2014/2020

– CNR-IBE, National Research Council - Istituto per la Bioeconomia. San Michele All'Adige (TN) ITALY. Thanks to the colleagues of IBE S. Fiorentino (FI), M. Nocetti and M. Brunetti who performed mechanical tests.

– Swedish University of Agricultural Sciences, Department of Forest Biomaterials and Technology, Uppsala, Sweden.

References

1. Romano, A., Cappellin, L., Cuccui, I., Bogialli, S., Khomenko, I., Tonezzer, M., Biasioli, F., Pastore, P., Allegretti, O.: Exploring volatile organic compound emission from thermally modified wood by PTR-ToF-MS. Analyst. **147**, 5138–5148 (2022). https://doi.org/10.1039/d2an01376b
2. Hill, C.A.S.: Wood Modification: Chemical, Thermal and Other Processes. John Wiley & Sons, Chichester/England.Hoboken, NJ (2006)
3. ISO 13061: Physical and mechanical properties of wood - Test methods for small clear wood specimens (2017)
4. ISO 11664-4: Colorimetry - Part 4: CIE 1976 L*a*b* Colour space (2019)
5. EN 1534: Wood floring - Determination of resistance to indentation -Test method (2010)
6. EN 113-2: Durability of wood and wood-based products -Test method against wood de stroying basidiomycetes - Part 2: Assessment of inherent or enhanced durability (2020)
7. EN 84: Wood preservatives. Accelerated ageing of preservative treated wood prior to bilogical testing. Leaching (1997)
8. EN 350: Durability of wood and wood-based products - Testing and classification of the durability to biological agents of wood and wood-based materials (2016)
9. Allegretti, O., Cuccui, I., Terziev, N., Sorini, L.: A model to predict the kinetics of mass loss in wood during thermo-vacuum modification. Holzforschung. **75**, 474–479 (2020). https://doi.org/10.1515/hf-2020-0127

Thermal Treatment as an Industrial Process to Produce Different Commercial Wood Products

Paola Cetera[1](✉) and Alessandro Porcu[2]

[1] Department of Agricultural Sciences, University of Sassari, Viale Italia 39/a, 07100 Sassari, Sardinia, Italy
pcetera@uniss.it
[2] Baschild S.r.l., Via degli Assini, 16, 24048 Treviolo, Italy
a.porcu@baschild.com

Abstract. Thermal treatment of sawn timber is carried out under carefully controlled conditions at high temperatures. In the processing process, only heat, water vapor and fresh water are used without the use of any chemical preparations or additives, which makes it completely ecologically friendly and limited harm to the environment. Albino Pozzi (Italy), Evolen (Croatia) and Sivalbp (France), are three European manufacturers that use high temperature of treatment to increase some properties of wood and preserve the material without using any chemical additive or fossil fuels according to the technology of Italian company Baschild. The positive effects of the high temperature on some wood properties, such as colour or durability, confirm that nowadays the thermal treatment is an important industrial process for commercial products wood in Europe. This contribution describes the experience of Baschild, specializing in the production of kilns for wood thermal modification. The thermally-treated wood products appeared suitable for various applications, including decking, balconies, facades, and similar outdoor projects, minimizing environmental impact.

Keywords: Wood · Thermal Treatment · Wood Paneling · Decking

1 Introduction

The thermal treatment of wood has gained significant attention in recent years due to its ability to enhance some properties of this natural material [1–3]. The most relevant kinds of the processes are summarized as Feuchte-Wärme-Druck and Oil Heat Treatment in Germany, Plato in The Netherlands, ThermoWood in Finland, Le Bois Perdure and Rectification in France and TermoVuoto in Italy [4, 5]. The process involves subjecting wood to high temperatures, typically ranging from 160 °C to 260 °C, in the absence of oxygen. This process, also known as thermo-wood or thermally modified wood, has been widely adopted by various industries [6], including construction, furniture manufacturing for outdoor uses, and architecture landscaping. The thermal treatment process alters the chemical composition of wood [7], reducing its hygroscopicity and

G. Goli et al. (Eds.): ECWM 2024, SPM 86, pp. 19–24, 2025.
https://doi.org/10.1007/978-3-031-99418-0_3

increasing its dimensional stability. This means that thermally treated wood is less prone to swelling and shrinking due to changes in moisture content, making it an ideal choice for outdoor applications such as decking, cladding, and garden furniture.

Moreover, the thermal treatment process improves the durability (generally at least 180 °C is for durability, but with reduced mechanical properties) of wood by reducing its susceptibility to fungal decay and insect attack [8]. This is particularly important in regions with high humidity or where wood is exposed to harsh environmental conditions. By enhancing the natural resistance of wood, the thermal treatment process reduces the need for chemical preservatives, making it an eco-friendly alternative. Baschild is an Italian manufacturer of kilns for wood modification and the innovative solutions of the company have helped to advance the field of wood thermal treatment, providing high quality products that meet the demands of various industries. This article will explore the benefits of the thermal treatment process for wood, highlighting its applications in different sectors and showcasing the work of Baschild in advancing this technology.

2 Materials and Methods

2.1 Thermal Treatment Process

The process of wood thermal modification adopted from Baschild Company involves heating wood at temperatures between 160 °C and 230 °C in order to change the aesthetic and functional characteristics of the wood itself. The treatment is performed within a closed kiln of different dimensions with capacity from 30 to 400 m^3 per drying cycle, made of stainless steel and other materials suitable for continuous use under the needs expressed from users of the machine. Usually, the kiln relies a thermal medium such as thermal oil for heating air through heat exchangers and for generating steam is used for boiling water contained in a special-designed tank. The kiln is thus able to operate at varying conditions of humidity and air temperature into loaded timber, and changes the speed of the air itself, on the basis of running schedules (cycles) that can be time to time and custom tailored to the needs of the case.

The ATK series ovens are machines designed and built to conduct the wood thermal modification process an automatic and safe mode. In general, the kiln for thermal modification can be operated at varying volumes of timber (from 10 to 60 m^3). Each treatment cycle runs through the sequence of 6 main stages:

- Step 1 - Heating - At this stage the loaded timber is brought to a safe temperature to allow free water to escape. The set temperature is set at 70–90 ° C.
- Step 2 - Climb to high temperature - Ascent set at a variable rate depending on type of wood and thickness; the duration is also dependent on the set temperature (which can be from 160 °C to 230 °C). Usually made in 2 or more steps.
- Step 3 - Treatment / Maintenance - For 3 h the temperature is maintained at the same set value in step 2.
- Step 4 - Cooling above 100 °C - At this stage sets a rate of temperature drop of about 7–10 °C/h up to a set of about 100 °C.
- Step 5 - Conditioning - This step allows the introduction into the timber of water content needed to make it workable again (the anhydrous wood is not). Temperature maintained at 80–100 °C under conditions of steam.

- Step 6 - Cooling below 100 °C - Phase is a gradual cooling of the wood before unloading, setting a rate of 8–12 °C/h up to a temperature set that is not above 30 °C above the outside temperature.

3 Results

3.1 Industry Applications of Thermally Treated Wood

Baschild has produced approximately 40 furnaces for the thermal treatment of various sizes. Of these, 25 are still in service: when combined, the annual productivity of thermally treated wood easily exceeds 30,000 cubic meters net. Products derived from thermally treated wood are primarily used for outdoor applications, such as flooring, fences, facades, garden furniture, the marine sector, and so on. However, there is also a decent offering of products for indoor use, such as engineered hardwood flooring, wall paneling, tables, and general furnishings.

The Albino Pozzi company located in Sondrio, Italy is specializes in "*turnkey*" projects and therefore offers a catalog of products that belong to both categories. Among these, the facades made of thermo-modified ash (Fig. 1) represent a consolidated product. The wood, of European origin, is treated at 212 °C and at the end of the Baschild HT process, it is ready for subsequent cutting, milling, and finishing. As a final step, a protective UV layer is applied to preserve the color: this operation can also be repeated during installation as a maintenance procedure.

Fig. 1. Thermal treaded facades installed by Albino POZZI, Italy.

According to UNI EN 13228:2011 "*Wood flooring. Solid wood overlay flooring elements including blocks with an interlocking system*" when regarding the characteristics of solid wood elements for flooring, in the Republic of Croatia, specifically in Novska, the company Evolen carries out the thermal modification of ash decking, achieving the highest resistance class. In conditions of wide variability in air humidity, wooden floors must ensure excellent dimensional stability. Evolen, through thermal modification and a proven installation system (Fig. 2), guarantees the performance and durability of the floors over time.

Fig. 2. Particular of installation system of wood thermal modified (a) and different uses of the thermal modified decking (b) by Evolen, Croatia.

Another example related to thermal modified facades (Fig. 3) comes from Thônes, France, where the Sivalbp plant is located. During the thermal process, Sivalbp uses exclusively wood supplies from sustainably managed french forests, certified PEFC (PEFC/10-31-1593) & FSC® (FSC-C118242). Selecting only the best wood species, such as western red cedar, larch, spruce, nordic pine or douglas fir, which are all different wood species, each with its own characteristics. The final product of Sivalbp is a range of wood cladding called Elegance, New Age, Vintage, Montagne and Colors. The latter certification, delivered by the Institut Technologique FCBA, guarantees the performance of the treatments used to durably protect cladding woods.

Fig. 3. Sivalbp wood thermal treated, France.

4 Conclusion

Wood thermal treatment, as highlighted in several studies, offers significant benefits that underscore its value in various applications. Baschild's expertise in this area ensures the delivery of high-quality heat-treated wood products with enhanced properties. Thus, is possible emphasize the following benefits of wood thermal treatment by Baschild:

- High durability: heat-treated wood exhibits improved resistance to decay, making it a durable for outdoor applications like porches and terraces.
- Dimensional stability: the thermal modification process enhances the stability of wood, reducing shrinkage and swelling, which is crucial for maintaining the structural integrity of wood products over time.
- Environmental sustainability: thermally modified wood is an eco-friendly alternative, as it does not require chemical additives and can be recycled without detrimental environmental impact, unlike chemically treated wood.

In conclusion, Baschild's wood thermal treatment process not only enhances the natural properties of wood but also ensures the production of durable, stable, and environmentally friendly wood products suitable for a wide range of applications in the wood industry in different parts of the Europe.

Baschild also ensures the quality of their products through a comprehensive quality assurance process by several key steps, such as standard definition, ensuring required specifications and customer expectations or process improving the processes to prevent defects.

References

1. Dvořák, O., Kvietková, M.S., Šimůnková, K., Machanec, O., Pánek, M., Pastierovič, F., Lin, C.F., Jones, D.: The Influence of the initial treatment of oak wood on increasing the durability of exterior transparent coating systems. Polymers. **15**(15), 3251 (2023)
2. Esteves, B.: Wood modification by heat treatment: a review. BioResources, 370–404 (2009)
3. Faraone, I., Russo, D., D'Auria, M., Bruno, M.R., Cetera, P., Todaro, L., Milella, L.: Influence of thermal modification and extraction techniques on yield, antioxidant capacity and phytochemical profile of chestnut (*Castanea sativa* Mill.) wood. Holzforschung. **75**, 260–268 (2021)
4. Sandberg, D., Kutnar, A., Mantanis, G.: Wood modification technologies-a review. iForest-Biogeosci. For. **10**(6), 895 (2017)
5. Hill, C.A.: Wood Modification: Chemical, Thermal and Other Processes. John Wiley & Sons (2007)
6. Militz, H., Altgen, M.: Processes and properties of thermally modified wood manufactured in Europe. In: Deterioration and Protection of Sustainable Biomaterials, pp. 269–285. American Chemical Society (2014)
7. Cetera, P., Russo, D., Milella, L., Todaro, L.: Thermo-treatment affects *Quercus cerris* L. wood properties and the antioxidant activity and chemical composition of its by-product extracts. Ind. Crops Prod. **130**, 380–388 (2019)
8. Candelier, K., Thevenon, M.F., Petrissans, A., Dumarcay, S., Gerardin, P., Petrissans, M.: Control of wood thermal treatment and its effects on decay resistance: a review. Ann. For. Sci. **73**, 571–583 (2016)
9. UNI EN 13228:2011 - Wood flooring. Solid wood overlay flooring elements including blocks with an interlocking system. European Committee for Standardization, Brussels, Belgium

Thermal Modification

Heat Treatment of Cryptomeria Japonica from Azores

Yurlet Mercado[1], Lina Nunes[2,3], Luisa Cruz-Lopes[1], and Bruno Esteves[1](✉)

[1] CERNAS Research Centre, Polytechnic University of Viseu, Viseu, Portugal
{lvalente,bruno}@estgv.ipv.pt

[2] Structures Department, National Laboratory for Civil Engineering, Lisbon, Portugal
linanunes@lnec.pt

[3] CE3C, Centre for Ecology, Evolution and Environmental Changes & CHANGE, Global Change and Sustainability Institute, University of the Azores, Angra do Heroísmo, Portugal

Abstract. Cryptomeria japonica has been introduced to various regions, including the Azores where it is an abundant resource. Its dimensional instability prompts exploration of heat-treatment for property enhancement. Chemical composition analysis involved determination of extractives, lignin, holocellulose, and α-cellulose. Mechanical properties were evaluated for untreated and heat-treated wood samples, including bending strength and stiffness. Shrinking and swelling behavior were assessed through cyclic exposure to extreme moisture conditions. The extractive composition of untreated and heat-treated Cryptomeria wood was analyzed, revealing predominant ethanol, water, and dichloromethane extractives in untreated samples. With heat treatment, a slight increase in ethanol extractives was observed, alike findings in other species. Additionally, the lignin percentage increased post-heat treatment, likely due to hemicellulose degradation and condensation reactions between lignin and polysaccharide degradation compounds. Despite variations in natural durability against subterranean termites, Cryptomeria heartwood remains susceptible to attack. While some antitermic extractives may be removed during heat treatment, no significant change in wood durability against termites was observed. Swelling decreased with heat treatment, particularly in the tangential direction. Mechanical strength, notably bending strength, decreased due to heat treatment, while modulus of elasticity (MOE) initially increased before declining with more severe treatments.

Keywords: Cryptomeria · Dimensional stability · Heat-treatment · Termites

1 Introduction

Cryptomeria japonica, commonly known as Japanese cedar or Sugi in Japanese, is a majestic and revered evergreen tree native to Japan. With its strikingly symmetrical form and towering height, Cryptomeria japonica has become an iconic symbol of Japanese landscapes and culture. In its natural habitat, Cryptomeria japonica can reach impressive heights of up to 50 m (164 feet) or more, with a straight trunk and dense foliage composed of flattened, needle-like leaves [1]. These leaves are arranged spirally around

G. Goli et al. (Eds.): ECWM 2024, SPM 86, pp. 27–33, 2025.
https://doi.org/10.1007/978-3-031-99418-0_4

the branches, creating a lush canopy that provides shade and shelter to various animals and plants beneath it. Renowned for its resilience and adaptability, Cryptomeria japonica thrives in a variety of soil types and climates, from humid coastal regions to mountainous terrains. It is often planted in Japan as a windbreak, erosion control measure, or as a timber resource. In addition to its practical uses, Cryptomeria japonica holds significant cultural importance in Japan, where it is often planted in temple gardens, shrines, and sacred groves. Beyond Japan, Cryptomeria japonica has been introduced to other parts of the world, including North America, Europe, and Asia, where it is valued for its ornamental beauty and timber qualities. Its wood, known as Japanese cedar, is highly prized for its durability, straight grain, and resistance to decay, making it a sought-after material for construction, furniture-making, and woodcraft. This species has been introduced to the Azores archipelago (Portugal) on the mid-nineteenth century and has been thriven since then.

Wood heat treatment is an environmentally friendly process used to enhance the properties of wood by improving its dimensional stability, durability, and resistance to biological degradation without the use of any harmful chemicals [2, 3]. Several processes have been developed throughout the years leading to well established commercial processes such as Thermowood, Plato Wood, Bois Perdure and Rectification. In the last decade, several new processes have emerged some of them using vacuum to protect wood from oxygen like for instance Termovuoto®, Moldrup-SSP® and TanWood®, others with superheated steam like ThermoTreat 2.0® or Firmolin® and others use modifications from traditional processes like Westwood® [4–8]. Among them, thermal modification by using water vapor as a shielding gas has gained attention due to its effectiveness and sustainability. In this method, wood is heated at elevated temperatures (typically between 160 °C and 260 °C) in an oxygen-poor environment, with water vapor acting as shielding gas [9]. The main function of the water vapor is to prevent combustion, minimize oxidation, and create a controlled atmosphere that influences the thermal degradation and chemical reactions within the wood. During this process, the structure and composition of the wood undergo significant changes, resulting in reduced moisture absorption, increased resistance to decay, and a darker appearance [10].

Water vapor plays a critical role in moderating the chemical transformation of wood components such as hemicellulose, cellulose, and lignin [11]. It acts as a heat carrier and limits oxygen availability, which helps to protect the wood from charring or burning. Heat treatment eliminates the need for chemical additives, making it a more sustainable and non-toxic alternative for improving wood's performance in applications like construction, furniture, and outdoor products. The main challenge with Cryptomeria wood is its high dimensional instability, which causes it to expand, contract, or warp due to changes in moisture content. Therefore, heat treatment emerges as a promising solution for improving the wood's dimensional stability and overall performance. Heat treatment can improve both the dimensional stability and durability of wood, making it more suitable for demanding applications such as outdoor structures and furniture.

2 Materials and Methods

Wood was treated in a Portuguese company known as Palser at 212 °C in accordance to Thermo D class from Thermowood®.

Chemical composition consisted in the determination of extractives, lignin, holocellulose and α-cellulose. The determination of wood extractives was done using Soxhlet extraction that began with the preparation of the wood samples. These samples were ground and sieved and the 40–60 mesh fraction was used. A Soxhlet extractor setup was assembled, comprising a round-bottom flask, a condenser, and a Soxhlet extractor thimble. The ground wood samples were placed into the Soxhlet thimbles and extracted with dichloromethane (8 h), ethanol (16 h) and water (16 h). Once the extraction was finished, the solvent containing the extracted compounds was evaporated off using a rotary evaporator. The remaining extract was collected and dried to remove any residual solvent. The dried extract was then weighed to determine the mass of the extracted compounds expressed as a percentage of the original wood sample mass. The Klason method was used to determine lignin content. Samples were hydrolyzed with 72% sulfuric acid, followed by a second hydrolyze 3% sulfuric acid. After filtration, washing, and drying, the residue was weighed. Results were reported as a percentage of the original dry sample weight. The chlorite method was used to determine the holocellulose content. Alpha-cellulose was quantified from dry holocellulose using a two-step hydrolysis process. Initially, the holocellulose was treated with 17.5% NaOH at 20 °C for 30 min, followed by a 15-min resting period outside the bath. Subsequently, 8.25 mL of water were added to the samples, which were then maintained for an additional hour at 20 °C in the thermal bath. The samples underwent filtration and washing with 25 mL of 8.3% NaOH solution, followed by distilled water rinse, and concluded with 3.75 mL of 10% acetic acid. Hemicelluloses were determined by difference.

Mechanical properties were determined in accordance to Esteves et al. [10]. Air-dry density was determined for untreated and heat-treated wood conditioned at 20 °C and 65% relative humidity by weighing 20 mm cubic samples, cut from described boards with faces oriented in three directions. Ten replicates were averaged. Bending strength and stiffness were determined on untreated and heat-treated wooden test specimens using a three-point bending test machine. Samples, sized 360 × 20 × 20 mm in transverse, radial, and tangential directions respectively, were conditioned at 20 °C and 65% relative humidity. Ten replicates were conducted for each test.

Shrinking, and swelling were assessed in treated and untreated wood cubic samples measuring approximately 20 mm edge, with faces oriented in three directions. One cycle of 0% (Oven at 100 °C) and 100% (Water at 20 °C) was conducted. Swelling was determined in relation to initial dry wood dimensions.

The resistance of the untreated and heat-treated wood samples (10 replicates; 50 × 25 × 15 mm) against the subterranean termite *Reticulitermes grassei* (Blattodea: Isoptera: Rhinotermitidae) was determined according to EN 117 [12].

3 Results and Discussion

Extractive composition of untreated and heat-treated cryptomeria wood is presented in Table 1. Untreated cryptomeria extractives are mainly composed of ethanol extractives (1.3%) followed by water (0.99%) and dichloromethane (0.6%). These results are not that different from the results presented before [13] that presented 1.81% of water extractives and 1.23% of ethanol-toluene extractives in wood coming from experimental plantings located in the Experimental Station of the Federal University of Paraná, in the city of Rio Negro in Brazil. Yinodotlgör and Kartal [14] presented 1.36% of cyclohexane/ethanol, 0.49% of ethanol extractives and 2.78% of extractives in water.

Table 1. Extractive composition of untreated and treated wood.

Sample	Extractives (%)			
	Dichloro.	Ethanol	Water	Total
Untreated	0.60 ± 0.00	1.30 ± 0.43	0.99 ± 0.11	2.89 ± 0.20
Heat-treated	0.55 ± 0.04	2.05 ± 0.13	0.99 ± 0.24	3.59 ± 0.13

With heat treatment there is a small increase in ethanol extractives similar to what was found before for other species like heat-treated Paulownia wood [10]. This increase, however, was not observed on some thermally modified tropical hardwoods like afrormosia (*Pericopsis elata*) and duka (*Tapirira guianensis*) probably due to the already higher ethanol extractives content in untreated wood [15].

The percentage of lignin increases from 32.5% to around 38.9% probably due to the higher degradation of hemicelluloses that decrease from 22.9% to 15.5% but also to the condensation reactions between lignin and degradation compounds from polysaccharides. This has been discussed by several authors throughout the years [10, 16–19].

The natural durability of C. japonica against subterranean termites has been shown to vary significantly even within trees of the same plantation though always susceptible to some level of attack [20–22]. The heartwood of Cryptomeria japonica has been studied as a potential source of natural termiticidal compounds. Recent studies have sought to identify the specific chemical constituents responsible for its termite-repelling properties [23].

In the present tests conducted according to the method described in EN117 (2012) [12], all samples, untreated or heat treated showed high susceptibility to subterranean termites (Table 2). Therefore, even if some antitermic extractives are removed from wood with heat-treatment, there is no significant change on the wood durability against termites.

Table 2. Untreated and treated wood resistance to termite (EN 117, 2012).

Material	Survival rate [%]	Moisture content [%]	Mass loss [%]	Attack level
Untreated	40.60	56.89	8.82	3.8
Heat-treated	31.30	64.42	10.48	4.0

Figure 1 presents the swelling differences between untreated and heat-treated cryptomeria. As expected, the swelling decreased with the treatment and the decrease was higher for tangential direction followed by radial and longitudinal directions (Fig. 1). The improvement was around 37% for the tangential direction changing from 8% to around 5% swelling. Similar results were presented before for several species and with different types of heat treatments [3, 4, 24–26].

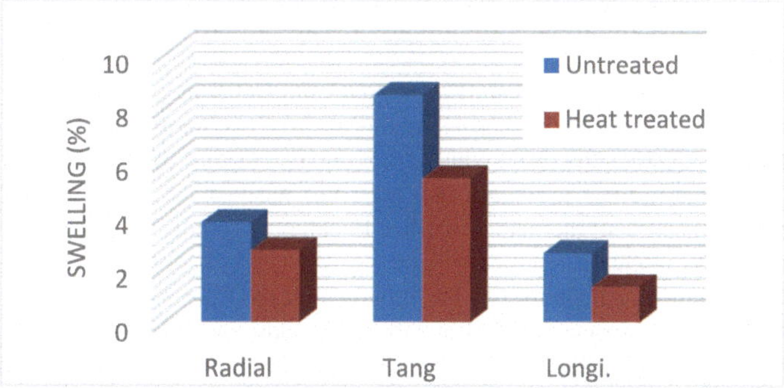

Fig. 1. Swelling of untreated and heat-treated wood.

Mechanical strength is known to decrease due to heat-treatment, especially bending strength [11, 27–29]. MOE has been known to increase in the beginning of the treatment, decreasing afterwards for more severe treatments. A similar pattern was observed here, where bending strength decreased from 52.6 MPa to 42.3 MPa corresponding to a 19.6% decrease in relation to untreated wood while MOE increased from 7268 MPa to 10,836 MPa (Table 3).

Table 3. MOE and Bending Strength.

Sample	MOE (MPa)		Bending strength (MPa)	
	Average	Std. Dev.	Average	Std. Dev.
Untreated	7268	1194	52.6	2.4
Heat-treated	10,836	771	42.5	6.0

4 Conclusion

Results show that cryptomeria chemical composition changes with heat-treatment, increasing lignin and ethanol extractives, improving dimensional stability, apparently with no significant difference on the resistance to subterranean termites and with relatively low mechanical degradation.

Acknowledgments. This work was supported by the FCT - Foundation for Science and Technology, I.P. Furthermore, we would like to thank the CERNAS (UIDB/00681/2020 DOI: 10.54499/UIDB/00681/2020), Research Centre and the Polytechnic University of Viseu for their support.

References

1. Cryptomeria japonica (Japanese Cedar, Japanese Cryptomeria) | North Carolina Extension Gardener Plant Toolbox, https://plants.ces.ncsu.edu/plants/cryptomeria-japonica/
2. Hill, C.A.: Wood Modification: Chemical, Thermal and Other Processes. John Wiley & Sons (2006)
3. Esteves, B., Pereira, H.: Wood modification by heat treatment: a review. Bioresources. **4**, 370–404 (2009)
4. Pockrandt, M., Jebrane, M., Cuccui, I., Allegretti, O., Uetimane, E., Terziev, N.: Industrial Thermowood® and Termovuoto thermal modification of two hardwoods from Mozambique. Holzforschung. **72**, 701–709 (2018). https://doi.org/10.1515/hf-2017-0153
5. Batista, D.C., Paes, J.B., de Muñiz, B., Ines, G., Nisgoski, S., da Silva Oliveira, J.T.: Microstructural aspects of thermally modified Eucalyptus grandis wood. Maderas. Ciencia y tecnología. **17**, 525–532 (2015)
6. Ferrari, S., Cuccui, I., Allegretti, O.: Thermo-vacuum modification of some European softwood and hardwood species treated at different conditions. Bioresources. **8**, 1100 1109 (2013). https://doi.org/10.15376/biores.8.1.1100-1109
7. Willems, W.: Patent NL2000405; US 8465845; WO 2008/079000 A1 (2006)
8. Dagbro, O., Torniainen, P., Karlsson, O., Morén, T.: Colour responses from wood, thermally modified in superheated steam and pressurized steam atmospheres. Wood Mater. Sci. Eng. **5**, 211–219 (2010). https://doi.org/10.1080/17480272.2010.520739
9. Mayes, D., Oksanen, O.: Thermowood Handbook, pp. 5–15. Finnforest, Finland (2002)
10. Esteves, B., Ferreira, H., Viana, H., Ferreira, J., Domingos, I., Cruz-Lopes, L., Jones, D., Nunes, L.: Termite resistance, chemical and mechanical characterization of Paulownia tomentosa wood before and after heat treatment. Forests. **12**, 1114 (2021). https://doi.org/10.3390/f12081114
11. Esteves, B., Nunes, L., Domingos, I., Pereira, H.: Comparison between heat treated sapwood and heartwood from Pinus pinaster. Eur. J. Wood Wood Prod. **72**, 53–60 (2014). https://doi.org/10.1007/s00107-013-0751-y
12. EN 117: Wood Preservatives - Determination of Toxic Values against Reticulitermes Species (European Termites) (Laboratory Method). CEN, Brussels (2012)
13. Fonte, A.P.N., Trianoski, R., Iwakiri, S., dos Anjos, R.A.M.: Physical and chemical properties of heartwood and sapwood of Cryptomeria japonica. Revista de Ciências Agroveterinárias. **16**, 277–285 (2017). https://doi.org/10.5965/223811711632017277
14. Yinodotlgör, N., Kartal, S.N.: Heat modification of wood: chemical properties and resistance to mold and decay fungi. For. Prod. J. **60**, 357–361 (2010). https://doi.org/10.13073/0015-7473-60.4.357
15. Esteves, B., Ayata, U., Cruz-Lopes, L., Brás, I., Ferreira, J., Domingos, I.: Changes in the content and composition of the extractives in thermally modified tropical hardwoods. Maderas-Cienc Tecnol. **24**, 1–14 (2022). https://doi.org/10.4067/s0718-221x2022000100422
16. Domingos, I., Ayata, U., Ferreira, J., Cruz-Lopes, L., Sen, A., Sahin, S., Esteves, B.: Calorific power improvement of wood by heat treatment and its relation to chemical composition. Energies. **13**, 5322 (2020). https://doi.org/10.3390/en13205322
17. Esteves, B., Graça, J., Pereira, H.: Extractive composition and summative chemical analysis of thermally treated eucalypt wood. Holzforschung. **62**, 344–351 (2008). https://doi.org/10.1515/HF.2008.057
18. Alén, R., Kotilainen, R., Zaman, A.: Thermochemical behavior of Norway spruce (Picea abies) at 180–225 C. Wood Sci. Technol. **36**, 163–171 (2002). https://doi.org/10.1007/s00226-001-0133-1

19. Windeisen, E., Strobel, C., Wegener, G.: Chemical changes during the production of thermo-treated beech wood. Wood Sci. Technol. **41**, 523–536 (2007). https://doi.org/10.1007/s00226-007-0146-5
20. Shibutani, S., Takata, K., Doi, S.: Quantitative comparisons of antitermite extractives in heartwood from the same clones of Cryptomeria japonica planted at two different sites. J. Wood Sci. **53**, 285–290 (2007). https://doi.org/10.1007/s10086-006-0866-4
21. Cheng, S.-S., Chang, H.-T., Wu, C.-L., Chang, S.-T.: Anti-termitic activities of essential oils from coniferous trees against *Coptotermes formosanus*. Bioresour. Technol. **98**, 456–459 (2007). https://doi.org/10.1016/j.biortech.2006.01.006
22. Xie, Y., Li, M., Huang, Q., Lei, C.: Chemical composition and termiticidal activity of essential oils from different tissues of Chinese cedar (Cryptomeria fortunei). Nat. Prod. Commun. **9**, 1934578X1400900536 (2014). https://doi.org/10.1177/1934578X1400900536
23. Arihara, S., Umeyama, A., Bando, S., Kobuke, S., Imoto, S., Ono, M., Yoshikawa, K., Amita, K., Hashimoto, S.: Termiticidal constituents of the black-heart of Cryptomeria japonica. J. Jpn. Wood Res. Soc. **50** (2004)
24. Unsal, O., Kartal, S.N., Candan, Z., Arango, R.A., Clausen, C.A., Green, F.: Decay and termite resistance, water absorption and swelling of thermally compressed wood panels. Int. Biodeterior. Biodegradation. **63**, 548–552 (2009). https://doi.org/10.1016/j.ibiod.2009.02.001
25. Behr, G., Bollmus, S., Gellerich, A., Militz, H.: Improvement of mechanical properties of thermally modified hardwood through melamine treatment. Wood Mater. Sci. Eng. **13**, 262–270 (2018). https://doi.org/10.1080/17480272.2017.1313313
26. Militz, H., Altgen, M.: Processes and properties of thermally modified wood manufactured in Europe. In: Deterioration and Protection of Sustainable Biomaterials, pp. 269–285. ACS Publications (2014)
27. Yildiz, S.: Physical, mechanical, technological and chemical properties of beech and spruce wood treated by heating [thesis]. Trabzon: Karadeniz Technical University (2002)
28. Yildiz, S., Gezer, E.D., Yildiz, U.C.: Mechanical and chemical behavior of spruce wood modified by heat. Build. Environ. **41**, 1762–1766 (2006). https://doi.org/10.1016/j.buildenv.2005.07.017
29. Boonstra, M.J., Van Acker, J., Tjeerdsma, B.F., Kegel, E.V.: Strength properties of thermally modified softwoods and its relation to polymeric structural wood constituents. Ann. For. Sci. **64**, 679–690 (2007). https://doi.org/10.1051/forest:2007048

Effects of Accelerated Weathering on Surface Hardness of Thermally Modified Woods (*Fagus Sylvatica L.* and *Pinus nigra*)

Holta Cota[1](✉), Entela Lato[1], Leonidha Peri[1], Hektor Thoma[1], Doklea Quku[1], Dritan Ajdinaj[1], Erald Kola[1], Ottaviano Allegretti[2], Marco Togni[3], and Giacomo Goli[3]

[1] Faculty of Forest Sciences, Department of Wood Industry, Agricultural University of Tirana, Tirana, Albania
{hcota,elato,leonidha.peri,hektor.thoma,dquku,d.ajdinaj, ekola}@ubt.edu.al

[2] Istituto per la BioEconomia, IBE, CNR, Trento, Italy
ottaviano.allegretti@ibe.cnr.it

[3] UniversitàdegliStudi di Firenze, DAGRI, Firenze, Italy
{marco.togni,giacomo.goli}@unifi.it

Abstract. Wood is an organic material used for structural, decorative and aesthetic applications. However, when it is used in external environment exposed to atmospheric agents a lot of its properties change. Thermal modification has become a conventional answer to improve some of the wood properties that makes the material suitable for external use. The aim of this study is to determine the effect of both thermal modification and accelerated weathering on the surface hardness of beech (*Fagus Sylvatica* L.) and pine (*Pinus nigra* J. F. Arnold). Five groups of 20 samples each are prepared for the modification process. The samples are modified at high temperatures (180 and 220 °C) with two different periods (2 h and 8 h). After that, the modified samples are exposed to the accelerated weathering conditions for two different time durations (120 and 240 h) according to the ASTM G154 standard. The hardness measurements of radial surface are made on natural (untreated), modified and accelerated weathered samples, according to the respective standards. The highest percentage of mass loss is detected at treatment with the higher temperature and the longer time. The hardness values are increased for samples modified at lower temperature and time (180 °C for 2 h), and decreased for treatment with higher temperature and longer time (the highest decrease is detected for treatment 220 °C for 8 h). The hardness of heat-treated wood samples exposed in accelerated weathering decreases compared to the control samples.

Keywords: thermal modification · accelerated weathering · mass loss · hardness · beech · pine

G. Goli et al. (Eds.): ECWM 2024, SPM 86, pp. 34–43, 2025.
https://doi.org/10.1007/978-3-031-99418-0_5

1 Introduction

Thermal modification of wood is a process that involves heating wood at high temperatures to alter its physical and chemical properties. This modification is typically done to improve wood durability, dimensional stability, and resistance to decay. Thermally modified (TM) wood has become common in the global market due to increasing customer demand for innovative wood products [17]. TM wood is more environmentally friendly while maintaining improved performance characteristics. Various methods of thermal wood modification exist, each with its parameters and variations.

Different methods used for industrial TM processes vary according to the medium in which the wood is treated, such as steam, water, nitrogen, or oil [1]. All these treatments increase wood value by reducing equilibrium moisture content, improving dimensional stability and durability, along with decreasing the heat transfer coefficient [2, 8, 10, 12]. Modified wood finds applications both indoors and outdoors, especially for decks, patios, outdoor claddings, facades, garden furniture, noise barriers, fences, ship decks, etc. Weathering resistance, which affects aesthetic service life, is crucial for users in these applications.

Artificial weathering of wood in UV (QUV Accelerated Weathering Tester) weathering chambers is a process used to simulate the effects of natural weathering on wood materials in a controlled environment. Weathering is the general term used to define the slow degradation of materials exposed to the weather [9, 18, 19]. The goal is to simulate the effects of sunlight, moisture, temperature fluctuations, and other environmental factors on wood surfaces. Particularly, exposure to UV light is the main factor that can lead to the degradation of surface components in wood, such as lignin [20]. Lignin contributes to the hardness and rigidity of wood. Significant degradation may affect the overall hardness of the wood surface.

Exposure to UV light can also change the colour of wood, making it brown or gray. This is more noticeable in lighter woods [18]. The breakdown of components like lignin causes these colour shifts. This process, called photo degradation, transforms the wood over time into a gray or silver hue, which might be appreciated for its appearance. Weathering, mainly due to wind and abrasion from particles, further erodes the wood surface.

Changes in wood's chemical, physical, and optical properties cause discoloration, loss of shine, and surface roughening. The alterations of the three main chemical components of wood also impact mechanical properties [7]. This process is commonly used in research, quality control, and product development to understand how durable, efficient, and resistant wood products are to aging.

The aim of this study is to investigate the impact of thermal modification and accelerated weathering using UV on the surface hardness of beech (*Fagus Sylvatica* L.) and black pine (*Pinus nigra* J. F. Arnold).

2 Material and Methods

Samples from beech wood (*Fagus Sylvatica* L.) and pine (*Pinus nigra* J. F. Arnold) were prepared in dimensions of 50 × 50 × 50 mm, selected randomly from planks without defects such as spiral grain and cracks. The samples displayed clear radial, tangential and cross-sections and were prepared according to the UNI ISO 2130 standard. In total, 100 samples were prepared and divided into five groups. Four groups were to be heat-treated at four different regimes, while one group was left untreated. Before treatment, all samples were dried at 103 ± 2 °C until reaching a moisture content of 0%. Subsequently, the oven-dried samples were weighed using a digital scale with an accuracy of 0.01 g.

The heat treatment was conducted in conventional ventilated oven at two different temperatures (180 °C and 220 °C) and two different durations (2 and 8 h) under atmospheric pressure. After heat treatment, the chamber's temperature was decreased to room temperature and the dry mass loss resulting from the heat treatment was determined referring the mass of the dry samples after the treatments to the dry mass measured before treatment.

Hardness was measured using a Mitutoyo Hardmatic Type D HH-334. For weathering tests, both heat-treated and natural wood samples were cut radially into dimensions of 50 × 50 × 8 mm for exposure in a UV accelerated weathering tester. The samples underwent accelerated weathering at two different time durations, 120 and 240 h, according to the ASTM G154 standard [3], which prescribes exposure cycles based on the material being evaluated. The exposure duration in the UV corresponds to 2.5 and 5 years of natural weathering. The accelerated weathering cycle consisted of three stages: a UV process (1 h at 60 °C with a light intensity of 0.71 w/m^2), water spray (10 min at 20 °C), and conditioning (4 h at 50 °C) respectively (Fig. 1) [15]. Also, the mass loss in % of the thermally modified and untreated samples was determined after the accelerated weathering process.

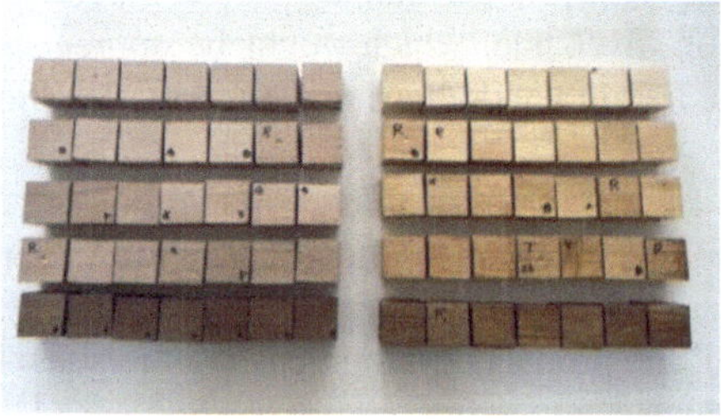

Fig. 1. UV treatment

After the procedure of accelerated weathering, the hardness of the samples was determined according to ASTM D 2240 standard (Fig. 2).

Fig. 2. Hardness measurement

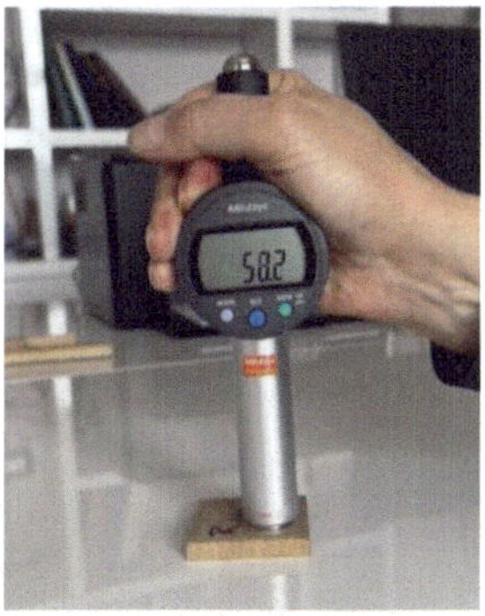

Fig. 3. Thermally modified samples

3 Result and Discussion

The heat treatment causes a colour darkness of samples (Fig. 3). The treatment of highest time and longest time duration cause the extreme colour darkness. During heat treatment as lignin breaks down, it produces dark-colored compounds, contributing to the overall darkening of the wood [2, 5, 11].

3.1 Mass Loss

The results showed that after the thermal modification occurs the mass loss. The extent of mass loss depends on factors such as treatment temperature, duration, and wood species.

The higher heat treatment temperature and longer duration caused greater weight losses in the samples [1].

ML varied between 2.56 – 2.92% at 180 °C to 10.77 – 11.37% at 220 °C. The minimum mass loss is detected during the 180 °C for 2 h (the minimum mass loss for pine 2.56% and for beech 2.92%). The maximum mass loss is detected during the 220 °C for 8 h (the maximum mass loss for pine 11.37% and for beech 10.77%) (Fig. 4).

The observed behaviour may be explained by degradation of chemical components [13].

While during accelerated weathering the minimum ML (2.73%) is detecting for pine samples treated in 220 °C/8 h and weathered for 120 h and the maximum ML (4.01%) is detecting for pine samples treated in 180 °C/2 h and weathered for 240 h. The minimum ML (2.85%) for beech samples is detected in samples treated 220 °C/2 h and weathered for 120 h and the maximum ML (4.87%) for beech samples is detected samples treated in 180 °C/2 h and weathered for 240 h (Fig. 5).

Such behaviour may be explained by wood chemical depolymeralisation, degrading wood extractives the leaching of cell wall extractives and partial hydrolysis of the hemicellulose and cellulose on the sample surface, which consequently leads to the reduction of wood density [13].

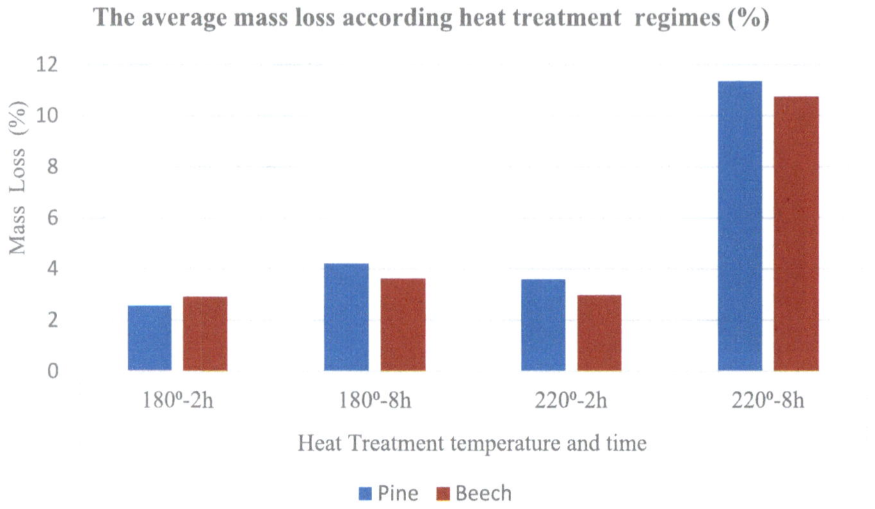

Fig. 4. The average mass loss according thermally modification regimes

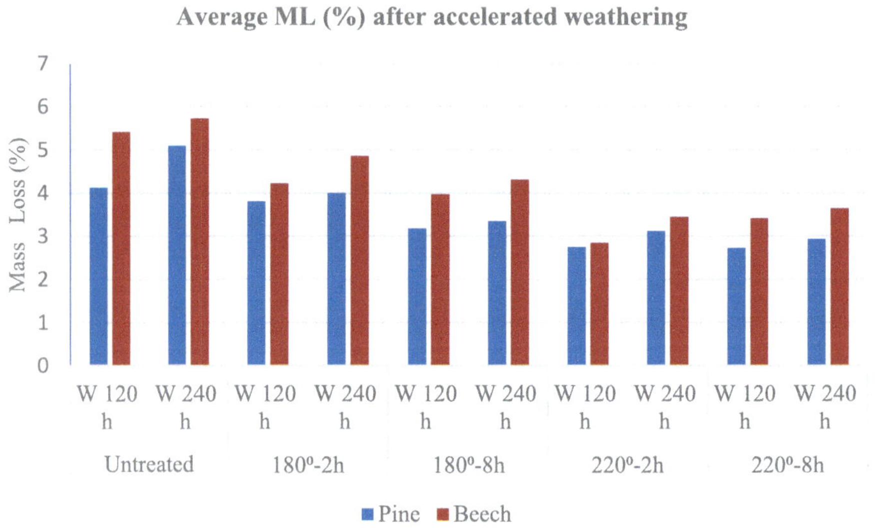

Fig. 5. The average mass loss in % after the accelerated weathering

3.2 Hardness

Regarding the hardness, the study's results show that hardness values increased at the heat treatment of 180 °C for 2 h and decreased at the heat treatment of 180 °C (8 h) and 230 °C for both times (2 and 8 h) (Fig. 6). The increase in hardness value at 180 °C for 2 h can be attributed to significant changes in the structure of lignin during thermal treatment [16]. While the exact effects on hardness may vary, alterations in lignin can contribute to the overall strengthening of the wood material.

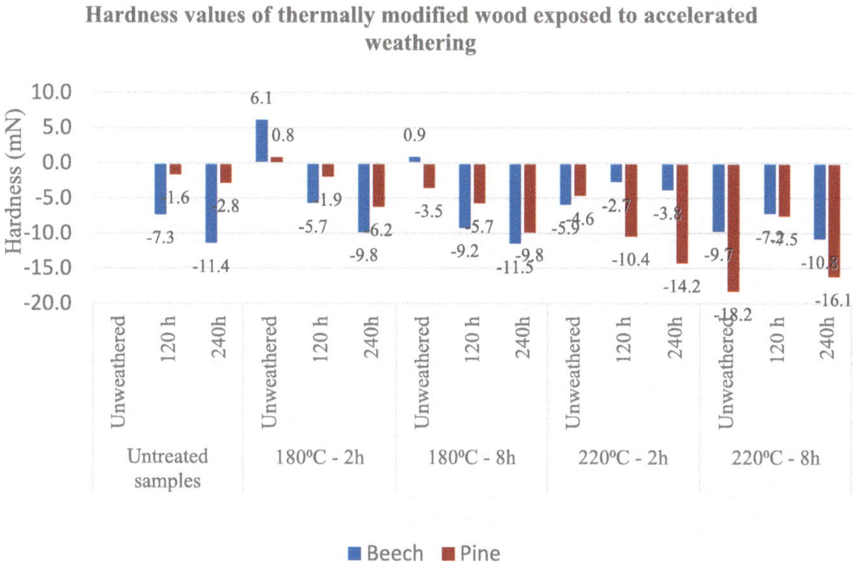

Fig. 6. The hardness change of heat-treated wood after accelerated weathering treatment

After the process of accelerated weathering, it was observed that the hardness values decreased in all groups of species compared to their respective control groups [4, 6, 14]. The hardness of two species decreased over time for both weathering durations of 120 and 240 h.

The maximum decrease in hardness for pine was recorded after 240 h of weathering following the 220 °C – 8 h heat treatment [15], with a reduction of 16.1%.

The maximum decrease in hardness for beech was observed after 240 h of weathering following the 180 °C – 8 h heat treatment, with a reduction of 11.5%. The minimum effect on hardness was noted after 120 h of weathering in pine samples treated at 180 °C for 2 h, with a decrease of 1.9%.

The hardness of black pine was most affected by the weathering process, with decreases ranging between −1.6% and 16.1%.

Regarding the hardness values of untreated samples after weathering, the maximum decrease was noted for beech after 240 h of weathering, with a reduction of −11.4%, while the minimum effect was observed for black pine after 120 h of weathering, with a decrease of −1.6% [15] (Table 1).

According to Jankowska and Kozakiewicz [13, 14], one of the main causes of the loss of mechanical performance of wood under aging treatments can be attributed to cyclic exposure conditions, which lead to the degradation of wood tissues and the formation of cracks. Numerous factors are associated with the loss of mechanical performance, including microscopic factors such as the distribution of cellular elements within the wood (parenchyma, fibers, vessels, etc.), macroscopic factors such as the presence of juvenile wood and latewood incidences, and the chemical composition of the wood (including extractives such as resins, oils, tannins, etc.) [13, 14].

Table 1. Hardness values of thermally modified wood exposed to accelerated weathering

Wood Species					Heat treatment temperature and time and weathering time											
		Untreated samples			180 °C – 2 h			180 °C – 8 h			220 °C – 2 h			220 °C – 8 h		
		Unweathered	120 h	240 h	Unweathered	120 h	240 h	Unweathered	120 h	240 h	Unweathered	120 h	240 h	Unweathered	120 h	240 h
Beech	Minimum	54.30	51.70	48.10	58.30	54.30	49.10	49.20	46.20	44.90	53.20	51.10	49.80	50.60	47.10	44.70
	Maximum	62.50	58.40	54.80	66.40	63.10	58.30	54.90	52.90	49.80	57.60	56.80	55.90	55.80	51.80	49.90
	Stand. Dev	1.28	1.68	1.27	0.81	1.23	1.12	0.88	2.10	0.69	1.23	0.80	1.59	1.01	1.30	1.12
	Average	58.80	54.51	52.10	62.39	58.83	53.07	53.80	48.85	47.61	55.33	53.84	53.23	53.10	49.27	47.36
	Change (%)		−7.30	−11.40	6.10	−5.70	−9.80	0.85	−9.20	−11.50	−5.9C	−2.70	−3.80	−9.70	−7.20	−10.80
Black Pine	Minimum	40.10	38.80	39.80	38.90	39.90	38.70	38.10	34.80	33.40	38.30	36.10	32.10	32.10	29.10	26.70
	Maximum	44.80	44.10	42.80	43.50	43.80	43.50	42.80	41.60	39.70	43.50	39.10	36.80	35.80	35.40	39.80
	Stand. Dev	1.32	0.89	1.01	1.35	0.78	0.98	0.86	1.12	1.24	0.09	1.16	1.18	1.54	1.65	1.28

(continued)

Table 1. *(continued)*

Wood Species		Heat treatment temperature and time and weathering time														
		Untreated samples			180 °C – 2 h			180 °C – 8 h			220 °C – 2 h			220 °C – 8 h		
		Unweathered	120 h	240 h	Unweathered	120 h	240 h	Unweathered	120 h	240 h	Unweathered	120 h	240 h	Unweathered	120 h	240 h
	Average	42.30	41.50	41.12	42.64	41.83	39.99	40.82	38.49	36.82	40.35	36.16	34.67	34.60	32.01	28.71
	Change (%)		−1.60	−2.80	0.80	−1.90	−6.20	−3.50	−5.70	−9.80	−4.60	−10.40	−14.20	−18.20	−7.50	−16.10

4 Conclusions

A small difference in hardness values after weathering was observed between untreated (natural) and heat-treated wood for both species. The mass loss values increased with higher heat treatment temperatures and longer durations. The mass loss is occurred after the weathering process too.

Regarding the hardness values for heat-treated wood the values slightly increased at lower heating temperatures and durations, and decreased with higher temperatures and durations.

During the weathering process, it was observed that the hardness values for both species decreased compared to the unweathered samples.

References

1. Allegretti, O., Brunetti, M., Cuccui, I., Ferrari, S., Novetti, M., Terziev, N.: Thermo-vacuum modification of spruce (Picea Abies Karst.) and fir (Abies Alba Mill.) wood. Bioresources. **7**(3), 3656–3669 (2012)
2. Ates, S., Akyildiz, M.H., Ozdemir, H.: Effects of heat treatment on calabrian pine (Pinus brutia ten.) wood. Bioresources. **4**(3), 1032–1043 (2009)
3. ASTM G154-06: Standard Practice for Operating Fluorescent Light Apparatus for UV Exposure of Nonmetallic Materials. ASTM International, West Conshohocken, PA (2006)
4. Baysal, E.: Surface characteristics of CCA treated Scots pine after accelerated weathering. Wood Res. **57**(3), 375–382 (2012)
5. Bekhta, P., Niemz, P.: Effect of high temperature on the change in color, dimensional stability and mechanical properties of Spruce wood. Holzforschung. **57**, 539–546 (2003). https://doi.org/10.1515/HF.2003.080
6. Cheng, L., Di, Y., Zhao, P., Dai, J., Yang, Z., Chang, Y.: Effects of accelerated weathering test on properties of Larch wood. Bioresources. **16**(4), 7400–7415 (2021)
7. Denes, A.R., Young, A.R.: Reduction of weathering degradation of wood through plasma-polymer coating. Holzforschung. **53**(6), 632–640 (1999)
8. Esteves, B.M., Pereira, H.M.: Wood modification by heat treatment: a review. Bioresources. **4**, 370–404 (2009)
9. Evans, P.D.: Review of the weathering and photo stability of modified wood. Wood Mater. Sci. Eng. **4**(1–2), 2–13 (2009)
10. Godinho, D., Araujo, S., Quilho, T., Diamantion, T., Gominho, J.: Thermally modified wood exposed to different weathering conditions: a review. Forests. **12**, 1400 (2021). https://doi.org/10.3390/f12101400
11. Goli, G., Negro, F., Emmerich, L., Militz, H.: Thermal and chemical modification of wood – a combined approach for exclusive, high-demanding performance products. Wood Mater. Sci. Eng. **18**(1), 58–66 (2023)
12. Hill, C.A.S.: Wood Modification: Chemical, Thermal and Other Processes. John Wiley and Sons, Chichester. ISBN 9780470021729 (2006)
13. Jankowska, A., Kozakiewicz, P.: Evaluation of wood resistance to artificial weathering factors using compressive properties. Drvnaindustrija [Internet]. **67**(1), 3–8 (2016) Available from: https://hrcak.srce.hr/154657
14. Jankowska, A., Kozakiewicz, P.: Comparison of outdoor and artificial weathering using compressive properties. Wood Res. **59**(2), 245–252 (2014)
15. Karamanoglu, M., Mehmet, A.H.: Colour, gloss and hardness properties of heattreated wood exposed to accelerated weathering. Pro Ligno. **9**(4), 729738 (2013)

16. Kocaefe, D., Poncsak, S., Boluk, Y.: Effect of thermal treatment on the chemical composition and mechanical properties of birch and aspen. Bioresources. **3**(2), 517537 (2008)
17. Militz, H.: Preface of the second European Conference on Wood Modification, in Proceedings for the 2nd European conference on wood modification, October 672005, Gottingen, Germany (2005)
18. Rowell, R.M.: Handbook of Wood Chemistry and Wood Composites. Taylor & Francis, Madison, WI (2005) ISBN 0849315883
19. Williams, R.: S "weathering of wood". In: Rowell, R.M. (ed.) Handbook of Wood Chemistry and Wood Composites, pp. 139–185. CRC Press, Boca Raton, FL (2005)
20. Tolvaj, L., Tsuchikawa, S., Inagaki, T., Varga, D.: Combined effects of UV light and elevated temperatures on wood discoloration. Wood Sci. Technol. **49**, 1225–1237 (2015)

Effect of Thermal Treatment on the Interaction of Wood with Liquid Water

Dace Cirule(✉), Edgars Kuka, Ingeborga Andersone, and Bruno Andersons

Latvian State Institute of Wood Chemistry, Dzerbenes 27, Riga, Latvia
dace.cirule@kki.lv

Abstract. The changes in the interaction of wood with liquid water caused by thermal modification (TM), which is a less studied effect of TM, was investigated. Birch (*Betula* spp.) wood and pine (*Pinus sylvestris* L.) sapwood were thermally modified and their properties compared with unmodified counterparts. The wettability was assessed by sessile drop contact angle measurements. The water permeability was measured by capillary absorption tests through the radial and tangential surfaces and indirectly evaluated from water distribution in cross-section of boards exposed on weathering rack to an intensive rain episode. The liability to rapid drying was evaluated by comparing the amount of the retained water in impregnated specimens after definite time period of drying in controlled environment. For both woods, TM caused not only a reduction in wettability, but also a reduction in drying rate, which could result in prolonged periods of high wetness during service. Lower liquid water permeability was observed for TM birch through both lateral surfaces which was supported by the results of water distribution in boards exposed to rain. Significant increase in water absorption was detected for TM pine through tangential surface in both tests used for evaluating water permeability of wood.

Keywords: Drying · Interaction with liquid water · Permeability · Thermally modified wood · Wettability

1 Introduction

Several thermal modification (TM) processes of wood differing in the treatment parameters have been developed with an aim to improve properties without application of chemicals. Intensive research has been carried out and a considerable number of results have been published in scientific and summarized in review articles on the changes in wood properties due to TM [e.g. 1–5]. However, among the extensively investigated properties of TM wood, the aspects related to changes in the interaction of wood with liquid water as a result of TM are less studied and the reported results are quite fragmentary. However, such knowledge is important for the proper use of materials in places where they may be exposed to liquid water [6]. Another aspect requiring the knowledge of these properties is the approach of enhancing the TM wood properties by combining TM and impregnation with biocides, which involves a leaching risk that can also be

G. Goli et al. (Eds.): ECWM 2024, SPM 86, pp. 44–51, 2025.
https://doi.org/10.1007/978-3-031-99418-0_6

influenced by water permeability of wood among other factors [7, 8]. The objective of the study was to evaluate the influence of TM on wood interaction with liquid water including wetting, absorption, and desorption.

2 Materials and Methods

Boards of birch *(Betula* spp.) wood and Scots pine (*Pinus sylvestris* L.) sapwood, which is of more interest for TM than heartwood due to its lower bioresistance, were used in the study. Boards measuring 25 × 100 × 700 mm were thermally modified in a water vapor environment under elevated pressure (0.6–0.8 MPa depending on the TM temperature) in a multifunctional pilot scale autoclave produced by Wood Treatment Technology (WTT, Denmark). Three TM peak temperatures were used for each species: 150, 160, 170 °C for birch and 160, 170, 180 °C for pine. The treatment time at the peak temperature was 1 h. The specimens of the TM and unmodified wood were conditioned (RH 65%; 20 °C) for at least 2 weeks both before and after preparing the test specimens in order to exclude the effect of the freshly mechanically processed wood surface, which is subjected to imminent aging processes that can change the interaction of the wood with water [9, 10]. When preparing the specimens to the required size, planning was used as the surface treatment method for all tests in the study.

The wettability of the surface was evaluated by sessile drop contact angle (CA) measurements (goniometer Data-physics OCA20, Germany) on the tangential surface of the specimens. CA of distilled water droplet of a volume of 10 μL was recorded during 30 s at a fixed interval of 1 s. Ten specimens were used for each TM temperature and CA measurements were performed at 10 locations on each specimen distancing at least 10 mm from each other. Constant wetting rate angle (CWRA) was calculated by differential method described by [11] and the penetration rate (Kvalue) was calculated for the CA values after reaching CWRA according to the equation:

$$\text{Kvalue} = \partial(\text{CA})/\partial t$$

Liquid water permeability was assessed by capillary water absorption tests through the radial (Rad) and tangential (Tang) surfaces. Cubic specimens (20 × 20 × 20 mm) with strict annual ring and grain orientations and water-tight sealed edges adjacent to the test surface were used for the tests. Modified floating tests were carried out with the specimens installed in frames that fix the test surface of the specimen 2 ± 0.2 mm under the water level throughout the experiment. The amount of the absorbed water during 24 h was used for calculation of water absorption (Aw). Ten replicates per each surface were used in the capillary water absorption tests.

Outdoor weathering tests were used for getting insights about the effect of TM wood on potential performance in a real service life. Unmodified and TM defect free flat sawn boards with sealed end grain surfaces were exposed on weathering racks faced south and inclined at an angle of 45° to the horizontal. After a prolonged rain episode, a 100 mm long section was cut from the center of the board and divided into four 5 mm thick layers aligned parallel to the exposed surface. Water content in each layer was assessed by the weight loss in drying the wood at 103 ± 2 °C. For this experiment, only TM boards modified at 160 °C for birch and 170 °C for pine were used.

The drying tests were carried out on specimens of two sizes measuring 20 × 20 × 100 mm and 20 × 100 × 350 mm with sealed end grain surfaces. The specimens were impregnated with water and exposed to drying in a controlled environment (RH 65%; 20 °C) with mass control after 180 h for the smallest specimens and 720 h (30 days) for the largest specimens.

Standard deviations were calculated for all results. A one-way analysis of variance (ANOVA) was used at a 5% level of significance to assess differences between the test groups. The test groups were compared separately for birch and pine.

3 Results and Discussion

3.1 Wettability

Different approaches have been used for evaluation of the wettability of wood surface because the phenomenon of water contacting wood is complex and involves several stages with permanent and dynamic changes in declining contact angle with time [12, 13]. In this study, the changes in surface wettability due to TM was evaluated according to the model proposed by Nusbaum [11] in which the wetting process is divided into spreading and penetration stages and the CWRA corresponds to the CA at which the penetration stage starts with an almost constant rate of change in the CA. Kvalue corresponds to the slope of the penetration stage of the changes in CA and is another characteristic of the wetting describing the penetration process. The results obtained from the CA measurements are presented in Fig. 1.

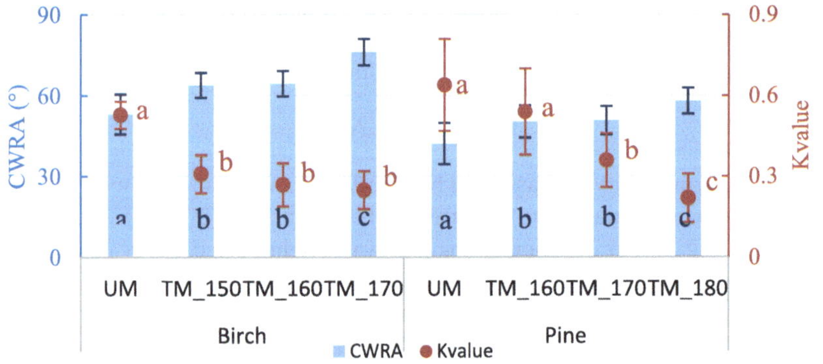

Fig. 1. Constant wetting rate angle (CWRA; primary axis) and penetration rate (Kvalue; secondary axis) of unmodified (UM) and thermally modified (TM) birch and pine wood. Different letters indicate on significant differences at 5% level; Differences were evaluated separately within each species and for each parameter.

The results of CWRA show that the surface wettability of both species decreased due to TM at all applied temperatures. However, since the threshold value of 90 °C was not reached, the TM wood included in this study had still surface of high wettability, even if it was more hydrophobic than that of the unmodified counterpart. No correlation was found between the changes in the CWRA and treatment temperature for the temperature ranges used in the study. Although the treatment at the highest used temperature resulted in the lowest wettability of the wood, which significantly differed ($p < 0.05$) from the effect of the other temperatures, no substantial difference was observed between the two lower treatment temperatures ($p > 0.05$). This trend was observed for both species. Similar results regarding the lack of dependence between TM intensity and changes in wettability have been also reported by others [14–16]. In some studies, only TM at higher temperatures had a significant effect on the change in wettability, while lower temperatures did not [10, 18].

Regarding the rate of CA decline, which in the Fig. 1 is described by Kvalue, the results show reduction in this parameter due to TM implying slower water penetration into wood at the initial contact. In general, decrease in the rate of CA decline with increase in TM temperature was observed for pine. However, no significant difference was observed in K-value between the pine treated at the lowest modification temperature (TM_150) and unmodified wood. For the birch, all applied temperatures considerably diminished the rate of the CA decline but the changes did not significantly differ depending on treatment temperature. Although changes in the rate of CA decline have been less evaluated for characterizing changes in wood surface wettability, reduction in this parameter due to TM has been reported [18] as well as can be seen from the graphs depicting CA vs. time [15].

3.2 Water Permeability

Decreased wettability caused by TM implies that liquid water penetration into the wood could be hindered due to the decrease in capillary forces. The results of the capillary water absorption tests are presented in Fig. 2.

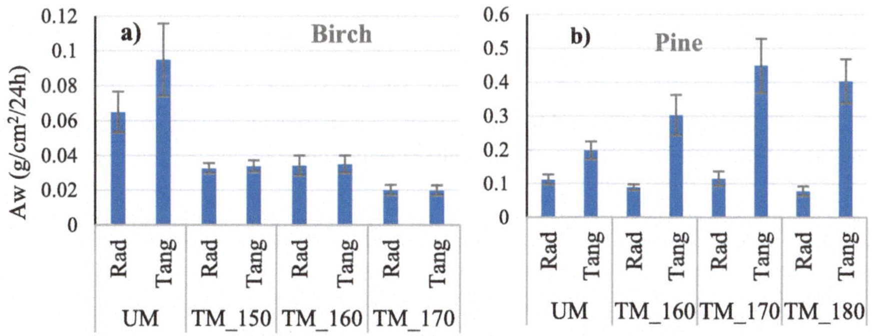

Fig. 2. Water absorption (Aw) through radial (Rad) and tangential (Tang) surfaces of unmodified (UM) and thermally modified (TM) birch (a) and pine (b) wood.

TM at all temperatures significantly but differently changed water permeability of both species. The higher water absorption through the Tang surfaces was similar for unmodified birch and pine, supporting the importance of the effect of the rays for the lateral penetration of liquid water into the wood [19–21]. The results show significantly reduced Aw for TM birch through both lateral surfaces (Tang and Rad) with more decrease through the Tang surface resulting in similar Aw values for water absorption through both surfaces. Similarly to the changes in CWRA (Fig. 1), the treatment at the highest used temperature resulted in the lowest Aw for the birch wood, while no substantial difference was observed between the two lower treatment temperatures. Consequently, no dependence on TM temperature was observed for changes in the water permeability of birch wood. Different trend can be seen for the changes in water permeability of TM pine. Regardless of the temperature, the TM did not cause significant changes in permeability through the Ra surface of pine. The results for the Tang surface show even a considerable increase in permeability, with the Aw values for TM_170 and TM_180 being more than twice as high as for unmodified pine. The difference in trends of changes in water permeability due to TM between wood species and lateral flow increase in *Pinus sylvestris* sapwood have been observed also in other studies [22–24]. Obviously, the changes in the wood structure of the pine due to TM and especially in the ray structure, which is the dominant pathway for water absorbed through Tang surface, was the determining factor in the Aw increase outweighing the effect of the chemical changes that imparted wood lower wettability. Different changes in wood structure have been found as a result of wood exposure to high temperature including degradation of pit apertures, disruption of pit membrane, formation of micro checks especially along the rays, destruction of tracheids [25–28]. However, slightly lower Aw observed for pine specimens TM_180 compared to TM_170, although difference was not significant ($p > 0.05$), may suggest that treatment at higher temperatures could reverse this outweighing effect of structural over chemical changes. This is supported by the results of other studies in which the increase in water absorption through Tang surface (in Rad direction) was followed by a decrease and even a lower Aw was reached than that observed for UM pine wood by increasing TM temperature above certain threshold temperature [22–24].

The opposite trend in the change in water absorption due to TM of birch and pine showed excellent agreement with the moisture distribution pattern in the cross-section of the boards exposed on outdoor weathering racks after prolonged rain event (Table 1). Since flat-sawn boards were used in this experiment, the Tang surfaces were exposed to rain and the water penetrated mainly radially into the wood.

Table 1. Moisture content (%) in cross-section of unmodified (UM) and thermally modified (TM) birch and pine boards depending on the distance from the surface.

Distance, mm	Birch_UM	Birch_TM	Pine_UM	Pine_TM
0–5	42.4 (6.0)	40.2 (6.9)	53.6 (4.8)	77.9 (10.7)
6–10	28.2 (1.8)	22.7 (5.7)	49.7 (8.0)	71.6 (10.4)
11–15	25.3 (1.1)	17.4 (1.7)	40.0 (5.2)	61.5 (9.3)
16–20	25.2 (0.7)	17.8 (1.1)	25.9 (1.0)	40.8 (8.8)

Standard deviations within parenthesis.

When comparing pine and birch, overall higher water uptake was found in all layers of both unmodified and TM pine. For birch, the water content in the top layer of unmodified and TM wood was similar, while a significant reduction in water content was observed for the deeper layers of TM birch compared to unmodified birch, which agrees with the lower Aw of the former (Fig. 2) and approves the positive effect of TM on liquid water uptake reduction for birch wood. Opposite effect of TM was observed for pine. Significantly more water in all layers of the TM pine boards compared to their unmodified counterparts indicates that TM facilitated the penetration of liquid water into pine wood through Tang surface.

3.3 Drying

Considering moisture dynamics of wood during its service life, easy drying is no less important than its indisposition to water absorption [6, 29]. The results of the absorbed water in impregnation and the retained water after conditioning are presented in Fig. 3.

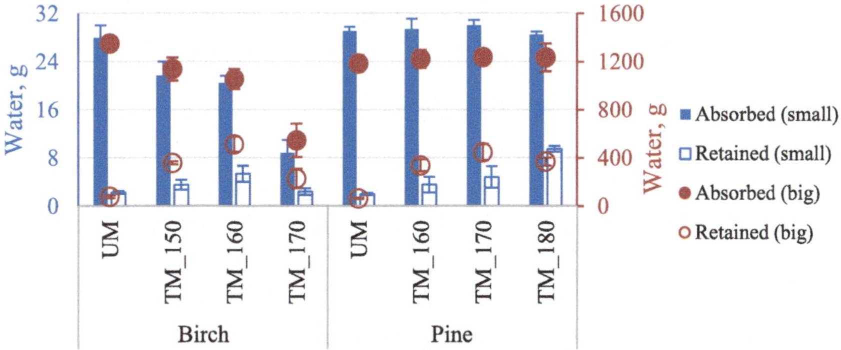

Fig. 3. Amount of water absorbed during impregnation (filled) and retained after conditioning (empty) in unmodified (UM) and thermally modified (TM) birch and pine wood: bars – small specimens (primary axes); dots – big specimens (secondary axes).

In both experiments, similar tendency of significantly delayed drying was observed for TM wood of both species compared to the unmodified wood with the drying rate slowing down with increased TM temperature. For birch, no significant difference in water content was observed between unmodified and TM_170 specimens at the end of drying experiments. However, this was because the slowed water release was balanced by lower water uptake in the TM_170 wood. TM of birch at the two other temperatures resulted in higher water content at the end of the tested drying periods. For pine, no significant difference was observed in the water uptake while significantly more water was retained in the TM specimens at the end of both drying experiments, clearly demonstrating extended desorption time for TM specimens. One of the reasons for the slowed drying of TM wood could be the clogging of the surface due to the increased amount of the water-soluble extractives that have been reported to migrate towards the surface during wood drying [30, 31]. These results indicate that exposure of unprotected TM wood of both species to liquid water can lead to a significantly prolonged state of increased moisture content due to the slowed release of water during the drying phase, which in the case of TM pine is coupled with increased wood permeability.

4 Conclusions

Influence of TM on wood interaction with liquid water was investigated for birch (*Betula* spp.) wood and pine (*Pinus sylvestris* L.) sapwood. Surface wettability, characterized by CWRA, and the penetration rate of liquid water into the wood at the initial contact, characterized by Kvalue, of birch and pine decreased due to TM at all applied temperatures, but no dependence on the temperature was observed. TM of birch wood resulted in significantly reduced capillary water absorption, which also ensured that less water was present in the board volume under the top layer for TM boards exposed to rain compared to their unmodified counterparts. Opposite results were observed for TM pine with increased both water absorption in capillary absorption tests and amount of water in boards after rain episode. Regarding water release, similar tendency of significantly delayed drying was observed for TM wood of both species compared to the unmodified wood with the drying rate slowing down by increasing TM temperature.

References

1. Esteves, B.M., Pereira, H.M.: Wood modification by heat treatment: a review. Bioresources. **4**, 370–404 (2009)
2. Sandberg, D., Kutnar, A.: Thermally modified timber: recent developments in Europe and North America. Wood Fiber Sci. **48**, 28–39 (2015)
3. Willems, W., Altgen, M.: Hygrothermolytic wood modification. Process description and treatment level characterisation. Wood Mater. Sci. Eng. **15**, 213–222 (2020)
4. Hill, C., Altgen, M., Rautkari, L.: Thermal modification of wood - a review: chemical changes and hygroscopicity. J. Mater. Sci. **56**, 6581–6614 (2021)
5. Zelinka, S.L., Altgen, M., Emmerich, L., Guigo, N., Keplinger, T., Kymäläinen, M., Thybring, E.E., Thygesen, L.G.: Review of wood modification and wood functionalization technologies. Forests. **13**, 1–46 (2022)
6. Van Acker, J., Van den Bulcke, J., De Windt, I., Colpaert, S., Li, W.: Moisture dynamics of modified wood and the relevance towards decay resistance. In: 8th European Conference on Wood Modification, pp. 44–55, Helsinki (2015)
7. Andersone, I., Cirule, D., Kurnosova, N., Andersons, B., Kuka, E.: Service properties of pine after combining an impregnation treatment with a thermo-hydro modification process. Wood Mater. Sci. Eng. **18**, 11–18 (2023)
8. Kuka, E., Cirule, D., Andersone, I., Andersons, B., Kurnosova, N., Verovkins, A., Puke, M.: Environmental performance of combined treated wood. Wood Mater. Sci. Eng. **18**, 88–96 (2023)
9. Gindl, M., Reiterer, A., Sinn, G., Stanzl-Tschegg, S.E.: Effects of surface ageing on wettability, surface chemistry, and adhesion of wood. Holz Roh Werkst. **62**, 273–280 (2004)
10. Altgen, M., Militz, H.: Thermally modified scots pine and Norway spruce wood as substrate for coating systems. J. Coatings Technol. Res. **14**, 531–541 (2017)
11. Nussbau, R.M.: Natural surface inactivation of scots pine and Norway spruce evaluated by contact angle measurements. Holz Roh Werkst. **57**, 419–424 (1999)

12. Liptáková, B.E., Kúdela, J.: Analysis of the wood-wetting process. Holzforschung. **48**, 139–144 (1994)
13. Qin, Z., Zhang, Q., Gao, Q., Zhang, S., Li, J.: Wettability of sanded and aged fast-growing poplar wood surfaces: II. Dynamic wetting models. Bioresources. **9**, 7176–7188 (2014)
14. Esteves, B., Marques, A.V., Domingos, I., Pereira, H.: Influence of steam heating on the properties of pine (Pinus pinaster) and eucalypt (Eucalyptus globulus) wood. Wood Sci. Technol. **41**, 193–207 (2007)
15. Kocaefe, D., Poncsak, S., Doré, G., Younsi, R.: Effect of heat treatment on the wettability of white ash and soft maple by water. Holz Roh Werkst. **66**, 355–361 (2008)
16. Humar, M., Kržišnik, D., Lesar, B., Thaler, N., Ugovšek, A., Zupančič, K., Žlahtič, M.: Thermal modification of wax-impregnated wood to enhance its physical, mechanical, and biological properties. Holzforschung. **71**, 57–64 (2017)
17. Metsä-Kortelainen, S., Viitanen, H.: Wettability of sapwood and heartwood of thermally modified Norway spruce and scots pine. Eur. J. Wood Wood Prod. **70**, 135–139 (2012)
18. Huang, X., Kocaefe, D., Boluk, Y., Kocaefe, Y., Pichette, A.: Effect of surface preparation on the wettability of heat-treated jack pine wood surface by different liquids. Eur. J. Wood Wood Prod. **70**, 711–717 (2012)
19. Murmanis, L., Chudnoff, M.: Lateral flow in beech and birch as revealed by the electron microscope. Wood Sci. Technol. **13**, 79–87 (1979)
20. Booker, R.E.: Changes in transverse wood permeability during the drying of Dacrydium cupressinum and Pinus radiata. New Zeal. J. For. Sci. **20**, 231–244 (1990)
21. Pfriem, A.: Alteration of water absorption coefficient of spruce (Picea abies (L.) Karst) due to thermal modification. Drv. Ind. **62**, 311–313 (2011)
22. Metsä-Kortelainen, S., Antikainen, T., Viitaniemi, P.: The water absorption of sapwood and heartwood of scots pine and Norway spruce heat-treated at 170°C, 190°C, 210°C and 230°C. Holz Roh Werkst. **64**, 192–197 (2006)
23. Johansson, D., Sehlstedt-Persson, M., Morėn, T.: Effect of heat treatment on capillary water absorption of heat-treated pine, spruce and birch. In: Kurjatko, S., Kúdela, J., Lagaňa, R. (eds.) Proceedings of the 5th IUFRO Symposium on Wood Structure and Properties '06, pp. 251–255. Arbora Publishers, Zvolen (2006)
24. Scheiding, W., Direske, M., Zauer, M.: Water absorption of untreated and thermally modified sapwood and heartwood of Pinus sylvestris L. Eur. J. Wood Wood Prod. **74**, 585–589 (2016)
25. Terziev, N., Daniel, G.: Industrial kiln drying and its effect on microstructure, impregnation and properties of scots pine timber impregnated for above ground use. Part 2. Effect of drying on microstructure and some mechanical properties of scots pine wood. Holzforschung. **56**, 434–439 (2002)
26. Awoyemi, L., Jones, I.P.: Anatomical explanations for the changes in properties of western red cedar (Thuja plicata) wood during heat treatment. Wood Sci. Technol. **45**, 261–267 (2011)
27. Ahmed, S.A., Hansson, L., Morén, T.: Distribution of preservatives in thermally modified scots pine and Norway spruce sapwood. Wood Sci. Technol. **47**, 499–513 (2013)
28. Altgen, M., Adamopoulos, S., Militz, H.: Wood defects during industrial-scale production of thermally modified Norway spruce and scots pine. Wood Mater. Sci. Eng. **12**, 14–23 (2017)
29. Rapp, A.O., Peek, R.D., Sailer, M.: Modelling the moisture induced risk of decay for treated and untreated wood above ground. Holzforschung. **54**, 111–118 (2000)
30. Terziev, N.: Migration of low-molecular sugars and nitrogenous compounds in Pinus sylvestris L. during kiln and air drying. Holzforschung. **49**, 565–574 (1995)
31. Cirule, D., Verovkins, A., Andersone, I., Kuka, E., Andersons, B.: Thermally modified birch wood interaction with liquids. Eur. J. Wood Wood Prod. **78**, 849–857 (2020)

Assessment of the Effect of Thermal Modification on the Colorimetric Properties of Three Amazonian Species

Kamilly da Silva Pereira[1](✉), Anna Clara Oliveira Rupf[1], Paulo Henrique dos Santos Silvares[1], Saulo José da Costa Lima[1], Djeison Cesar Batista[1], and Victor Fassina Brocco[2]

[1] Department of Forest and Wood Sciences, Federal University of Espírito Santo, Jeronimo Monteiro, Brazil
kamillysilv4p@gmail.com

[2] Center for Higher Studies of Itacoatiara, University of the State of Amazonas, Itacoatiara, Brazil

Abstract. Thermal modification is a process that aims to improve some properties of wood, such as biological resistance, dimensional stability and color change, enhancing the value of the product. Additionally, it is an eco-friendly process that does not use toxic chemicals. The main objective of this work was to evaluate the effect of thermal modification on the wood color of two species, jatobá (*Hymenaea* spp.) and caxinguba (*Ficus* sp.). The secondary objective was to compare two systems of color measurement. The wood samples were obtained from sawmills in the municipality of Itacoatiara, Amazonas State, Brazil. Thermal modification was carried out in a laboratory oven with forced air circulation, at 180 °C for 150 min. We calculated the mass loss and measured the wood color before and after thermal modification using the Munsell and the CIE-L*a*b* systems. There was a significant darkening of all the wood samples tested. The use of the Munsell and the CIE-L*a*b* systems returned different colors for all species. The CIE-L*a*b* was more effective in detecting the total color variation of *Ficus* sp.

Keywords: Caxinguba · CIE-L*a*b* · Colorimetry · Jatobá · Munsell Color Chart

1 Introduction

Among the important factors affecting the marketing of tropical hardwood for uses such furniture making are its aesthetic attributes, conferred by organoleptic properties such as color, luster, texture, and figure pattern. Silva et al. (2017) observed that Amazonian wood presented a broad spectrum of variation of basic colorimetric parameters. The genus *Hymenaea* has 16 species distributed from Mexico to South America, 13 of which are found in Brazil and known as "jatobá" (Lewis 2005). These species have high density and good mechanical properties, medium to high natural durability, and good dimensional stability (Klitzke et al. 2008). Jatobá wood is described as being difficult to machine.

G. Goli et al. (Eds.): ECWM 2024, SPM 86, pp. 52–58, 2025.
https://doi.org/10.1007/978-3-031-99418-0_7

It has thick yellowish-white sapwood that is easily distinguished from the heartwood (Lorenzi 2008). It has both internal and external uses, such as flooring, moldings, fencing, furniture and railroad sleepers (Mainieri and Chimelo 1989).

The genus *Ficus* (Moraceae) occurs in Neotropical regions (Pereira 2015), and one of its species is known in Brazil as caxinguba (Rios and Pastore Junior 2011). It is a large tree, found mainly in floodplains (Lima et al. 1995). Correa and Bernal (1995) described some of its wood characteristics: whitish color, with indistinct sapwood, thick texture, and straight grain; medium density and mechanical resistance, easy machining, and low natural durability. It is used for packaging, firewood, door panel cores, and veneers.

Thermal modification darkens wood and enlarges its possible uses (Farias and Silveira 2014). Heat treatment can, therefore, improve the qualities of commercial wood, producing an alternative product for use in external environments. According to Carvalho et al. (2019), this process is spreading due to its ease of application, low cost, and low environmental impact. Because of the variable improvement of some wood parameters, thermal modification is attracting research aimed at making better commercial use of wood products. Regarding tropical wood, thermal modification is appropriate for species with low natural durability and large sapwood volume. Thermal treatment makes it less susceptible to biological agents and expands its range of uses, as is the case of caxinguba and jatobá sapwood.

The main objective of this work was to evaluate the effect of thermal modification on the color of jatobá (*Hymenaea* spp.) and caxinguba (*Ficus* sp.) woods. The secondary objective was to compare two systems of color measurement.

2 Materials and Methods

2.1 Material Collection and Identification

Three wood species from the Brazilian Amazon region were used, two from the genus *Hymenaea* (jatobá) – (mixture) with a basic density of 0.82 g/cm^3 and one from the genus *Ficus* (caxinguba) – (softwood) with a basic density of 0.65 g/cm^3. The wood samples were obtained from sawmills in the municipality of Itacoatiara, Amazonas State, Brazil. The identification of the species was carried out based on macroscopic characteristics following the instructions of the Pan American Commission of Technical Standards (1972). Macrophotographs were taken from ring-oriented samples (30 $\times$ 30 $\times$ 30 mm) using a magnifying lens (24$\times$) coupled to a smartphone camera (12-megapixel, aperture of f/1:8) as well as with a stereomicroscope with a photographic system and specific software for image analysis (Figs. 1, 2 and 3).

2.2 Thermal Modification

For thermal modification, five samples were employed per species, measuring 50 $\times$ 100 $\times$ 120 mm (radial $\times$ tangential $\times$ longitudinal), which were manually sanded (grits of 60 and 100). After that, they were thermally-modified in an open system in a laboratory oven with forced air circulation using a three-step schedule: (i) drying at 103 $\pm$ 2 °C for 24 h, followed by weighing (nearest 0.01 g) (M1); (ii) heating to 180 °C; (iii) maintenance of the target temperature for 150 min, followed by a last weighing (M2). The mass loss (ML) was calculated with the following formula: ML (%) = [(M1 – M2)/M1] $\times$ 100.

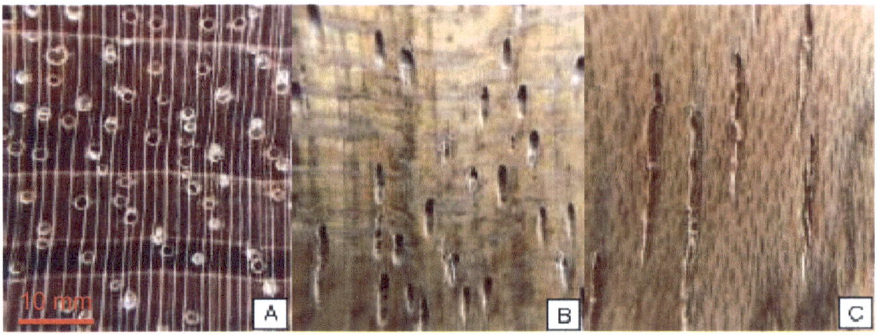

Fig. 1. – Macrophotographs of *Hymenaea* sp.1: transversal (a), radial (b), and tangential (c) sections.

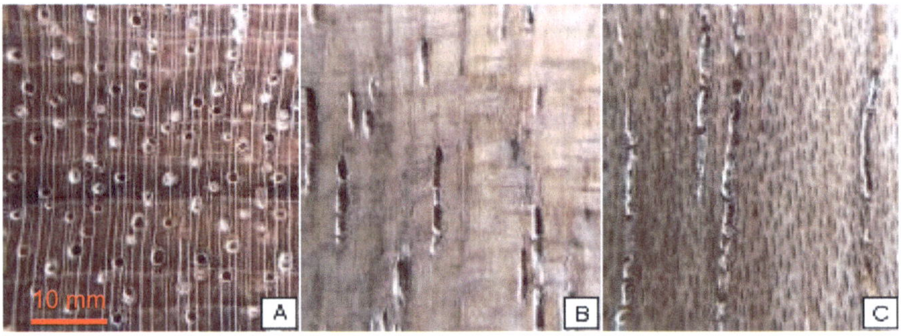

Fig. 2. – Macrophotographs of *Hymenaea* sp.2: transversal (a), radial (b), and tangential (c) sections.

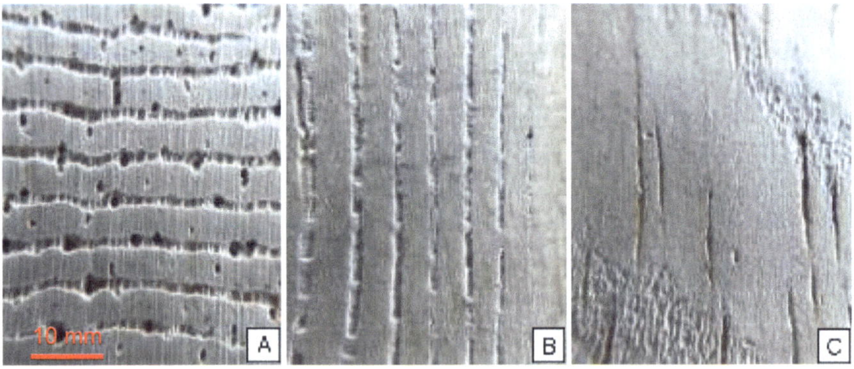

Fig. 3. – Macrophotographs of *Ficus* sp.: transversal (a), radial (b), and tangential (c) sections.

2.3 Colorimetric Characterization

The wood color was determined by three evaluators using the Munsell Soil Color Charts (Munsell Color 2000), based on the color attributes hue (basic color), chroma (color intensity), and value (lightness). The measurements were taken on the tangential surfaces of the samples before and after thermal modification. The measurements before thermal modification represented the control of every species.

These Munsell system data were transformed into CIE-L*a*b* parameters according to the conversion table proposed by Vodyanitskii and Kirillova (2016), resulting in the colorimetric parameters lightness (L*), red-green coordinate (a*), and yellow-blue coordinate (b*). The color variations before and after thermal modification were calculated for the L*, a*, and b* parameters, as well as the total color variation (ΔE, Eq. 1). ΔE was classified according to Hikita et al. (2001).

$$\Delta \mathrm{E} = \sqrt{\Delta L^2 + \Delta \mathrm{a}^2 + \Delta b^2} \tag{1}$$

Where: ΔE: total color variation; ΔL^2: variation of lightness; Δa^2: variation of the colorimetric parameter a* (red-green coordinate); and Δb^2: variation of the colorimetric parameter b* (yellow-blue coordinate).

Table 1. Classification of total color variation (ΔE).

Color variation (ΔE)	Classification
0.0–0.5	Negligible
0.5–1.5	Slightly noticeable
1.5–3.0	Remarkable
3.0–6.0	Appreciable
6.0–12.0	Very appreciable

Source: Hikita et al. (2001)

3 Results and Discussion

The mass losses were low for *Hymenaea* sp. (1) (0.77%), *Hymenaea* sp. (2) (0.72%) and *Ficus* sp. (0.23%), which was expected due to the low thermal treatment temperature. These results were similar to those reported by Ronsoni (2015), who studied the thermal modification of the tropical species *Trattinnickia rhoifolia* (0.70%) and *Apuleia leiocarpa* (0.50%). According to the literature, at 180 °C the main degraded chemical constituents are hemicelluloses. This occurs alongside the volatilization of some extractives (Czajkowski et al. 2020).

The results using the Munsell system are presented in Table 1, while samples before (Control) and after thermal modification can be seen in Fig. 4. During the stages of collection and identification, we noticed two different species of *Hymenaea*, although it was not possible to reach the species level. Possibilities were, for example, *H. courbaril, H. stigonocarpa* or *H. martiana.* The color classification reinforced the previous identification, where the *Hymenaea* species had different colors (reddish- yellow and reddish-brown). Silva et al. (2017) classified *Hymenaea courbaril* wood as reddish-brown and this was the same classification as our *Hymenaea* sp. 2.

Table 2. Wood colors according to the Munsell system.

Species	Treatment	Hue	Value	Chroma	Color
Hymenaea sp. 1	Control	7.5YR	6	6	Reddish-yellow
	180 °C	5YR	3	2	Dark reddish-brown
Hymenaea sp. 2	Control	5YR	5	4	Reddish-brown
	180 °C	2.5YR	3	1	Dark reddish-gray
Ficus sp.	Control	2.5Y	8	2	Light grayish-yellow
	180 °C	2.5Y	7	4	Light grayish-yellow

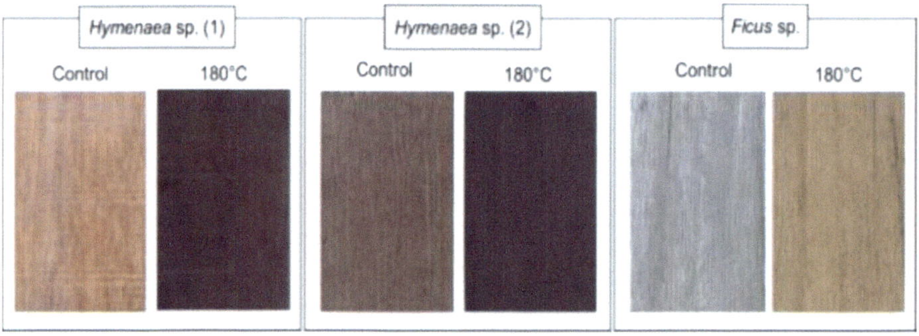

Fig. 4. Color changes in thermally modified wood.

For *Ficus* sp., the color assigned was "light grayish-yellow" and it was the only species in which the thermal modification had little effect on color. Pereira (2015) classified *Ficus* spp. wood as grayish-white.

The results using the CIE-L*a*b* system are presented in Table 2. After thermal modification, the L* parameter was reduced for all species, meaning the wood darkened. The highest reduction was of *Hymenaea* sp. 1 while the opposite was observed for *Ficus* sp.

Table 3. Wood colors according to the CIE-L*a*b* system.

Species	Treatment	L[a]	ΔL[a]	a[a]	Δa[a]	b[a]	Δb[a]	ΔE[a]	Color[a]
Hymenaea sp. 1	Control	61.7	–	12.5	–	34.4	–	–	Orange-yellow
	180 °C	30.8	−30.9	6.9	−5.6	10.1	−24.3	39.71	Purplish-brown
Hymenaea sp. 2	Control	51.6	–	11.2	–	20.1	–	–	Reddish-brown
	180 °C	30.8	−20.8	4.4	−6.8	4.3	−15.8	26.99	Reddish-black
Ficus sp.	Control	81.4	–	−0.4	–	15	–	–	Grayish-white
	180 °C	71.6	−9.8	1.2	1.6	27.8	12.8	16.20	Light-olive

[a] Camargos and Gonçalez (2001)

As discussed previously regarding the results using the Munsell system, the *Hymenaea* species also were classified with different colors according to the CIE-L*a*b* system. However, the different systems returned different colors. For example, *Hymenaea* sp. 1 was classified as reddish-yellow and orange-yellow in the different systems.

A different pattern was noticed for the coordinates a* and b* among the species. For the *Hymenaea* species, the variation of these coordinates was negative, indicating that the color shifted from red to green and yellow to blue, respectively for a* and b*. The opposite occurred for *Ficus* sp., where these variations were positive.

The total color variation was classified as "very appreciable" (Table 3) for all species, indicating that thermal treatment significantly changed the wood color. Comparing both color systems, the ΔE based on the CIE-L*a*b* system was more effective to detect the color change of *Ficus* sp. (Fig. 4) than the Munsell system. Table 1 shows that the hue did not change after thermal modification (2.5Y) and the value changed slightly (from 8 to 7), indicating slight darkening. The chroma increased slightly (from 2 to 4), meaning the color got more intense or saturated.

We highlight that these color changings were achieved even with low mass losses (less than 1%). This is positive, because it means low degradation of the cell wall constituents and maintenance of the original mechanical resistance. Changes in the natural color of wood imparted by thermal modification can result in a "new product" that can be recommended to replace other wood species that are difficult and/or expensive to obtain (Griebeler 2013).

4 Conclusion

Even at low thermal modification temperature (180 °C), there was a significant darkening of all the wood samples tested. Greater changes in color were observed for *Hymenaea* species than for *Ficus* sp.

The use of the Munsell system and the CIE-L*a*b* system returned different colors for all species. The CIE-L*a*b* system was more effective to detect the total color variation of *Ficus* sp. than the Munsell system.

References

Camargos, J.A.A., Gonçalez, J.C.: A colorimetria aplicada como instrumento na elaboração de uma tabela de cores de madeira. Brasil Florestal. **71**, 30–41 (2001)

Carvalho, D.E., Juizo, C.G.F., França, M.C., Loiola, P.L., da Rocha, M.P.: Effect of thermal modification in the natural resistance of *Eucalyptus grandis* and *Pinus taeda* woods. Revista Brasileira de Ciências Agrarias. **14**(1), 1–7 (2019)

Comissão Panamericana de Normas Técnicas: COPANT- 459: métodos de ensayos físicos y mecânicos. [S.l] (1972)

Correa, J.E.Q., Bernal, H.Y.M.: Especies vegetales promisorias de los países del Convenio Andrés Bello. Bogotá: SECAB, p. 515. Tomo 11, letra M. (PREVECAB. Serie Ciencia y Tecnologia, 54) (1995)

Czajkowski, Ł., Olek, W., Weres, J.: Effects of heat treatment on thermal properties of European beech wood. Eur. J. Wood Wood Prod. **78**, 301–312 (2020)

Griebeler, C.G.O.: Colorimetria da madeira de *Eucalyptus grandis* W. Hill ex Maiden modificada termicamente. In: 66 f. Dissertação (Mestrado em Engenharia Florestal) – Setor de Ciências Agrárias. Universidade Federal do Paraná, Curitiba (2013)

Hikita, Y., Toyoda, T., Azuma, M.: Weathering testing of timber: discoloration. In: Imamura, Y. (ed.) High Performance Utilization of Wood for Outdooor Uses. Press-Net, Kyoto (2001)

Klitzke, R.J., et al.: Caracterização dos lenhos de cerne, alburno e transição de jatobá (*Hymenaea* sp.) visando ao agrupamento para fins de secagem convencional. Scientia Forestalis. **36**(80), 279–284 (2008)

Lewis, G.P.: Tribe Cassieae. In: Lewis, G.P., Schrire, B., Mackinder, B., Lock, M. (eds.) Legumes of the World, pp. 111–125. Kew Publishing, Londres (2005)

Lorenzi, H.: Árvores brasileiras: manual de identificação e cultivo de plantas arbóreas nativas do Brasil, vol. 1, 5th edn, p. 368. Instituto Plantarum, Nova Odessa (2008)

Mainieri, C., Chimelo, J.P.: Fichas características das madeiras do Brasil, p. 418. IPT, São Paulo (1989)

Munsell Color: Munsell Soil Color Charts. Koelmorgen, Baltimore, MD (2000)

Pereira, K.D.N.: Qualidade da superfície da lâmina de madeira de figueira (*Ficus* spp.) submetida a tratamentos de acabamentos. In: Monografia (Engenharia Florestal). Universidade de Brasília, Brasília – DF (2015)

Rios, M.N.D.S., Pastore Junior, F.: Plantas da Amazônia: 450 espécies de uso geral, p. 3140. Universidade de Brasília, Biblioteca Central, Brasília (2011)

Ronsoni, T.: Influência do tratamento térmico nas propriedades tecnológicas da madeira de três espécies amazônicas. Trabalho de Conclusão de Curso (Graduação em Engenharia Florestal) – Universidade Federal de Mato Grosso, Instituto de Ciências Agrárias e Ambientais, Sinop (2015)

Silva, R.A.F., Setter, C., Mazette, S.S., Melo, R.R., Stangerlin, D.M.: Colorimetria da madeira de trinta espécies tropicais. Ciência da Madeira. **8**(1), 36–41 (2017)

Vodyanitskii, Y.N., Kirillova, N.P.: Conversion of Munsell color coordinates to the Cie- L* a* b* system: tables and calculation examples. Moscow Univ. Soil Sci. Bull. **71**(4), 139–146 (2016)

Zanuncio, A.J.V., Farias, E.D.S., Da Silveira, T.A.: Tratamento térmico e colorimetria da madeira de *Eucalyptus grandis*. Floresta e Ambiente. **21**(1), 85–90 (2014)

Effect of Paraffin-Thermal Modification on Water Absorption and Dimensional Stability of Louro-Preto Wood (*Nectandra dioica* Mez.)

Saulo José da Costa Lima[1](✉), Anna Clara Oliveira Rupf[1], Kamilly da Silva Pereira[1], Paulo Henrique dos Santos Silvares[1], Djeison Cesar Batista[1], Victor Hugo Pereira Moutinho[2], and Fernando Wallase Carvalho Andrade[2]

[1] Graduate Program in Forest Sciences, Center for Agrarian Sciences and Engineering, Federal University of Espírito Santo, Jerônimo Monteiro, Espírito Santo, Brazil
saulo.j.lima@edu.ufes.br

[2] Institute of Biodiversity and Forests, Federal University of Western Pará, Santarém, Brazil

Abstract. It is important the spread of thermal modification as a process that aims to improve undesirable properties of woods with low commercial value. Better results can be obtained by combining processes and adding hydrophobic agents such as oils and waxes. The objective of this study was to evaluate the effect of paraffin-thermal modification on the hygroscopicity and dimensional stability of *Nectandra dioica* Mez wood. The combined treatment was applied at two temperatures (190 °C and 210 °C) for 150 min. The samples were subjected to three immersion-drying cycles to evaluate the water absorption (WA) and the anti-swelling efficiency (ASE). Dimensional stability increased proportionally to the temperature intensity, increasing with the addition of paraffin. The opposite behavior was observed for WA, recording results of only 17.03% in the most intense treatment (210 °C + paraffin). In general, the best results were for paraffin-thermally modified wood at 210 °C. Thermal modification with paraffin is recommended, as it will result in some advantages in terms of dimensional stability and water repellency, making it an option for wood in various applications.

Keywords: Hygroscopicity · Swelling · Thermally modified wood · Tropical wood · Water-wood relations

1 Introduction

Wood is a porous and structurally capillary material, which directly influences its performance in environments with considerable amounts of humidity. The chemical constitution of wood is composed of cellulose, hemicelluloses, and lignin, which are polymers responsible for its hygroscopic characteristics. Among these structural components, hemicelluloses are considered the most hydrophilic, having a major influence on water sorption and, consequently, on the dimensional changes of wood (Willems et al. 2015).

G. Goli et al. (Eds.): ECWM 2024, SPM 86, pp. 59–66, 2025.
https://doi.org/10.1007/978-3-031-99418-0_8

Depending on the magnitude, dimensional changes can compromise the structural integrity of the material, due to the occurrence of warping and cracks (Borges and Quirino 2004). The application of sustainable methods or technologies aimed at improving these characteristics can promote the appreciation of wood and a reduced need for maintenance, reducing the cost of wooden structures and products.

Thermal modification is an alternative for minimizing problems related to wood, which is commonly carried out between 160 °C and 220 °C, under different environments such as water vapor, nitrogen, vacuum, or plant oils (Boonstra and Tjeerdsma 2006; Fengel and Wegener 1989). It is an environmentally friendly process that has been developed and improved for a long time to improve the durability, hygroscopicity, and physical-mechanical properties of wood, which are the result of changes in its chemical composition (Modes et al. 2013; Conte et al. 2014; Przybysz et al. 2014; Wentzel et al. 2019; Pratiwi et al. 2019; Altgen et al. 2020; Brito et al. 2023).

In addition to thermal modification, wax impregnation is an option for protecting wood against moisture, due to its great hydrophobicity. When treated in this way, wood has its water sorption kinetics reduced and, consequently, its dimensional stability is improved by the deposition of wax in the cell lumens and in the S3 layer of the cell wall (Scholz et al. 2009; Scholz et al. 2010; Esteves et al. 2014). The use of thermal modification with paraffin impregnation can result in a synergistic effect, providing a more efficient product, mainly on wood species that are more susceptible to water absorption.

Louro-preto (*Nectandra dioica* Mez, Lauraceae, Fig. 1) is a tree species with phytogeographical distribution in the Amazon Region. Its wood has medium mechanical resistance, and a mean basic density of 0.54 g cm^{-3} and, based on wettability analyses (contact angle), it was classified as a wood with rapid water absorption (Cardoso et al. 2012; Amorim et al. 2013). The objective of this study was to evaluate the effect of paraffin-thermal modification on the hygroscopicity and dimensional stability of louro-preto wood.

Fig. 1. Tangential section of *Nectandra dioica* Mez wood (Luchtemberg 2013).

2 Materials and Methods

2.1 Material Characterization

The tests were carried out on samples of louro-preto wood from trees planted in an experimental stand located in the Western Region of the state of Pará, Brazil, coordinates 02°41′16, 1″S and 54°31′56, 0″W. Five samples measuring 20 × 20 × 20 mm^3 (tangential × radial × longitudinal) were used per treatment for hygroscopicity and anti-swelling efficiency (ASE) tests.

2.2 Thermal Modification and Treatments

The thermal modification of the wood was carried out according to adaptations of the method by Reinprecht and Repák (2019). The samples were thermally modified with and without paraffin (Table 1) in a laboratory electric oven (Fabbe, Model 070, São Paulo), at atmospheric pressure, via hot air irradiation. The thermally modified wood at the same temperature (with and without paraffin) was processed in the same batch.

Table 1. Treatments.

Treatment	Temperature (°C)	Time (min)	Paraffin
Untreated	–	–	–
190	190	150	No
190P	190	150	Yes
210	210	150	No
210P	210	150	Yes

The thermal modification schedule was divided into five stages: (i) the wood samples and the granulated paraffin were arranged in a stainless steel tray and exposed to a temperature of 100 °C for 1 h, to allow the paraffin to melt; (ii) the samples were immersed in liquid paraffin for 1 h at 100 °C; (iii) the temperature was raised until the target temperature (Table 1); (iv) maintenance of the target temperature for 150 min; (v) the temperature was reduced to 100 °C, and the samples were removed from the paraffin and placed on a steel screen for 1 h, allowing the unabsorbed paraffin to drain. Finally, the oven was turned off and cooled naturally, allowing the paraffin to solidify.

2.3 Hygroscopicity and Dimensional Stability

The effect of the process on the hygroscopicity and dimensional stability of wood was evaluated by determining the water absorption (WA) and anti-swelling efficiency (ASES), as adapted from the method by Andrade et al. (2023). The samples were immersion in distilled water at 20 °C for 24 h, obtaining their dimensions with a digital caliper (Mitutoyo, Model Absolute, Brazil) to the nearest 0.001 mm and mass with a digital scale (Shimadzu, Model AUX-320, Brazil) to the nearest 0.001 g. The material was dried in an oven at 40 °C for 24 h and at 103 ± 2 °C for an extra 24 h (to avoid the occurrence of cracks due to moisture content gradients), measured, and weighed again, according to the previous procedure, finishing the first immersion-drying cycle. In total, we carried out three cycles to evaluate the ASE (Eq. 1) and the WA (Eq. 2). Additionally, we calculated the anisotropy coefficient (Eq. 3).

$$ASE(\%) = \frac{Su - \text{Sm}i}{Su} \times 100 \quad (1)$$

ASE is anti-swelling efficiency (%); S_u is the mean swelling of the untreated wood (%, third cycle) and Sm*i* is the mean swelling of the thermally modified treatment *i* (%, third cycle).

$$WA(\%) = \frac{AA1 - AA3}{AA1} \times 100 \quad (2)$$

WA is the water absorption (%); AA1 and AA3 are the mean water absorptions in the first and third cycles, respectively.

$$\text{CA} = \frac{\alpha t}{\alpha r} \quad (3)$$

The Coefficient of Anisotropy (CA) is the ratio between the tangential swelling (αt) and the radial swelling (αr).

2.4 Data Analysis

The data were analyzed with the analysis of variance, and in cases of the rejection of the null hypothesis ($p < 0.05$), we used the test of Tukey for the means differentiation.

3 Results and Discussion

Figure 2 shows the mean linear and volumetric ASE per treatment. Regarding dimensional stability, an increase in the anti-swelling efficiency of treatments with paraffin (190P and 210P) was observed, when compared to treatments without paraffin (190 and 210), respectively. The best results occurred with the combination of using paraffin at the highest temperature (210P), in which the ASE was statistically different from the other treatments (Fig. 2), with a mean above 70% in the tangential direction.

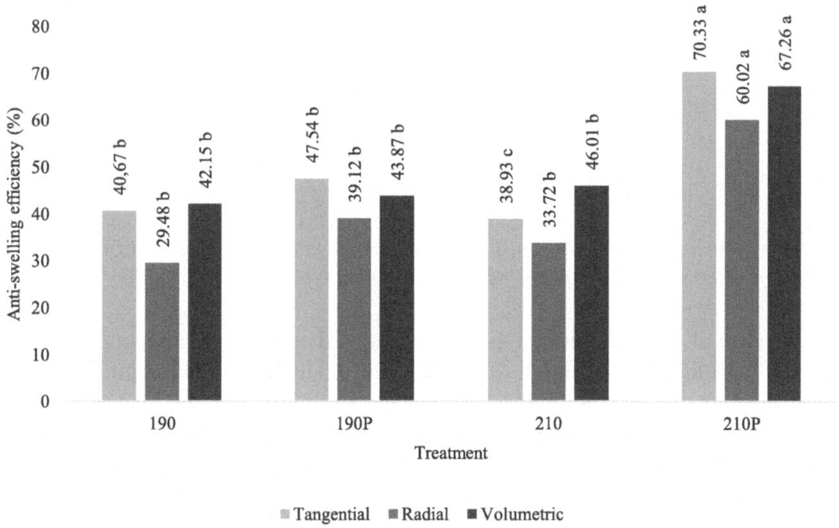

Fig. 2. Means of linear and volumetric anti-swelling efficiency.

Reinprecht and Repák (2019) analyzed the efficiency of thermal modification with paraffin on European beech wood (*Fagus sylvatica* L.) and achieved similar results, with a reduction in wood swelling by approximately 60% at 210 °C. In a similar work with *Pinus* spp. from, Wang et al. (2015) concluded that the combined treatment with paraffin was able to improve some negative effects left by thermal modification (alone), leading to better dimensional stability.

In the most intense phase of thermal modification (210 °C), cleavage reactions occur in the bonds in the lignin complex, increasing the concentration of phenolic groups. This leads to self-condensation and polycondensation reactions, forming cross-links in the lignin-lignin and lignin-carbohydrate complexes, causing an insulation of the cellulose microfibrils. This process increases the lignin content in the wood, providing greater dimensional stability and reducing its hygroscopicity (Kollmann and Fengel 1965; Nuopponen et al. 2005).

The WA was satisfactorily reduced with the increase in the thermal modification temperature associated with the paraffin (Table 2), with the best results for the treatments 190P and 210P. Among the treatments without paraffin, only 210 differed significantly from the untreated wood, indicating the effect of higher temperature on the hygroscopicity of the wood.

Table 2. Means of water absorption (WA) and coefficient of anisotropy ($\alpha t/\alpha r$).

Treatments	WA [%]	$\alpha t/\alpha r$
Untreated	62.30 c (0.23)	2.11 c (0.24)
190	56.25 c (0.21)	1.95 bc (0.21)

(*continued*)

Table 2. *(continued)*

Treatments	WA [%]	$\alpha t/\alpha r$
190P	25.42 a (0.13)	1.81 ab (0.36)
210	34.63 b (0.37)	1.78 ab (0.19)
210P	17.03 a (0.26)	1.57 a (0.50)

Means followed in a column by the same lowercase letter do not differ statistically from each other (Tukey test, $p < 0.05$); Values in parentheses correspond to the standard deviation.

Przybysz et al. (2014) reported that there is a predisposition to an increase in water absorption with the intensification of the thermal modification temperature, due to an increase in pore dimensions, which would lead to more space for water. The same was reported by Wang et al. (2015) when they analyzed thermally modified wood. In general, thermal modification reduces the hygroscopicity of wood, due to the hydrolysis of hemicelluloses, in addition to the plasticization of lignin and the restructuring of its polymeric matrix, reactions that promote hydrophobic characteristics (Korkut and Hiziroglu 2009).

Table 2 also contains the mean coefficient of anisotropy, which is an indication of materials more prone to drying defects, such as warping and cracking. The means ranged from 2.11 (Untreated) to 1.57 (210P), implying greater efficiency of the combined treatment.

Huller et al. (2017) evaluated thermally modified *Eucalyptus cloeziana* wood and found reductions up to 17% in the coefficient of anisotropy, reaching 1.45 at 210 °C, while the result of the Untreated wood was 1.75. For "Louro-preto" wood (Table 2), this reduction at the same modification temperature was 25%, but considering the combined treatment.

These results are related to the immediate degradation of accessible hydroxyl groups present in the cell wall, caused by thermal modification. As these groups are responsible for absorbing moisture, the wood becomes less hygroscopic (Poubel et al. 2013). Additionally, the paraffin present in cell lumens slows down vapor diffusion in the modified wood structure (Wang et al. 2015). Thus, for louro-preto wood, it was clear the effectiveness of the combined treatment and its ability to produce a material with good dimensional stability and a lower tendency to defects resulting from dimensional changes.

4 Conclusions

The combination of thermal modification with paraffin was effective in increasing the repellency of *Nectandra dioica* wood to water and increasing its dimensional stability. The positive effects were verified even after successive cycles of water immersion and drying. It is suggested that this work be continued with more tests, such as biological durability and weathering, as well as image analysis to verify the paraffin penetration into wood.

References

Andrade, F.W.C., Moutinho, V.C.P., Severo, E.T.D., Ferreira, J.S.S., Ballarin, A.W.: Impacto da modificação térmica no inchamento e comportamento mecânico de *Couratari* spp. Bioresources. **18**(3), 5242–5252 (2023)

Altgen, M., Kyyrö, S., Paajanen, O., Rautkari, L.: Resistance of thermally modified and pressurized hot water extracted scots pine sapwood against decay by the brown-rot fungus *Rhodonia placenta*. Eur. J. Wood Wood Prod. **78**, 161–171 (2020)

Amorim, M.R.S., Ribeiro, P.G., Martins, S.A., Del Menezzi, C.H.S., Souza, M.R.D.: Surface wettability and roughness of 11 Amazonian tropical hardwoods. Floresta e Ambiente. **20**, 99–109 (2013)

Boonstra, M., Tjeerdsma, B.: Chemical analysis of heat-treated softwoods. Holz Roh Werkst. **64**(3), 204–211 (2006)

Borges, L.M., Quirino, W.F.: Higroscopicidade da madeira de *Pinus caribaea* var. *hondurensis* tratada termicamente. Revista Biomassa Energia. **1**(2), 173–182 (2004)

Brito, T.M., Ferreira, J.S.S., Ferreira, G., da Silva, J.G.M., de Mendonça, A.R., Fantuzzi Neto, H., et al.: Physical and mechanical properties of thermally modified short-rotation wood in a closed system. Int. Wood Prod. J. **14**(2), 94–101 (2023)

Cardoso, C.C., Moutinho, V.H.P., Melo, L.O., dos Santos Sousa, L.K.V., de Souza, M.R.: Caracterização físico-mecânica de madeiras amazônicas com aptidão tecnológica para comercialização. Revista de Ciências Agrárias. **55**(3), 176–183 (2012)

Conte, B., Missio, A.L., Pertuzzatti, A., Cademartori, P.H.G., Gatto, D.A.: Propriedades físicas e colorimétricas da madeira termorretificada de Pinus elliottii var. elliottii. Sci. For. Piracicaba. **42**(104), 555–563 (2014)

Esteves, B., Nunes, L., Domingos, I., Pereira, H.: Improvement of termite resistance, dimensional stability and mechanical properties of pine wood by paraffin impregnation. Eur. J. Wood Wood Prod. **72**(5), 609–615 (2014)

Fengel, D., Wegener, G.: Wood: chemistry, ultrastructure, reactions. Walter de Gruyter, Berlin (1989)

Huller, L.A.S., Haselein, C.R., da Silveira, A.G., de Menezes, W.M., Talgatti, M., de Souza, J.T., Santini, E.J.: Modificação térmica e propriedades tecnológicas da madeira de *Eucalyptus cloeziana*. Pesquisa Florestal Brasileira. **37**(90), 183–188 (2017)

Kollmann, F., Fengel, D.: Changes in the chemical composition of wood by thermal treatment. Holz als Roh-und Werkstoff. **23**, 461–468 (1965)

Korkut, S., Hiziroglu, S.: Effect of heat treatment on mechanical properties of hazelnut wood (*Corylus colurna* L.). Mater. Des. **30**(5), 1853–1858 (2009)

Luchtemberg, P.H.Q.: Resistência natural de dez espécies de madeiras amazônicas submetidas ao ataque de fungos apodrecedores em ensaio de laboratório. 45 f., il. Monografia (Bacharelado em Engenharia Florestal)—Universidade de Brasília, Brasília (2013)

Modes, K.S., Santini, E.J., Vivian, M.A.: Higroscopicidade da madeira de Eucalyptus grandis e Pinus taeda tratadas termicamente. Cerne. **19**, 19–25 (2013)

Nuopponen, M., Vuorinen, T., Jämsä, S., Viitaniemi, P.: Thermal modifications in softwood studied by FT-IR and UV resonance Raman spectroscopies. J. Wood Chem. Technol. **24**(1), 13–26 (2005)

Pratiwi, L.A., Darmawan, W., Priadi, T., George, B., Merlin, A., Gérardin, C., et al.: Characterization of thermally modified short and long rotation teaks and the effects on coatings performance. Maderas. Ciencia y tecnología. **21**(2), 209–222 (2019)

Reinprecht, L., Repák, M.: The impact of paraffin-thermal modification of beech wood on its biological, physical and mechanical properties. Forests. **10**(12), 1–15 (2019)

Scholz, G., Krause, A., Militz, H.: Capillary water uptake and mechanical properties of wax-soaked Scots pine. In Proceedings of the European Conference on Wood Modification, Stockholm, Sweden, 27–29. SP Technical Research Institute of Sweden, Stockholm (2009)

Scholz, G., Krause, A., Militz, H.: Exploratory study on the impregnation of Scots pine sapwood (*Pinus sylvestris* L.) and European beech (*Fagus sylvatica* L.) with different hot melting waxes. Wood Sci. Technol. **44**, 379–388 (2010)

Poubel, D.D.S., Garcia, R.A., Santos, W.A.D., Oliveira, G.D.L., Abreu, H.D.S.: Efeito da termorretificação nas propriedades físicas e químicas da madeira de Pinus caribaea. Cerne. **19**, 391–398 (2013)

Przybysz, M., de Oliveira Machado, G., Christoforo, A.L., da Silva, M.R., Junior, C.C.: Estudo da Higroscopicidade de madeira de *Pinus oocarpa* Termorretificada. Madeira: arquitetura e engenharia. **14**(34), 37–46 (2014)

Wang, W., Zhu, Y., Cao, J., Guo, X.: Thermal modification of Southern pine combined with wax emulsion preimpregnation: effect on hydrophobicity and dimensional stability. Holzforschung. **69**(4), 405–413 (2015)

Wentzel, M., Fleckenstein, M., Hofmann, T., Militz, H.: Relation of chemical and mechanical properties of *Eucalyptus nitens* wood thermally modified in open and closed systems. Wood Mater. Sci. Eng. **14**(3), 165–173 (2019)

Willems, W., Lykidis, C., Altgen, M., Clauder, L.: Quality control methods for thermally modified wood: a review. Holzforschung. **1–10** (2015)

Exploring the Mechano-Sorptive Behavior of Thermally Modified Wood

Claude Feldman Pambou Nziengui[1](✉), Goli Giacomo[2], and Rostand Moutou Pitti[3]

[1] URMM, Ecole Polytechnique de Masuku (EPM), Université des Sciences et Techniques de Masuku (USTM), Franceville, Gabon
pclaudefeldman@gmail.com

[2] DAGRI – Dipartimento di Scienze e Tecnologie Agrarie, Alimentari, Ambientali e Forestali, Universita degli di Firenze, Florence, Italy

[3] Institut Pascal, Clermont Auvergne INP, CNRSS, Université Clermont Auvergne, Clermont Ferrand, France

Abstract. Mechano-sorptive tests in climatic chamber have been performed on unmodified and thermally modified wood by using 3-points bending test. The specimens were cut by splitting to get a straight grain. A geometrical configuration, with a small thickness was chosen to speed-up the moisture steady state (moisture equilibrium) within the sample. Then the samples were loaded at 10% of their average of rupture load. Average elastic modulus of the samples, as well as the average bending strength were determined on another set of samples coming from the same board. The tests were performed in a climate chamber, where the relative humidity can be changed between 35% and 80%. Three level of relative humidity have been tested (45%, 55%, and 65%) following three cycles. The average initial relative humidity was 55%, and the samples were loaded after equilibrium at their relative humidity level is reached. After 19 h of creep the relative humidity was changed to 65% (during two cycles), then 45% after 65 h. The results obtained show a more pronounced impact of humidity on non-thermally modified wood compared to thermally modified wood. Additionally, the measured elastic displacement is greater in non-thermally modified wood than in thermally modified wood. These results demonstrate that thermally modified wood exhibits significantly better physico-mechanical characteristics than unmodified wood.

Keywords: Poplar Wood · Thermally modified wood · Hydric behavior · Impact of relative humidity

1 Introduction

In structures, wood is subjected to the effects of mechanical and environmental loadings. The mechanical loading could be defined by the effects of weight or wind forces on the structures. On the other hand, environmental loads could be characterized by the effects of snow or rain by variations of humidity within the wood. Those two-phenomena coupled affect the mechanical behavior of wood by their interactions [1–3]. For these reasons, it seems very important to investigate the coupled effects of these two phenomena by

G. Goli et al. (Eds.): ECWM 2024, SPM 86, pp. 67–72, 2025.
https://doi.org/10.1007/978-3-031-99418-0_9

proposing some experimental or analytical models to predict the mechanical wood's behavior in structures. Until today, the tests widely considered as the best for these studies, are viscoelastic and mechano-sorptive tests. The tests performed during a creep test have shown for viscoelastic or mechano-sorption a mechanical loading going from 10% or 40% of maximal loads used [4–7]. Exceptionally, it exists some works as those of [8] which have been done specifically to show the coupled impact of loads, crack propagations and environmental changes in the time where the loads used are going from 50% to 65% of the maximal load. All these tests have in common the fact that they are conducted on non-thermally modified wood. In the context of this work, we conduct a comparative study of the mechano-sorptive behavior of thermally and non-thermally modified poplar wood. The objective is ultimately to identify which type of specimens may exhibit the best mechanical characteristics in structures but above all to propose an experimental model allowing to predict the structural behavior of thermally or non-thermally modified woods during a mechano-sorptive test.

2 Materials and Methods

Mechano-sorptive tests in a Climate-Controlled Chamber performed in the case of this study, involved conducting 3-point bending tests on both unmodified and thermally modified poplar wood samples, as depicted in Fig. 1. To ensure a straight grain, the specimens were obtained through a split-cutting method and sized at $15 \times 5 \times 160\ mm^3$ (width $\times$ height $\times$ length) as presented in Fig. 1a.

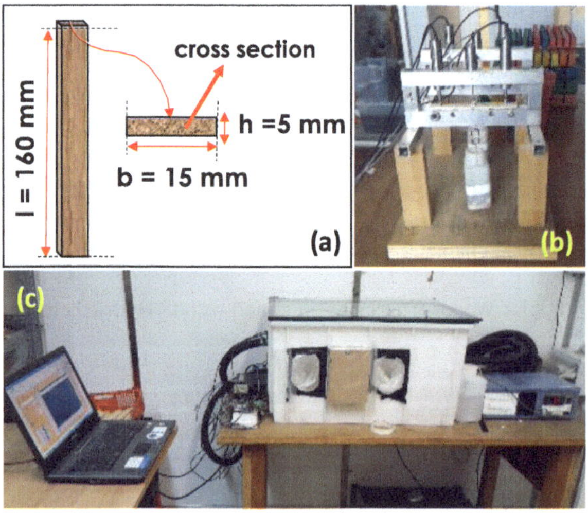

Fig. 1. experimental devices used: (a) typical samples of thermally modified popular wood tested; (b) typical 3-points bending test used for mechanosorptive test; (c) box of environmental control chamber.

This specific geometric configuration, with a reduced thickness, was chosen to facilitate the time diffusion of moisture within the sample. Subsequently, the samples were loaded at 10% of the average breaking loads obtained on another samples set (see Fig. 1b), following the approach of [9]. The elastic modulus and bending strength were determined using a separate set of samples sourced from the same board.

The tests were conducted within a controlled climate chamber capable of adjusting the relative humidity within the range of 35–80% (see Fig. 1c). In our study, three relative humidity levels were specifically targeted (45%, 55%, and 65%). Initially, the relative humidity was set at 55%, and the sample was subjected to loading under these conditions. After 19 h of creep, the relative humidity was then elevated to 65%, which was maintained for 5 h before returning to 55%. Subsequently, another 19-h period lapsed, following which a second humidity cycle was initiated, reducing the environment's humidity to 45% over 5 h. Finally, the chamber's humidity was restored to 55%. The load was removed after 19 h following the conclusion of the second cycle, marking the commencement of a new test.

In the case of this study, to compare the results obtained from the different samples, the specimens were tested in pairs (modified and unmodified) under the conditions previously presented. Two samples of unmodified (PM7 and PM8) and two samples of thermally modified poplar wood (PT8 and PT10) have been loaded. As presented by [10] in its woks, Fig. 2 presents the typical representation of the loading system performed on the samples. The device is a sample loaded in 3-points bending test compound with three (3) transducers T_1, T_2 and T_3 (see Fig. 2).

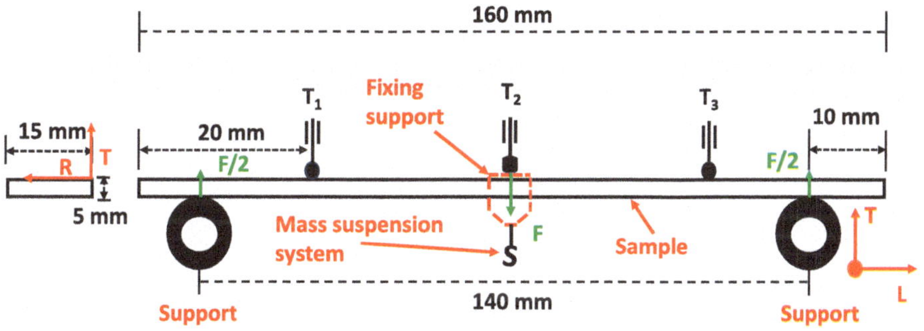

Fig. 2. Experimental loading dispositive sample as presented by [10].

The thermal modification was conducted exposing the samples to a saturated steam environment with 150 °C of temperature for 3 h with an extra heating ramp of 54 min with an expected dry mass loss of about 5% [11, 12].

The principal central relative displacement (U) considered in the case of this study, where determined by using the Eq. (1):

$$U = T_2 - \frac{(T_1 + T_3)}{2} \tag{1}$$

This configuration of the transducers allowed the automatic subtraction of the hygroscopic deformations because the shrinkage and swelling were acting simultaneously on the three transducers.

3 Results and Discussions

Figure 3 shows typical curves of mechano-sorptive tests obtained for all the samples loaded during the creep tests (Fig. 3a) and all cycles of the relative humidity (RH) conducted in the climatic chamber (Fig. 3b). Hence, the corresponding mechano-sorptive curves (C(t)) have been obtained from the Eq. (2):

$$C(t) = \frac{(U - U_0)}{U_0} \tag{2}$$

Where U represents the central deformation measured in Eq. (1) and $U_0$the elastic deformation of the sample.

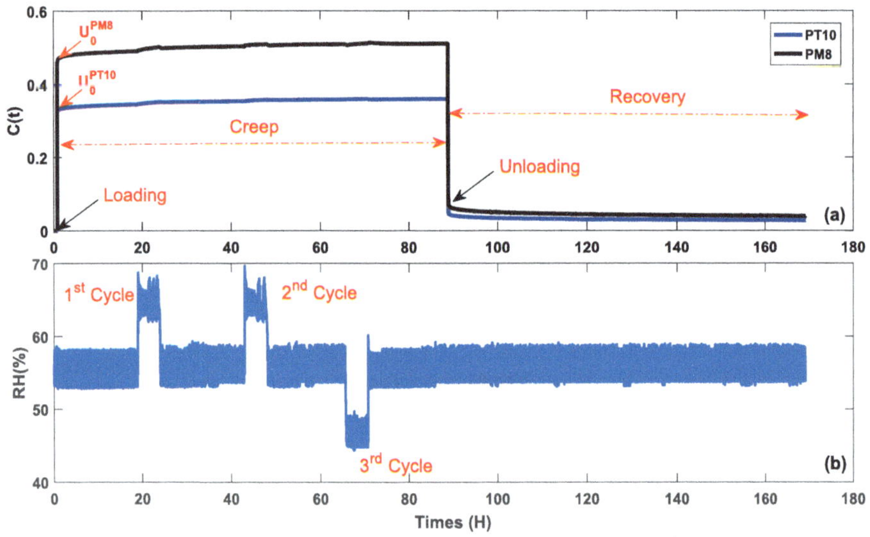

Fig. 3. Typical experimental loading dispositive sample

The curves show that for PM8 (unmodified sample) its initial relative creep ($U_0{}^{PM8}$) is more important (in the sense that it exhibits a higher deflection) compared to this obtained on PT10 (thermally modified wood $U_0{}^{PT10}$). This result shows that thermally modified wood exhibit better mechanical performances during mechano-sorptive loading compared to unmodified wood. Indeed, we can remark by this result that the mechanical load used has a greater impact on unmodified wood than on modified wood. A reduction of elastic deformation of about 30% is obtained because of thermal modification. The literature [13] shows that these results are mainly due to the difference of moisture content in the sample.

A specific study was conducted to evaluate the impact of moisture content (MC) variation in the samples, influenced by relative humidity (RH), on the mechanical behavior of each type of loaded specimen. Table 1 shows the results obtained, where δ represents the time taken by RH before impacting the sample (see Fig. 4) and C_{RH} the relative deformation (see Fig. 4).

Table 1. Values of relative deformation (C_{RH}) and time diffusion δ.

	Thermally modified sample	Unmodified sample
C_{RH}	0.008	0.059
(h)	0.82	0.72

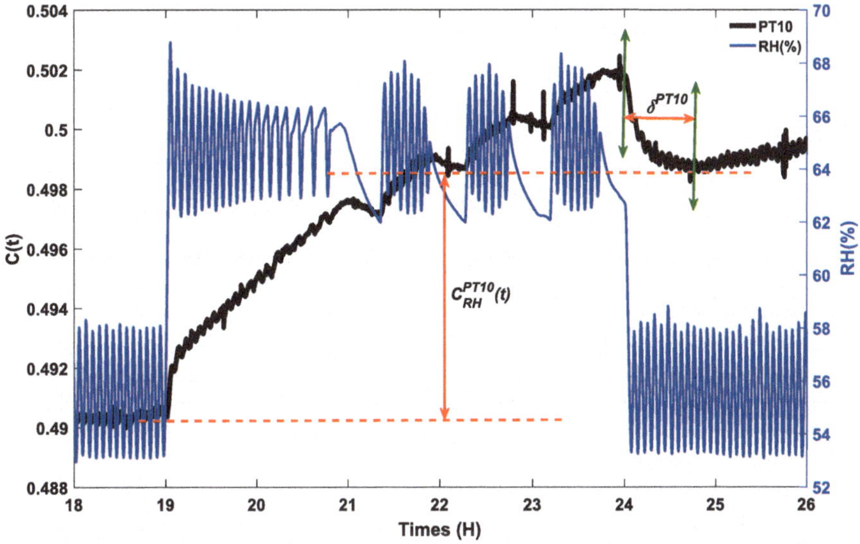

Fig. 4. Typical zoom performed on PT10 sample to evaluated viscous elastic deformation and diffusion time of RH in wood sample.

A lower tendency to viscous deformation of thermally modified sample has also been observed, 2.42% for thermally modified wood and 12.04% for unmodified wood of their respective elastic deformation. Figure 4 shows the typical analysis performed on the sample.

By considering all the results previously presented, thermally modified wood shows better characteristics against impact of humidity and mechanical load compared to unmodified thermally wood.

4 Conclusion

A comparative study on the mechano-sorptive behavior between thermally modified poplar wood and unmodified poplar wood has been done. The results display that, for the rates of modification applied on the samples, the mechanical behavior of thermally modified wood exhibits a better performance to long term loads and to environmental humidity variations. Indeed, in terms of mechanical behavior, it is noted that for the low modification rates applied (5% ML) in this work thermally modified wood present as higher stiffness compared to un-modified wood. Furthermore, the environmental study shows that moisture diffusion is slower in thermally modified wood. This could improve the stability conditions in wooden structures and ensure their safety. In perspective, to propose a mechano-sorptive analytical model of thermally modified wood, a more extensive experimental campaign, increasing the number of samples, needs to be considered in the development of new hygromechanical models for thermally modified wood.

Acknowledgements. The authors would like to acknowledge the DAGRI (University of Florence) for the provision of the material used for the tests carried out and the COST Action FP1407 for the financial support during the STSM.

References

1. Grossman, P.U.A., Kingston, R.S.T.: Creep and stress relaxation in wood during bending. pp. 403–17 (1954)
2. Randriambololona, H., Dubois, F., Petit, C.: Modelling the deferred mechanical behavior of wood in a variable environment. French J. Civ. Eng. **6**(7–8), 1333–1357 (2002). https://doi.org/10.1080/12795119.2002.9692412
3. Dubois, F., Husson, J.M., Sauvat, N., Manfoumbi, N.: Modeling of the viscoelastic mechano-sorptive behavior in wood. Mech. Time-Depend. Mater. **16**, 439–460 (2012)
4. Mukudai, J., Yata, S.: Further modeling and simulation of viscoelastic behavior (bending deflection) of wood under moisture change. Wood Sci. Technol. **21**(1), 49–63 (1987)
5. Hoyle Jr., R.J., Itani, R.Y., Eckard, J.J.: Creep of Douglas fir beams due to cyclic humidity fluctuations. Wood Fiber Sci., 490–497 (1986)
6. Navi, P., Heger, F.: Thermo-Hydro-Mechanical Behaviour of Wood: Technical and Structural Applications, 1st edn, pp. 119–120. In French. Presses Polytechniques et Universitaires Romandes, Lausanne, CH (2005)
7. Nguyen, S.L., Saifouni, O., Destrebecq, J.F.: Incremental formulation for coupled viscoelasticity and Hydrolock effect in softwood. In: Ralph, C., Silberstein, M., Thakre, P., Singh, R. (eds.) Mechanics of Composite and Multi-Functional Materials, Conference Proceedings of the Society for Experimental Mechanics Series. Cham, Springer (2016)
8. Pambou Nziengui, C.F., Moutou Pitti, R., Fournely, E., Gril, J., Godi, G., Ikogou, S.: Notched-beam creep of Douglas fir and white fir in outdoor conditions: experimental study. Constr. Build. Mater. **196**, 659–671 (2019)
9. Saifouni, O., Destrebecq, J.F., Froidevaux, J., Navi, P.: Experimental study of the mechanosorptive behaviour of softwood in relaxation. Wood Sci. Technol. **50**(4), 789–805 (2016). https://doi.org/10.1007/s00226-016-0816-2
10. Asseko Ella, M., Goli, G., Pambou Nziengui, C.F., Gril, J., Fournely, E., Moutou Pitti, R.: Comparative investigation on viscoelastic and mechanosorptive creep behavior of two tropical hardwoods and one temperate softwood. Eur. J. Wood Wood Prod. **82**(1), 35–52 (2024)
11. Hill, C., Altgen, M., Rautkari, L.: Thermal modification of wood—a review: chemical changes and hygroscopicity. J. Mater. Sci. **56**, 6581–6614 (2021)
12. Godinho, D., Ferreira, C., Lourenço, A., de Oliveira Araújo, S., Quilhó, T., Diamantino, T.C., Gominho, J.: The behavior of thermally modified wood after exposure in maritime/industrial and urban environments. Heliyon. **10**(3) (2024)
13. Marcon, B., Goli, G., Matsuo-Ueda, M., Denaud, L., Umemura, K., Gril, J., Kawai, S.: Kinetic analysis of poplar wood properties by thermal modification in conventional oven. IForest. **11**, 131–139 (2018)

Wood Modification as an Opportunity for Local Wood Species in Musical Instrument Making

Mario Zauer(✉), Robert Krüger, Tobias Dietrich, Herwig Hackenberg, and André Wagenführ

TU Dresden University of Technology, Dresden 01062, Germany
mario.zauer@tu-dresden.de

Abstract. This paper addresses wood modification as a tool for the use of European native wood species in high quality musical instruments as a brief review of a case study carried out. Regional woods were thermally modified and tested for their suitability as a tropical wood substitute in classical guitars. Thermal modification was developed in laboratory scale using a patented procedure. The reached properties due to the thermal modification of the wood were always compared with the tropical wood assortments so far to be used as well as to be replaced. Component-specific preferred variants were then manufactured on a pilot scale. Finally, entire test instruments, classical guitars, were made using the modified wood and compared with identical reference guitars by means of subjective testing methods with professional musicians. As a result, it was found that it is possible to replace the tropical woods previously used with some thermally modified regional woods without any significant loss of sound quality.

Keywords: Classical guitars · dynamic properties · hardness · swelling coefficient · thermal treatment

1 Introduction

The use of wood in high-quality musical instruments requires high-quality material. These mainly include specially selected, sometimes rare, European and non-European as well as tropical wood species. Depending on the component or function in the instrument, the wood must have i. a. high dimensional stabilities, sound velocities, stiffness's and hardness's. An essential evaluation criterion for customer acceptance is, first and foremost, the color and grain of the wood. A wood that does not have the right look, color or haptic may be rejected purely for these reasons, even if their performance is similar or even superior to other materials [1]. In addition to a comparatively low dimensional stability, most regional types of wood in Europe also only have a slight, correspondingly fascinating optical quality. Furthermore, the hardness and vibration properties are sometimes significantly weaker compared to the tonewood ranges currently used.

In recent years, a number of studies have been carried out to determine the effectiveness of various wood modification processes in changing the sorptive, optical, mechanical and acoustic properties of different types of wood and thus to investigate the suitability of specially modified woods for the construction of musical instruments [2].

G. Goli et al. (Eds.): ECWM 2024, SPM 86, pp. 73–80, 2025.
https://doi.org/10.1007/978-3-031-99418-0_10

Based on the available literature, it can be concluded that some treatment processes appear to be suitable for significantly improving the quality of individual types of wood in terms of musical instrument construction if the technical parameters (temperature, humidity, pressure, time) are applied appropriately. Unfortunately, most of the research work only describes the change in the physical property profile of the specially treated wood compared to that of untreated wood. Finally, the all-important reference in the sense of a comparison of characteristic values with the wood assortments to be replaced is often missing. There is also a lack of component-specific analyses on the fundamental suitability of the modified material for the entire musical instrument.

Therefore, two specific studies were carried out in this context [3] and excerpts from one study are presented in this paper. In this case, thermal modification was used to provide regional woods to replace tropical woods in classical guitars.

The following aspects are important to know about the individual components and woods commonly used in classical guitar making.

Cedro (*Cedrela odorata* L.) is often used as a neck material for classical guitars. The neck has the task of maintaining the string tensions and damping the string vibrations as less as possible. It should have a relatively low mass in order to avoid top-heaviness of the entire instrument. Suitable properties would therefore be comparatively high stiffness, especially under torsional stress, with low bulk density.

Indian rosewood (*Dalbergia latifolia* Roxb.) is often used as a material for the back, sides and bridge of classical guitars. The main task of the back and sides is to reflect the sound waves emitted by the guitar top with low loss. Ultimately, these reflected sound waves are emitted into the surroundings through the sound hole in the top. The bridge is responsible of both the fixation of the strings and transferring the vibrational energy generated by the strings to the top with as little loss of energy as possible. Suitable properties of wood for use in the three components mentioned would therefore be a comparatively high hardness, stiffness, especially under torsional stress, and speed of sound.

Ebony (*Diospyros crassiflora* Hiern.) is often used as a fingerboard material for classical guitars. Rosewood is also used for fingerboards in acoustic guitars and especially in electric guitars and electric basses. The main task of this component is to resist the mechanical stresses while playing the instrument. The most important property is therefore a relatively high hardness of the material.

The top of a classical guitar is usually made of high-quality spruce. As a soundboard, the top of the guitar is responsible for the conduction, radiation and transmission of the sound generated by the strings. Suitable properties would therefore be comparatively high stiffness with low bulk density and correspondingly high speed of sound.

As mentioned above, all instrument components must have a high dimensional stability under changing climatic conditions and an individual, fascinating appearance (similar to tropical wood).

The aim of the thermal modification of this paper was primarily to improve the dimensional stability, the acoustic properties and the visual aspects with regard to darker color nuances of the regional wood species. In this context, significant changes due to thermal treatment have already been demonstrated and discussed in numerous studies.

2 Materials and Methods

Untreated (UT) and thermally modified (TM) heartwood of spruce (*Picea abies* (L.) Karst.), cherry tree (*Prunus avium* L.), service tree (*Sorbus torminalis* (L.) Crantz) and plum tree (*Prunus domestica* L.) as well as untreated cedro, Indian rosewood and ebony (Latin names see above) as reference material were investigated. The last three tropical wood species mentioned should be replaced with the first-mentioned native species (untreated or treated) wherever possible.

All wood assortments were provided by a classical guitar manufacturer (Hanika Gitarren, Baiersdorf, Germany). The dimensions were on average up to: 1 m long, 30 mm thick and ca. 250 mm width. The boards for the wood to be modified were cut into two halves. One half was used as untreated reference. The other one was used for thermal modification (TM). TM for the pure material tests was conducted at 180 °C for 4 h in a laboratory oven using a patented treatment method [4]. The boards to be treated were conditioned at 20 °C (±2 °C) and 60% (±5%) relative humidity (RH) and subsequently sealed using a thermally stable film to realize an exclusion of oxygen. In addition, the sealed wood packages were vacuumed before being placed in the oven. The heating rate to reach the treatment temperature was 5 K h^{-1}. Likewise, after thermal treatment the temperature was gradually lowered by 5 K h^{-1}.

Specimens were cut from untreated and treated boards after conditioning to equilibrium at 20 °C (±2 °C) and 65% (±5%) RH with dimensions of 28 × 6.5 × 190 mm^3 (R × T × L). They continued to be stored in the aforementioned climate until the respective test. Twenty replicate specimens were used for testing density, swelling behavior, hardness, and dynamical properties.

Density were conducted according to DIN 52182 (1976). The swelling coefficient was measured in accordance with DIN 52184 (1979) in the range between 45% and 85% RH at 20 °C, because this range is of practical use for classical guitars. Brinell hardness was determined perpendicular to the grain in tangential direction in accordance with EN 1534 (2011). In the same test direction, dynamical properties were investigated using the experimental modal analysis (EMA) method [5]. In this connection, dynamic modulus of elasticity and dynamic shear modulus (torsion) were determined. The speed of sound was calculated from the former and the density [1]. A detailed description and calculation of all tests performed was given in [3].

Based on this, tropical wood-free test instruments were finally built from local wood species and compared with guitars of identical construction made from the usual tropical woods. For example, the following material substitutions were made (see Fig. 1).

- Neck: untreated cherry wood vs. cedro
- Fingerboard: TM plum wood treated at 195 °C for 2 h vs. ebony
- Back/sides: TM service tree wood treated at 195 °C for 2 h vs. rosewood
- Top: TM spruce treated at 140 °C for 2 h vs. untreated spruce
- Bridge: TM plum wood treated at 195 °C for 2 h vs. rosewood

In contrast to the pure material tests, a higher treatment temperature was chosen for service tree and plum wood in order to achieve a darker coloration. With spruce, the objective was only to increase the dimensional stability without discoloring the wood to dark shades. The risk of confusion with red cedar (*Thuja plicata* DONN EX D.), which is also used as a top material, should be avoided here.

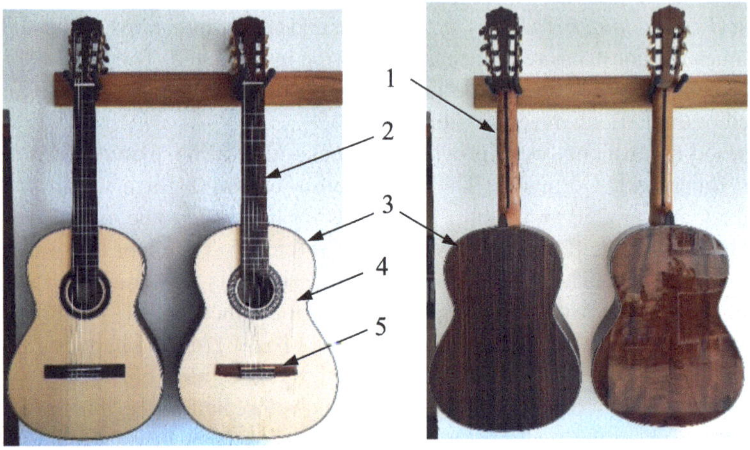

Fig. 1. Reference guitar (on the left in the photos) and tropical wood-free demonstrator guitars (on the right in the photos). 1 - neck, 2 - fingerboard, 3 - back/ sides, 4 - top, 5 - bridge.

The evaluation of the guitars was carried out with four professional guitarists. Two guitars were available for each series. The test subjects wore eye coverings (e.g. covered glasses, sleep masks) during the sound and haptic evaluation of the guitars in order to initially rule out any fundamental psychological optical influence. The assessment of each individual guitar in relation to individual evaluation criteria was based on absolute school grades from 1 (very good) to 5 (very poor). For this purpose, an evaluation scheme was created based on a special study [6].

This initially comprised the overall judgment with regard to the general sound impression of the instrument as well as the basic psychoacoustic characteristics: Timbre and loudness. Furthermore, the characteristics: Response (attack), dynamics and decay behavior (sustain) were also included, which to a certain extent reflect the transition between the acoustic and technical playing characteristics [6]. The general overall impression on the one hand and the balance between the individual positions (frets on the fingerboard) and strings (bass and treble strings) on the other were to be assessed. A well-balanced set of characteristics with differentiated pitches and ranges is evidence of a high-quality instrument that is easy to play. Due to the separate global and detailed query of the characteristics, the test person had to (or could) repeatedly deal with the corresponding criterion.

The handling of the guitars was also evaluated. Finally, the visual assessment of both the entire instrument and individual components was carried out with regard to the color and structure of the materials used.

The first step in the evaluation was to average the individual scores from the respective evaluation criteria for each instrument as well as the individual guitar players. The averaged school grades of the test guitars were then compared directly with those of the reference guitars. This ultimately made it possible to show the extent to which the test guitars were rated better or worse than the reference guitars.

3 Results and Discussion

A visual comparison of the individual woods used clearly shows the visual advantages of tropical woods compared to untreated domestic wood species (see Fig. 2). Rosewood in particular has a fantastic grain and variety of colors. Ebony, on the other hand, is characterized by its black color and simple texture. Cedro offers a slightly reddish color with a slight ripple grained and open-pored structure.

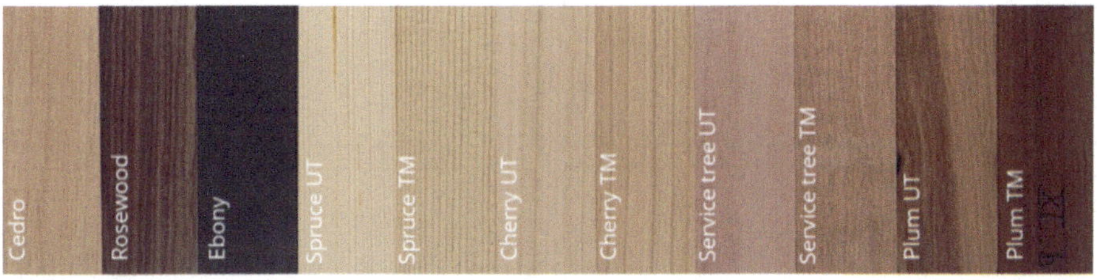

Fig. 2. Visual comparison of different types of wood used: cedro, rosewood, ebony, spruce, cherry, service tree and plum wood, each untreated (UT) and thermally modified (TM at 180 °C/4 h).

In the course of the thermal change, the woods discolored to varying degrees to light brown tones, as in TM spruce, TM cherry and TM service tree, or to dark brown tones, as in TM plum. In TM cherry and TM service tree, the reddish parts visually increased, although it can be assumed that these decrease again at higher treatment intensities and the gray parts increase [7]. It is clear that the color of ebony could not be achieved with the TM. However, there are certain similarities to cedar through TM cherry and TM spruce as well as to rosewood with TM plum.

The physical properties investigated were almost analogous to this (see Fig. 3). The density of spruce and cherry wood is almost equal to that of cedro (neck material) and lowest compared to the other woods examined. Furthermore, the densities of service tree and plum wood are almost the same as those of rosewood (material for back, sides and bridge) and, compared to the three aforementioned wood species, are significantly higher but also significantly lower than those of ebony (fingerboard material). In analogy to this, the material hardness's also adjusted according to the material densities. No significant changes in density or hardness were achieved as a result of the TM of any of the woods tested.

In contrast, the swelling coefficients of all woods decreased significantly as a result of the TM. In addition to the coloring to darker shades, this is another enormous advantage of the TM wood for suitability as "tonewood". It should also be noted that ebony exhibited the highest swelling coefficients compared to all other woods examined.

The sound velocities were highest in spruce. Those in the other woods examined were within a certain level, except for plum wood. As a result of TM, the sound velocities increased in all treated woods, except for cherry wood. Analogous to the densities and

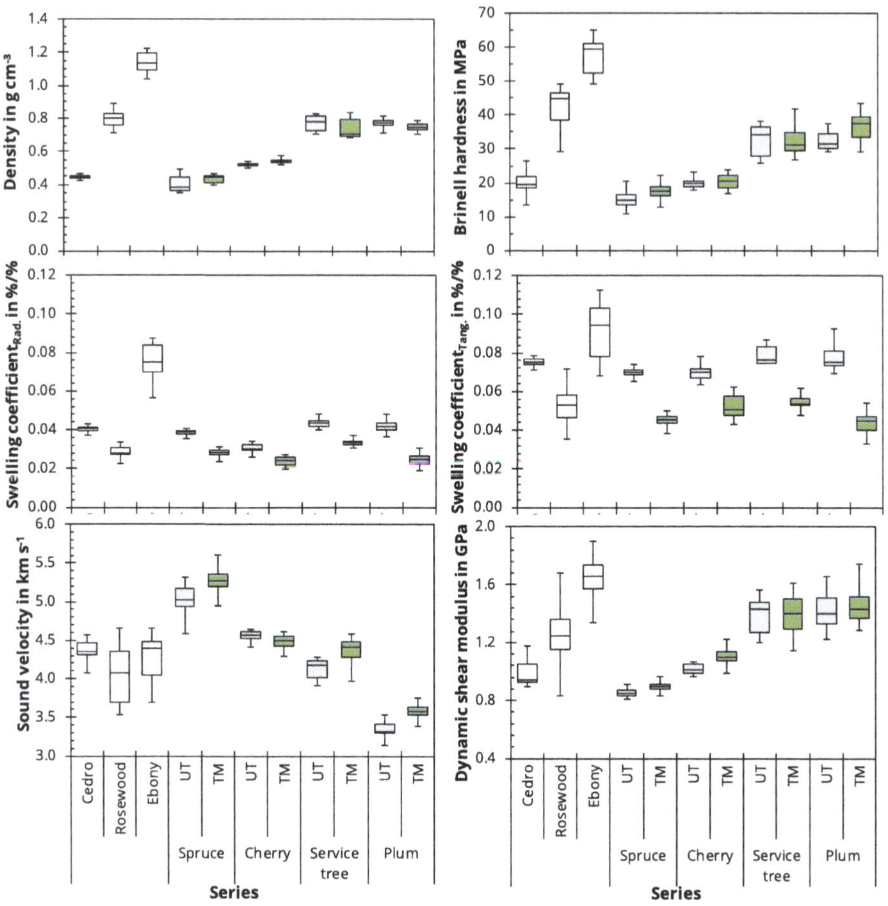

Fig. 3. Comparison of the determined densities, Brinell hardness's, swelling coefficients (radial and tangential direction), sound velocities and dynamic shear modules of all wood series investigated.

hardness's, similar value ranges of the dynamic shear moduli were measured for cedro (neck material), spruce and cherry wood on the one hand and for rosewood (material for back, sides and bridge), service tree and plum wood on the other. In this context, ebony again exhibited the highest dynamic shear stiffness.

From the results listed, it can be summarized that the woods examined can be divided into three groups: Cedro (1) and as possible substitute woods TM spruce and TM cherry wood as neck material, rosewood (2) and as possible substitute woods TM service tree and TM plum wood as material for back, sides and bridge in classical guitars. No equivalent wood could be found for ebony (3) with properties as similar as possible. However, the question arises as to whether the fingerboard material of a classical guitar has to be as hard as ebony. After all, rosewood is often used for many other acoustic and electric guitars and bass guitars. This means that TM service tree or TM plum wood could possibly be used as a substitute. Otherwise, a combination of mechanically densified and then thermally modified wood would be another alternative for using native woods with a lower natural density as fingerboard material. Mechanical densification leads to a higher density and thus hardness of the regional wood species. The second process step leads to a darker color and a reduction or even compensation of the set recovery effect of the densified wood [3]. However, it should be noted that compared to completely untreated wood, the swelling and shrinkage of the combined modified wood is significantly higher in the direction of densification [3].

The comparison of the averaged scores resulting from the guitar play tests between the test and reference guitars (see Fig. 4) clearly shows that no significant differences were found in terms of sound behavior and handling.

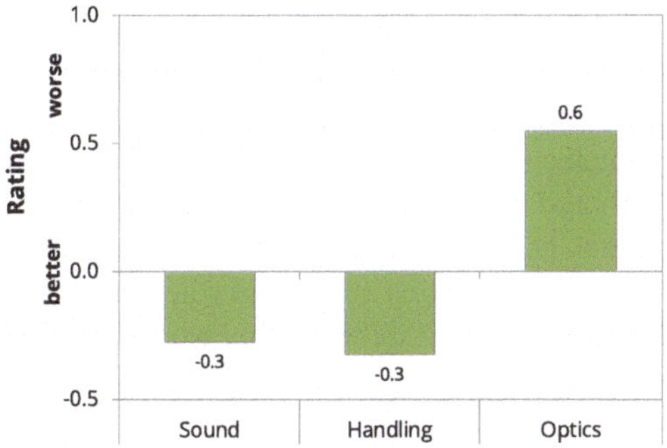

Fig. 4. Results from the guitar playing tests (the extent to which the test guitars were rated better or worse than the reference guitars).

In terms of handling, the reference guitars were always felt to be somewhat heavier and slightly top-heavy compared to the test guitars. This even resulted in a slightly better assessment of the tropical wood-free reference guitars. In terms of visual aspects, however, the standard reference guitars were rated slightly better. This could possibly have a negative effect on the decision to buy a tropical wood-free version, as classical guitars are a very traditional area.

Due to the thermal modification, it is not possible to achieve ebony-like shades or rosewood-like grains (see Fig. 1). It is known that at higher temperatures the wood can be discolored to black-brown, but the wood becomes extremely brittle, which is accompanied by a decrease in mechanical properties. In general, the color stability of the TM wood must be examined.

After all, everyone - manufacturers, dealers and customers - must be aware that new materials, such as TM wood, from regional areas look somewhat different from the tropical wood species used to date, but sound the same and can have similarly high dimensional stability. The company Hanika Gitarren (Baiersdorf, Germany) currently produces various models of classical guitars without tropical wood from different regional woods, which are mainly thermally modified.

Finally, it should be noted that the well-known initial smoky smell of the thermally modified wood disappears. None of the test subjects in this study were able to give negative ratings in this context.

4 Conclusion

In the context of this case study, it was possible to prove in principle that some regional woods in a thermally modified state can replace some tropical woods in individual components of classical guitars. The acoustic and mechanical properties improved only slightly. However, the treatment gives the wood significantly improved dimensional stability and a color enhancement almost analogous to the wood to be replaced. Specifically, it has been shown that cedro can be replaced by thermally modified spruce or cherry wood and rosewood by thermally modified service tree wood or plum wood. Cedro is commonly used as a neck material in classical guitars. Rosewood is used as a material for the back, sides and bridge in classical guitars and also as a fingerboard in some acoustic guitars and mostly in electric guitars and bass guitars. In classical guitars, ebony is often used as the fingerboard material. Neither the plain black color nor the hardness of this wood could be matched by the regional woods in the untreated and thermally modified state in this study. The question arose as to whether the high hardness of ebony is necessary at all. The dimensional stability of this wood was clearly the lowest of all the woods tested.

Acknowledgements. The research project was financially supported by the Federal Ministry of Economic Affairs and Energy of Germany (Grant reference 16KN035325).

References

1. Wegst, U.G.K.: Wood for sound. Am. J. Bot. **93**(10), 1439–1448 (2006)
2. Zauer, M., Pfriem, A.: Tropical and traditional wood species in musical instruments and case studies of their substitution with modified wood. In: Pérez, A., Marconi, E. (eds.) Wooden Musical Instruments - Different Forms of Knowledge, Book of End of WoodMusICK, COST Action FP1302, pp. 81–97, Paris (2018)
3. Zauer, M.: Die Holzmodifikation als Chance für einheimische Holzarten im Musikinstrumentenbau. Habilitation thesis. Technische Universität Dresden, Dresden (2023)
4. Zauer, M., Sproßmann, R.: Verfahren zur Herstellung von Tonholz. EP 2 977 156 B1 (2018)
5. Möser, M.: Messtechnik der Akustik. Springer, Berlin (2000)
6. Ziegenhals, G.: Subjektive und objektive Beurteilung von Musikinstrumenten. In: Eine Untersuchung anhand von Fallstudien. Dissertation. TUD Press, Dresden (2010)
7. González-Peña, M.M., Hale, M.D.C.: Colour in thermally modified wood of beech, Norway spruce and scots pine. Part 1: colour evolution and colour changes. Holzforschung. **63**(4), 385–393 (2009)

Correlation Between Color and Biodeterioration of Short-Rotation Thermally Modified Teak Wood

Anna Clara Oliveira Rupf[1](✉), Kamilly da Silva Pereira[1], Saulo José da Costa Lima[1], Paulo Henrique dos Santos Silvares[1], Jessica Sabrina da Silva Ferreira[1], Jaqueline Rocha de Medeiros[2], Adriano Ribeiro de Mendonça[1], Juarez Benigno Paes[1], and Djeison Cesar Batista[1]

[1] Graduate Program in Forest Sciences, Center for Agrarian Sciences and Engineering, Federal University of Espírito Santo, Jeronimo Monteiro, Espirito Santo, Brazil
annac.rupf@gmail.com

[2] Graduate Program in Forest and Environmental Sciences, Institute of Forests, Federal Rural University of Rio de Janeiro, Seropedica, Brazil

Abstract. Short-rotation teak wood (*Tectona grandis* L.f.) is composed of a higher proportion of sapwood than heartwood, giving the species heterogeneous color and natural durability. Thermal modification can be an alternative for improving these properties of teak wood from fast-growing forests. This study aimed to verify how colorimetric parameters correlate with mass loss caused by fungi, as an exploratory analysis for the development of wood durability prediction models based on colorimetry. Untreated and thermally modified wood were tested in closed system at 160 °C and in open system at 185 °C and 210 °C. For the fungi test, the white-rot *Trametes versicolor* and the brown-rot *Coniophora puteana* were used. Colorimetric analyses were carried out using the CIEL*a*b* system, using a portable spectrophotometer. Measurements were taken on the cross-section and the tangential surfaces of the specimens, both before and after the fungi test. The colorimetric coordinates L*, a*, and b* were measured and the parameters C* and h° were calculated. The correlation between the colorimetric parameters and mass loss was calculated considering the data obtained before and after the fungi test. A significance test was carried out for the correlation coefficient (r) and a scale was adopted for the qualitative assessment of the r. There were significant "strong" and "very strong" correlations between all the colorimetric parameters for both fungi. However, it was not possible to establish a method for developing biodeterioration prediction models based on colorimetry.

Keywords: Colorimetry · CIELab · *Tectona grandis* · *Coniophora puteana* · *Trametes versicolor*

G. Goli et al. (Eds.): ECWM 2024, SPM 86, pp. 81–88, 2025.
https://doi.org/10.1007/978-3-031-99418-0_11

1 Introduction

Teak wood (*Tectona grandis* L.f.) is a relevant wood species in the world market due to its attractive aesthetic characteristics, easy workability and high natural durability, resulting in suitable material for high-value uses, such as shipbuilding, construction, and furniture. On the other hand, trees from short-rotation plantations have a high proportion of juvenile wood, which has lower quality than adult wood. For instance, juvenile wood is composed of a higher proportion of sapwood than heartwood, giving it heterogeneous characteristics in terms of color and natural durability (Bellon 2013; Menezes 2017).

The benefits that thermal modification brings to wood properties are widely known. These include reduced hygroscopicity, increased dimensional stability, increased resistance to rot fungi, and wood darkening (Hill 2006; Jones and Sandberg 2020; Zelinka et al. 2022). Thermal modification is therefore an alternative to improve the properties of short-rotation teak wood, diversifying its use and making it more economically attractive.

Many studies have been carried out to evaluate the effect of thermal modification on the color and resistance to biodeterioration of teak wood (Bellon 2013; Motta et al. 2013; Lopes 2012; Gasparik et al. 2019). However, biodeterioration tests are time-consuming and expensive, while the measurement of colorimetric parameters can be carried out quickly and cheaply.

Therefore, the general objective of this work was to verify how colorimetric parameters correlate with mass loss caused by rot fungi, in an exploratory analysis for the development of wood durability prediction models based on colorimetry.

2 Materials and Methods

2.1 Statistical Treatments and Characterization of the Material

This study used short-rotation teak wood (*Tectona grandis* L.f.) and tested four treatments: untreated wood; CS160: thermally modified wood (TMW) at 160 °C in a closed system (pressurized) – corrected mass loss (CML) = 18.5%; OS185: TMW at 185 °C in an open system (non-pressurized) – CML = 0.33%; OS210: TMW at 210 °C in an open system – CML: 2.44%. The CML was calculated in another study by our research group, based on the total extractives content (Gomes et al. 2023).

Fifteen tangential slats (25 × 90 × 2250 mm) were collected from TWBrazil's stock (Ponta Grossa, Paraná), where the closed thermal modification process was carried out. Later, every slat was trimmed into four 500 mm long slats, which were homogeneously distributed among the treatments. The wood came from the first thinning (6 years old), acquired from a producer in the municipality of Cáceres, Mato Grosso, Brazil.

2.2 Thermal Modification of Wood

The closed thermal modification process adopted in this work is classified as hygrothermal (Hill 2006). The entire process lasted 12 h and was carried out in an autoclave with a nominal capacity of 6 m^3.

The temperature of 160 °C was adopted since it is commonly used by the company in its regular processing, which consists of seven phases: initial heating from 20 °C (room temperature) to 110 °C, at a heating rate of 1.30 °C min^{-1}; constant temperature for 30 min; heating to 140 °C at a rate of 0.50 °C min^{-1}; constant temperature for 30 min; heating to 160 °C, at a rate of 0.60 °C min^{-1}; constant target temperature for 90 min; and cooling to 60 °C, at a rate of 0.25 °C min^{-1}.

The open thermal modification process was carried out at the Department of Forest and Wood Sciences of Federal University of Espírito Santo (DCFM/UFES), located in Jerônimo Monteiro, Espírito Santo, Brazil. An electric furnace (Linn Elektro Therm, Bad Frankenhausen, Germany) was used, with internal dimensions of 600 × 600 × 700 mm, equipped with a time *x* temperature programmable controller.

The process schedule was similar to that of Lopes et al. (2014), who worked with the same species and equipment of the same model. It consisted of three phases: initial heating from 20 °C (room temperature) to 100 °C, with a heating rate of 0.50 °C min^{-1}; heating from 100 °C to the thermal modification temperature (185 °C or 210 °C) at a rate of 2.60 °C min^{-1}; and maintaining the target temperature for 180 min. At the end of phase 3, the equipment was immediately switched off and the slats were removed.

The entire process lasted 6 h and 2 min at 185 °C, and 6 h and 12 min at 210 °C. The temperatures chosen for the open system process, 185 °C and 210 °C, correspond respectively to the Thermo-S (tangential swelling/contraction between 6% and 8%, durability class 3) and Thermo-D (tangential swelling/contraction between 5% and 6%, durability class 2) standards of the ThermoWood® process for hardwoods (International Thermowood Association 2023).

2.3 Colorimetry

Colorimetric analyses were carried out using the CIEL*a*b* system, with a CM 2600d portable spectrophotometer, version 1.41, from Konica Minolta Sensing, Inc. (Tokyo, Japan). The standard illuminant D65 was used, with an observation angle of 10° and specular light included. The device was previously calibrated with the black and white standard, using the CM-A145 plate. The colorimetric parameters L* (lightness), a* (red-green chromaticity coordinate) and b* (yellow-blue chromaticity coordinate) were measured. The parameters C* (saturation) and h° (hue angle) were calculated using Eqs. (1) and (2), respectively.

$$C^* = \left[\left(a^{*2} + b^{*2}\right)\right]^{1/2} \tag{1}$$

$$h^\circ = \tan^{-1}\left(\frac{b^*}{a^*}\right) \tag{2}$$

This test used the same specimens as the biodeterioration test (more details in the next section). Measurements were taken at two points, one on the cross-section and the other on the tangential face, through direct contact between the surface of the specimens and the lighting area of the device. The color of the wood was analyzed again after the biodeterioration test, using this same procedure.

2.4 Fungi Test

The fungi test was conducted in accordance with standard E-30-16 (American Wood Protection Association 2016), with the rot fungi *Trametes versicolor* (white-rot) and *Coniophora puteana* (brown-rot). Ten specimens measuring 25 × 25 × 9 mm (tangential × radial × longitudinal) were used for each treatment *x* fungus combination. *Gmelina arborea* and *Pinus elliottii* specimens were used as a reference for the *T. versicolor* and *C. puteana* fungi, respectively.

Initially, the specimens were placed in an oven at 103 ± 2 °C until they reached constant mass and were weighed on an analytical scale (0.0001 g) to determine the initial anhydrous mass (A). The test was carried out in a climate-controlled room (27 ± 2 °C and 70 ± 5% relative humidity) for 12 weeks.

At the end of the test, the specimens were thoroughly cleaned. After removing the fungal mycelium, the final anhydrous mass (B) of the specimens was obtained using the same initial procedure, and the mass loss (ML, %) was calculated (Eq. 3).

$$\mathrm{ML} = \frac{(A - \mathrm{B})}{A} \times 100 \tag{3}$$

2.5 Statistical Analysis

The correlation between the colorimetric parameters and mass loss was calculated using Eq. (4). The combination of the colorimetric parameters (five), the two measurement surfaces (cross-section and tangential) of the specimens, the two measurement periods (before and after the fungi test), and the four treatments resulted in 80 correlation coefficient measurements (r) per fungus (two species).

A significance test (Student's t-test, Eq. 5) was carried out for r. A scale proposed by Appolinário (2006) was adopted for the qualitative assessment of r (Table 1).

$$\mathrm{r} = \frac{\sum \mathrm{XY} - \frac{\sum \mathrm{X} \cdot \sum \mathrm{Y}}{\mathrm{n}}}{\sqrt{\left[\sum \mathrm{X}^2 - \frac{(\sum \mathrm{X})^2}{\mathrm{n}}\right] \cdot \left[\sum \mathrm{Y}^2 - \frac{(\sum \mathrm{Y})^2}{\mathrm{n}}\right]}} \tag{4}$$

Where - r: correlation coefficient; X: given colorimetric parameter; Y: mass loss - fungi test (%); n: number of specimens.

$$\mathrm{t} = \frac{\mathrm{r} \cdot \sqrt{\mathrm{n} - 2}}{\sqrt{1 - \mathrm{r}^2}} \tag{5}$$

Table 1. Qualitative classification of the correlation coefficient.

Correlation coefficient	Classification
0.00	Null
0.01–0.10	Very weak
0.11–0.30	Weak
0.31–0.59	Moderate
0.60–0.80	Strong
0.81–0.99	Very strong
1.00	Absolute

3 Results and Discussion

Since this study involved exploratory analysis, correlations were made before and after the fungi test. However, with a view to modeling, the main focus was on the correlations before biodeterioration.

Since 160 correlation coefficients (r) were determined and tested for significance, for reasons of space limitation and to simplify the discussion, we focus only on those that were significant ($p < 0.05$) and "strong" – $\geq$0.60 (Table 1).

The OS210 was the only sample in which all the correlations were not significant (Tables 2 and 3). The same occurred for OS185, considering the correlations of the colorimetric data before the test with *C. puteana* (Table 3).

Table 2. Number of significant correlation coefficients - r ($p < 0.05$) - per treatment before and after the fungi test with *Trametes versicolor*, and their qualitative classification.

Treatments	Before		After	
	"Very strong"	"Strong"	"Very strong"	"Strong"
Control	2	5	5	1
CS160	0	3	0	4
OS185	2	5	3	3
OS210	0	0	0	0

Table 3. Number of significant correlation coefficients - r ($p < 0.05$) - per treatment before and after the fungi test with *Coniophora puteana*, and their qualitative classification.

Treatments	Before	After
	"Strong"	"Strong"
Untreated	1	3
CS160	6	8
OS185	0	4
OS210	0	0

Overall, better correlations (Tables 2 and 3) were found for *T. versicolor* (12 "very strong" and 21 "strong" correlations) than for *C. puteana*, which had no "very strong" and 22 "strong" correlations.

For *T. versicolor*, lightness (L*) was the colorimetric parameter that better correlated with mass loss (Table 4), with six "very strong" and four "strong" correlations, followed by the hue angle (h°). The highest correlation (0.97**) with mass loss was seen in OS185, for L* obtained on the cross-section before the fungi test. In general, the best correlations for both before and after the fungi test were verified for the untreated wood (Table 2).

Table 4. Number of significant correlation coefficients - r ($p < 0.05$) - per colorimetric parameter and fungus, and their qualitative classification.

Colorimetric parameters	*Trametes versicolor*		*Coniophora puteana*
	"Very Strong"	"Strong"	"Strong"
L*	6	4	2
a*	1	2	6
b*	1	5	6
C*	1	5	5
h°	3	5	3

For *C. puteana*, the colorimetric parameters a* and b* were the ones that best correlated with mass loss, both with six "strong" correlations (Table 4). However, the highest correlation (0.79**) with mass loss was seen in 160CS for the hue angle (h°) obtained on the tangential surface. In general, the best correlations for both before and after the fungi test were verified for the CS160 sample (Table 3).

Regarding the effect of the measurement surface of the colorimetric parameters (Table 5), it was possible to identify some patterns.

Table 5. Number of significant correlation coefficients - r ($p < 0.05$) - by treatment and measurement surface of the colorimetric parameters, and their qualitative classification.

Treatments	Cross-section		Tangential	
	"Very strong"	"Strong"	"Very strong"	"Strong"
Untreated	3	2	4	8
CS160	0	10	0	11
OS185	3	6	2	6
OS210	0	0	0	0

For untreated wood, regardless of the fungal species and the period of analysis (before or after the fungi test), the best correlations were observed for the tangential surface. The same occurred for CS160, only for the correlations after the fungi tests; for the correlations before the fungi test, the best results were found for the cross-section. For OS185, the best correlations were obtained for cross-section for *T. versicolor* (before and after the fungi test). On the other hand, for *C. puteana*, there were no significant correlations before the fungi test and there was no definition of the best analysis surface (x or t) in the correlations after the fungi test.

4 Conclusions

There were significant "strong" and "very strong" correlations between the colorimetric parameters L*, a*, b*, C*, and h° for both fungal species. However, it was not possible to establish a method for developing models for predicting mass loss based on colorimetry, either with colorimetric parameter results obtained before or after the fungi test.

For all the conditions tested, there were no significant correlations for thermally modified wood in the open system at 210 °C. The correlations were also not significant in the open system at 185 °C before the fungi test with *Coniophora puteana*.

References

American Wood Protection Association: AWPA. E30-16: Standard method for evaluating natural decay resistance of woods using laboratory decay tests. In: AWPA Book of Standards, pp. 494–499, Birmingham (2016)

Appolinário, F.: Methodology of Science: Philosophy and Practice of Research. Thomson, São Paulo, Brasil (2006)

Bellon, K.R.R.: Thermal Modification of Wood from Three Species of Planted Forests Using the Holzsysteme® VAP Process. Master's thesis,. Federal University of Paraná, Curitiba (2013)

Gasparik, M., Gaff, M., Kacik, F., Sikora, A.: Color and chemical changes in teak (*Tectona grandis* L. f.) and meranti (*Shorea* spp.) wood after thermal treatment. Bioresources. **14**(2), 2667–2683 (2019)

Gomes, F.D.S., Reis, M.D.S., Rupf, A.C.O., Silvares, P.H.D.S., Ferreira, J.S.D.S., Dias, L.D.C., Madi, J.P.S., Cadermartori, P.H.G.D., Fantuzzi Neto, H., Paes, J.B., Batista, D.C.: Chemical composition of short-rotation teak wood thermally modified in closed and open systems. Appl. Sci. **13**, 1–15 (2023)

Hill, C.: Wood Modification: Chemical, Thermal and Other Processes. John Wiley & Sons, West Sussex (2006)
International Thermowood Association: ThermoWood® Handbook, Helsinki (2023)
Jones, D., Sandberg, D.: A review of wood modification globally—updated findings from COST FP1407. Interdisciplinary perspectives on the. Built Environ. **v 1** (2020)
Lopes, J.O.: Color uniformity and stability of thermally rectified wood from *Tectona grandis* L.f. Master's thesis,. Federal Rural University of Rio de Janeiro, Seropédica, Brasil (2012)
Lopes, J.O., Garcia, R.A., Latorraca, J.V.F., Nascimento, A.M.D.: Heat treatment changes the color of teak wood. Floresta e Ambiente, Seropédica. **21**, 521–534 (2014)
Menezes, W.M.: Effect of thermal modification on an industrial scale on the quality of *Tectona grandis* Linn. F. Doctoral thesis,. Federal University of, Santa Maria, Santa Maria (2017)
Motta, J.P., Oliveira, J.T.S., Paes, J.B., Alves, R.C., Dambroz, G.B.V.: Natural resistance of *Tectona grandis* wood in laboratory tests. Ciência Rural. **43**(8), 1393–1398 (2013)
Zelinka, S.L., Altgen, M., Emmerich, L., Guigo, N., Keplinger, T., Kymäläinen, M., Thybring, E.E., Thygesen, L.G.: Review of wood modification and wood functionalization technologies. Forests. **13**, 1004 (2022)

Physical Properties of Thermally Modified *Gmelina arborea* Wood Under Different Process Conditions

Samuel Olaniran(✉) and Holger Militz

Wood Biology and Wood Products, Georg-August-Universität Göttingen, Gottingen, Germany
samuel.olaniran@uni-goettingen.de

Abstract. This study investigated the physical properties of *Gmelina arborea* wood following thermal modification using the open and closed processes. The aim of examining these processes on gmelina wood is to find a suitable processing condition that will offer acceptable reduction of moisture ingress, deliver good quality of thermally modified (TM) wood, while enhancing other important technological properties based on target end uses. For this study, thermal treatment using the open process covered three temperature regimes; 180 °C, 200 °C, and 220 °C. Modification in the closed process was done under temperature regimes including 160 °C, 170 °C and 180 °C, and under low pressure, as use of high pressure have been reported to cause more severe degradation to wood polymers. The outcome of the above modification methods showed that corrected mass loss of samples modified under both processes increased with increasing temperature; but higher losses was recorded in the open process. Mass loss significantly influenced the final moisture content, and slightly density through the open process. Volumetric swelling was also reduced from 9.86% to 3.8% for samples modified at 220 °C in the open process, while similar volumetric swelling of 5.41% and 5.26% was recorded for samples modified at 180 °C in open and low-pressure closed process respectively. From these results, gmelina wood showed improved resistance to moisture ingress in the open process at higher temperatures between 200 °C to 220 °C compared to the low-pressure closed process. However, thermal modification of gmelina in the open process negatively influenced its final quality and may have further implications for its strength properties.

Keywords: Gmelina wood · Thermal modification · open and low-pressure closed processes

1 Introduction

Gmelina arborea wood is grown in tropical and sub-tropical countries of the world and its wood is sold in different markets in Europe and the United States due to its versatility for various applications [1]. In most plantations where they are initially established either for pulp or furniture production. Recent investigations have shown that gmelina wood can be channeled to other uses due to its excellent material properties, which

G. Goli et al. (Eds.): ECWM 2024, SPM 86, pp. 89–95, 2025.
https://doi.org/10.1007/978-3-031-99418-0_12

competes well with European spruce and pine [2, 3]. One major downside to the utilization of gmelina wood is its low durability and dimensionally unstable; and for this, treatments that enhances its properties are required. However, due to low treatability of gmelina, modification via chemical treatment is not feasible [4, 5], hence the choice of thermal modification processes. Modifications through thermal processes following ThermoWood® and Firmolin® have been used in modifying species such as *Eucalyptus nitens* (shining gum), *Fagus sylvatica* (European beech), *Pinus sylvestris* (Scots pine) sapwood and *Picea abies* (Norway spruce) [6, 7]. Both processes differ in their effects on modified wood products due to variation in processing parameters such as treatment temperature, pressure and relative humidity of the system. The influence of both processes on the physical properties of gmelina wood is yet to be compared to determine a suitable process. The nature of process used in thermal modification of gmelina wood, which may impact its end-use application, is hereby investigated and presented in this study.

2 Materials and Methods

Gmelina arborea wood was procured from mixed Teak and Gmelina plantations aged 25 years at the Federal University of Technology Akure, Nigeria; processed into large planks (1500 × 340 × 71 mm), pre-dried and shipped to University of Goettingen, Germany. Samples of dimension 650 × 50 × 20 mm (L-R-T) were prepared for thermal treatment. For each process (i.e., open and closed process), a total of fifteen slats were modified under three temperature regimes; open process temperature include: 180 °C, 200 °C and 220 °C, while the closed process includes 160 °C, 170 °C and 180 °C under low pressure. Modification under open process occur at atmospheric pressure, and gradual increase in temperature of the reactor until the target temperature is attained and a holding time of 180 min was applied followed by gradual cooling of the system. In the closed process, modification was done under low pressure of 0.18 MPa (160 °C), 0.23 (170 °C), and 0.29 MPa (180 °C); a stepwise increase in temperature and pressure of the system until the target was attained and a holding time of 180 min was applied, followed by cooling of the system as in the open process. Mass loss due to thermal modification under both processes were corrected to obtain a corrected mass loss (MLc) by deducting extractives from dry weight of the slats. After determining mass loss from the slats, 30 samples of dimensions 25 × 25 × 10 mm (R-T-L) each were prepared from slats modified from both open and low-pressure closed processes, and also from unmodified gmelina wood from same boards as those thermally modified.

2.1 Determination of Physical Properties

The physical properties of thermally modified gmelina wood investigated include; moisture content, density and volumetric swelling. To determine the moisture content, density and volumetric swelling of the modified and unmodified samples, they were initially weighed, measured in the radial, tangential and longitudinal directions, and then oven-dried at 103 °C until constant weight. Thereafter, samples were conditioned at 65%

relative humidity and then immersed completely in distilled water for ten (10) days. Volumetric swelling was calculated according to the equation:

$$VS(\%) = \frac{V_2 - V_1}{V_1} \times 100 \quad (1)$$

where V_1 and V_2 represents the volume of samples with conditioning at 65% RH and saturation in water and the volume of oven-dried samples, respectively.

3 Results and Discussion

3.1 Mass Loss Due to TM of Gmelina Under Open and Low-Pressure Closed Processes

Gmelina wood modified by two major thermal processes showed differing responses in terms of mass loss and corrected mass loss, which resulted in some changes physical properties. While it is evident that under both processes, mass loss increases with increase in treatment temperature (Fig. 1a, b); samples modified through closed process under low temperature and pressure showed minimal losses, signifying low degradation of wood polymers. However, comparing samples modified at 180 °C under open and closed process, it will be observed that mean corrected mass loss in the former is 7.86% and in the latter 9.25%. This is an indication that degradation is higher at for samples modified in the closed process due to increased pressure and relative humidity of the process compared to the open process where temperature is the sole factor. The goal of this study is to keep pressure at the minimum and limit the extent of wood-polymer degradation that may occur during modification under high pressure in the closed process. Higher corrected mass losses in the closed process than that in open process have been previously reported in wood species modified under high pressure and at low temperature [8].

While comparing the physical characteristics and quality of gmelina wood modified under both processes, Olaniran and Militz [9] reported that, subsequent to cross-cutting of modified gmelina, cracks were observed for samples modified with the open process; especially as temperature increased from 200 °C to 220 °C, whereas, in samples modified in the closed process no crack was present. This points to the susceptibility of gmelina wood to cracking in the process of thermal modification under high temperature and its poor drying characteristics. Hence, modification under humidified conditions as in the close process may be more advantageous for gmelina wood compared to modifications under extremely dry conditions as is the case in the open process. It should be noted that, thermal modification under low temperature and high pressure leads to accelerated degradation of hemicelluloses as previously reported in the literature, and will translate to higher losses in mechanical properties; although with an advantage of improved dimensional stability.

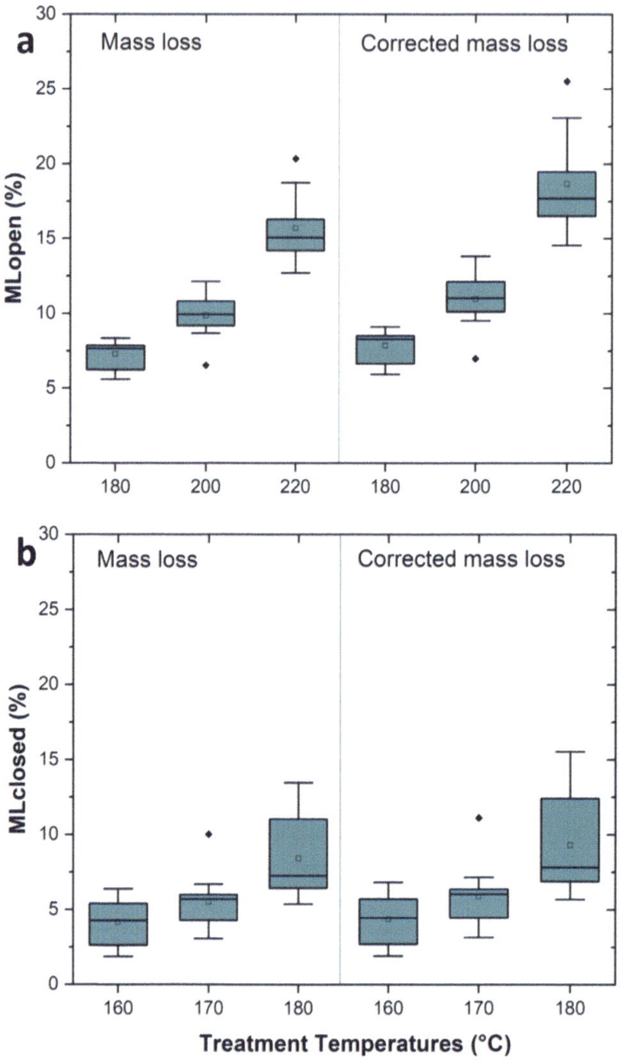

Fig. 1. Mass loss (ML) of Gmelina wood modified through open process (a) and low-pressure closed processes (b)

3.2 Physical Properties of TM Gmelina

The relationship between mass losses, moisture content and density from the open process and closed process (at low-pressure) are shown in Fig. 2a, b. For moisture content, both processes showed a decrease in moisture content as mass loss increases, but the relationship appeared stronger for samples modified through the open process ($R^2 = 0.93$). This indicates that the degradation of hemicelluloses may be higher in the samples modified in the open process, making them less hygroscopic due to limited amount of hydroxyl groups available for sorption [10] when compared to samples modified through the low-pressure closed process ($R^2 = 0.70$). The relationship between mass loss and density under both processes was quite low, but higher loss in density was evident in samples modified through the open process ($R^2 = 0.25$) when compared to samples modified in the low-pressure closed process ($R^2 = 0.03$). At modification temperature of 180 °C in the open process and low-pressure closed process; density seem to be comparable, but increasing pressure may lead to further reduction in density for the closed process. It is evident that lower density of TM wood under open process indicated higher degradation in cell wall polymers, especially hemicelluloses which may have been converted to volatile products and are evaporated alongside extractives as previously reported for heat treatment of other wood species by Boonstra and other coworkers [11].

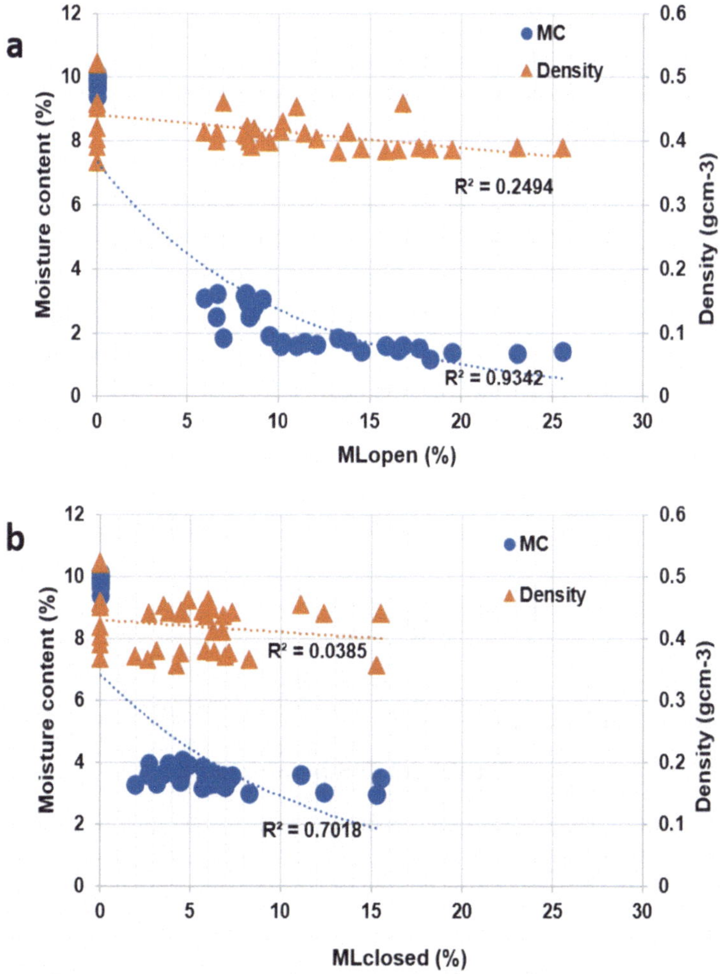

Fig. 2. Influence of mass loss on moisture content and density of gmelina wood thermally modified in open (a) and closed process (b).

Results of volumetric swelling of modified gmelina wood showed a similar pattern of decrease in volumetric swelling with treatment temperature under both processes (Fig. 2). In general, when compared to the control (unmodified) samples, volumetric swelling of gmelina wood is reduced significantly under both processes. However, reduction in volumetric swelling is more pronounced in samples modified through the open process. This could be a further confirmation of a higher degradation and loss of hemicelluloses in samples modified through the open process than in closed process when modification is done at low-pressure. It should be noted that samples modified at 180 °C under both processes showed comparable effect on volumetric swelling, although a higher mass loss (7.86%) was observed in samples modified at 180 °C under the low-pressure closed process (9.25%). There is then a possibility that as further degradation of wood polymers occur when thermal modification is done under (at temperature of 180 °C as in open process) a higher in the closed process, reduction in swelling may occur. This decrease in swelling could arise due to removal of water-soluble degradation process degradation products, leading to possible aggregation of cellulose microfibrils [12] and consequent reduction of capacity of cell wall of wood samples modified under high pressure for water uptake [13] (Fig. 3).

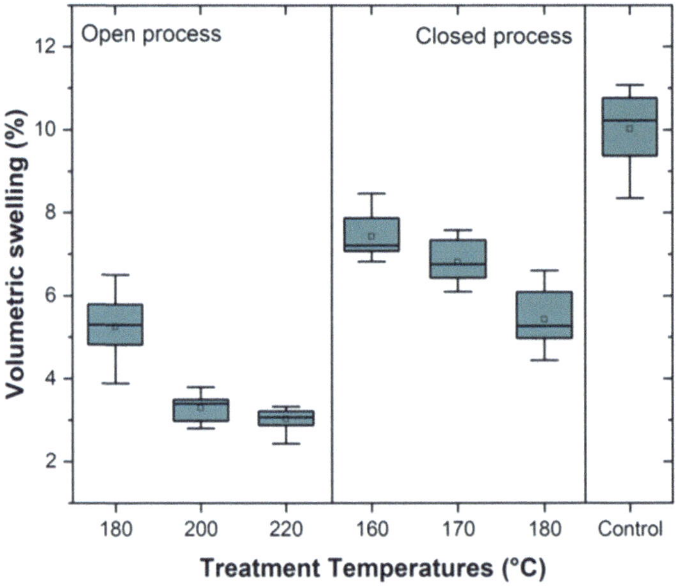

Fig. 3. Volumetric swelling of thermally modified Gmelina wood from both open and low-pressure closed processes

4 Conclusion

This study has shown that gmelina wood modified through the open process and closed process under low pressure showed distinct variations in terms of their physical properties. Samples modified in the low-pressure closed process at 180 °C and a higher pressure than 180 °C in the open process showed a higher corrected mass loss. Samples in this category showed comparable effect on volumetric swelling. It is however possible that thermal treatment under high pressure in closed process yield products with low volumetric swelling due to expected higher degradation of hemicelluloses than we now observed for the low-pressure treatment in this study. A closer look into effect of both processes on other properties (mechanical and durability) will provide further understanding and insight into selecting the appropriate processing conditions for gmelina wood. Future studies will focus on the influence of the open process and low-pressure closed process on the resistance of Gmelina wood to decay fungi. Furthermore, it is imperative to find a suitable recipe of temperature-pressure mix in the closed process TM for gmelina wood.

References

1. Dvorak, W.S.: World view of *Gmelina arborea*: opportunities and challenges. New For. **28**, 111–126 (2004)
2. Iwuoha, S.E., Seim, W., Onyekwelu, J.C.: Mechanical properties of *Gmelina arborea* for engineering design. Constr. Build. Mater. **288**, 123123 (2021)
3. Iwuoha, S.E., Seim, W., Olaniran, S.O.: Statistical distributions and their influence on the material property values of tropical timber: case study of *Gmelina arborea*. In: Structures, vol. 53, pp. 205–213. Elsevier (2023)
4. Ahmed, S.A., Chun, S.K.: Descriptions of the wood anatomy and safranine impregnation in *Gmelina arborea* Roxb. from Bangladesh. J. Korea Furnit. Soc. **18**(2), 100–105 (2007)
5. Olaniran, S.O., Löning, S., Buschalsky, A., Militz, H.: Impregnation proper-ties of Nigerian-grown *Gmelina arborea* Roxb. Wood. For. **13**(12), 2036 (2022)
6. Altgen, M.: Impact of Process Conditions in Open and Closed Reactor Systems on the Properties of Thermally Modified Wood. Cuvillier Verlag (2016)
7. Wentzel, M.: Process optimization of thermal modification of Chilean Eucalyptus nitens plantation wood (Doctoral dissertation, Niedersächsische Staats-und Universitätsbibliothek Göttingen (2019)
8. Wentzel, M., Fleckenstein, M., Hofmann, T., Militz, H.: Relation of chemical and mechanical properties of Eucalyptus nitens wood thermally modified in open and closed systems. Wood Mater. Sci. Eng. **14**(3), 165–173 (2019)
9. Olaniran, S.O., Militz, H.: Modification of *Gmelina arborea* wood for utilization in Nigeria. In: Proceedings of the 54th Annual Meeting of the International Research Group on Wood Protection. Held at Cairns Australia 28th May – 1st June, 2023. Document No: IRG 23–40986 (2023)
10. Kocaefe, D., Poncsak, S., Doré, G., Younsi, R.: Effect of heat treatment on the wettability of white ash and soft maple by water. Holz Roh Werkst. **66**, 355–336 (2008)
11. Boonstra, M., Van Acker, J., Tjeerdsma, B., Kegel, E.: Strength properties of thermally modified softwoods and its relation to polymeric structural wood constituents. Ann. For. Sci. **64**, 679–690 (2007)
12. Andersson, S., Serimaa, R., Vaananen, T., Paakkari, T., Jamsa, S., Viitaniemi, P.: X-ray scattering studies of thermally modified Scots pine (Pinus sylvestris L.). Holzforschung. **59**, 422–427 (2005). https://doi.org/10.1515/hf.2005.069
13. Wentzel, M., Altgen, M., Militz, H.: Analyzing reversible changes in hygroscopicity of thermally modified eucalypt wood from open and closed reactor systems. Wood Sci. Technol. **52**, 889–907 (2018)

Direct Evaluation of the Effect of Thermal Treatment on the Parallel Compression Strength of Wood

Rossana da Rosa[1](✉), Annie Cavalcante[1], Isabella de Sá[1], Bento Viana[1], Paula Dornelles[1], Lucia Garcia[2], and Cláudio Del Menezzi[1]

[1] Department of Forest Engineering, University of Brasilia, Brasilia, Brazil
ro.cortelini@hotmail.com
[2] Forest Product Laboratory, Brasilia, Brazil

Abstract. Due to the hygroscopicity, wood dimensional instability which limits its applications, especially in structural uses. A viable and well-studied solution to counteract this problem is thermal modification treatments. This study aims to evaluate the effect of thermal treatment on the parallel compression strength of Ipe wood and the chemical and physical changes. For this purpose, the thermal treatment was performed using wood specimens of 150 × 50 × 50 mm (l × w × t) and an average moisture content of 10%. The set temperatures and residence time in each stage were determined considering the preestablished drying parameters and the high density of Ipe wood. The treatment was divided into three phases. The equilibrium moisture content (EMC), density, dimensional stability, and compressive strength (fc,0) were determined before and after the thermal treatment. The results of this study indicated that the thermal treatment of Ipe wood resulted in a significant weight loss (WL), averaging a 15.7% difference. The EMC of Ipe wood was statistically different (p-value > 0.0001) before treatment were 10.1% and 2.7% after heat treatment. Before the treatment, the Ipe samples presented an average compressive strength of 102 MPa and 124 MPa afterwards, which means 22% increase. From the results obtained, it is evident that heat treatment has a positive effect on the mechanical properties of wood.

Keywords: Heat Treatment · Mechanical Properties · Wood Modification

1 Introduction

Wood is a hygroscopic material that has the capacity to absorb and lose water to the environment. This occurs in two main zones. In the first zone, water undergoes interaction between the hydroxyl groups and the wood constituents (cellulose, hemicellulose, and lignin) in the cell wall through hydrogen bonding, whereas in the second zone, absorption occurs through capillarity in the lumen region of the cells [1, 2]. Due to this characteristic, wood suffers dimensional instability, which negatively interferes with its final use, especially in structural applications. A possible solution to this problem is the use of heat treatments [3].

G. Goli et al. (Eds.): ECWM 2024, SPM 86, pp. 96–100, 2025.
https://doi.org/10.1007/978-3-031-99418-0_13

Knowing the variability of wood properties, to mitigate undesirable effects on products, transformation techniques and processes are often applied, such as thermal modification, which aims to preserve or improve the physical, chemical, and mechanical characteristics of wood. Thermal modification is a process of partial degradation of the main constituents of wood, using high temperatures, ranging between 160 °C and 260 °C with or without the presence of an inert atmosphere or not for a certain period, varying between 30 and 72 h [4–7]. Characteristics such as porosity, lumen size, pits, quantity and type of macromolecules, and the arrangement of the cell wall affect the success of heat treatment and determine other parameters such as heating rate, residence time, and temperature range [8].

The intensity, temperature, and duration of heat treatment cause changes in the chemical structure of wood, which are responsible for favorable changes in its physical structure: reduced swelling and shrinkage, low equilibrium moisture content, greater resistance to weathering, and biological deterioration, in addition to a better esthetic appearance [9–11]. Despite improvements in the physical and biological properties of heat-treated wood, several studies have reported a reduction in its mechanical properties [3, 12–15].

Considering the use of wood in Brazilian civil construction, it is possible to observe that its use has increased, driven mainly by the environmental movement of green constructions; however, its main application is still in temporary elements such as shoring and finished products such as floors and frames [16].

Ipe wood (*Handroanthus* spp.) is known for its high mechanical resistance, good dimensional stability, and high durability, and is widely used in the production of flooring, strung musical instruments, and wooden decks [17–19]. Therefore, the present study aims to direct evaluation of the effect of thermal treatment on the parallel compression strength of wood.

2 Materials and Methods

Ipe wood (*Handroanthus* sp.) measuring 100 × 50 × 50 mm with an average moisture content of 10% were used. The heat treatment of the wood was performed using a Binder oven with forced air circulation. The temperatures and duration of each level were determined by considering the high basic density of Ipe wood. To avoid defects such as cracking and collapse of the wood, three temperature levels were used: (I) 50 °C for 3 h; (II) 120 °C for 48 h, and (III) 180 °C for 48 h with the treatment totaling 99 h (Fig. 1).

The equilibrium moisture content (EMC), density, and compressive strength of the specimens were determined before and after heat treatment. The mass loss (ML) of the specimens was determined by weighing the samples before and after heat treatment and expressed as percentages. To evaluate variations in density before and at the end of treatment, the average density of the samples was determined using X-ray densitometry (Fergus-GreCon).

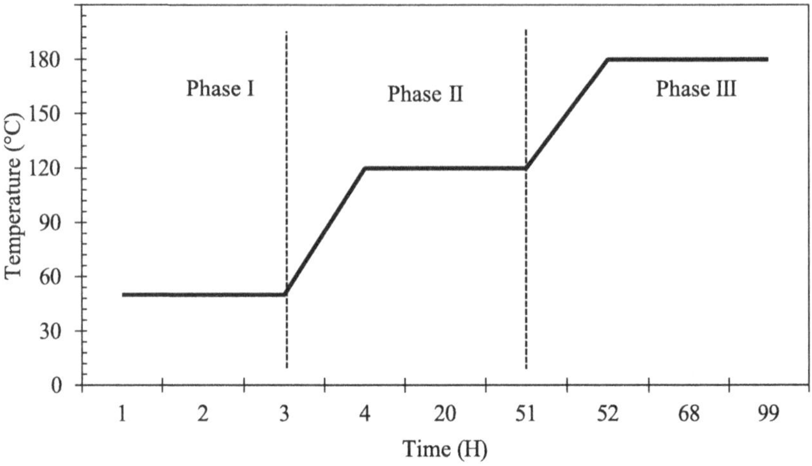

Fig. 1. Temperature thresholds and treatment duration at each stage of heat treatment.

The dimensional stability and parallel compression resistance were determined in accordance with the ABNT 7190:2022 and ASTM D143-22 standards. The compressive stress ($^{\partial}0$) was calculated from the ratio between the rupture and the cross-sectional area parallel to the loading. The values of the untreated specimens were corrected for 12% moisture content. The effect of thermal treatment on the parallel compression strength and physical properties of Ipe wood were calculated using T-test with a probability of 0.05.

3 Results and Discussion

The results of this study indicated that the thermal treatment of Ipe wood resulted in a significant weight loss (WL), averaging a 15.7% difference. The WL is most likely caused by the degradation of wood chemical components, such as the hemicellulose. The mains changes in mass were observed in phases II (150 °C) and III (180 °C) of the treatment, which were performed at the highest temperatures. The Table 1 presents the results of density, EMC, and compression strength. The EMC of Ipe wood was statistically different (p-value > 0.0001) before were 10.1% and 2.7% after heat treatment. This indicates that the hemicellulose was heavily degraded since they are the most sensitive wood polymer to thermal degradation and their degradation directly affects wood water adsorption ability.

For the results of fc,0, the t-test showed a significant difference before and after the thermal treatment (p-value = 0.000664). Before the treatment, the Ipe samples presented an average compressive strength of 102 MPa and 124 MPa afterwards, which means 22% increase. Studies have shown that thermal treatments with maximum temperatures of 190 °C positively impact mechanical properties [20–22]. In this temperature range, the degradation of low molecular weight sugars in wood and a proportional increase in the amount of crystalline cellulose and lignin occur due to the degradation of components that are more fragile to high temperatures, such as hemicellulose and the non-crystalline

Table 1. Heat treatment effects on properties of Ipe wood (*Handroanthus* sp.)

Ipe wood	EMC [%]	Density [kg/m^3]	fc,0 [MPa]
Non treated	10.1 (0.18)*	1016.2 (72.67)ns	102.0 (5.53)*
Heat treated	2.7 (0.44)*	961.0 (76.85)ns	124.1 (11.62)*
p-value	<0.0001	0.161	<0.001

Note: values between parentheses indicate the standard deviation; *, *ns* significant and no-significant in T-test at 0.05 level.

fraction of cellulose. From the results obtained, it is evident that heat treatment has a positive effect on the compressive properties of Ipe wood.

4 Conclusions

Based on the parallel compression strength results obtained in this study, it is evident that heat-treated Ipe wood has potential for structural applications. However, it is necessary to consider factors such as the distribution and application of forces in certain applications since the type of wood failure undergoes changes.

Acknowledgements. The authors thank the Coordenação de Aperfeiçoamento de Pessoal de Nível Superior (CAPES) for granting the Scholarship to the first and second authors.

References

1. Fredriksson, M.: On wood-water interactions in the over-hygroscopic moisture range-mechanisms, methods, and influence of wood modification. Forests. **10**(9) (2019)
2. Riza, M.A., Go, Y., Maier, R.R.J., Harun, S.W., Anas, S.B.: Hygroscopic materials and characterization techniques for fiber sensing applications: a review. Sens. Mater. **32**(11), 3755–3772 (2020)
3. Ferreira, M.D., Melo, R.R., Zaque, L.A.M., Stangerlin, D.M.: Propriedades Físicas e Mecânicas da Madeira de Angelim-Pedra submetida a Tratamento Térmico. Tecnologia Em Metalurgia Materiais e Mineração. **16**(1), 3–7 (2019)
4. Godinho, D., Araújo, S.O., Quilhó, T., Diamantino, T., Gominho, J.: Thermally modified wood exposed to different weathering conditions: a review. Forests. **12**(10) (2021)
5. Hill, C., Altgen, M., Rautkari, L.: Thermal modification of wood—a review: chemical changes and hygroscopicity. J. Mater. Sci. **56**(11), 6581–6614 (2021)
6. Modes, K.S., Santini, E.J., Vivian, M.A., Garlet, A.: Influência da termorretificação na resistência a degradação biológica das madeiras de pinus taeda e Eucalyptus Grandis. Ciencia Florestal. **27**(3), 993–1002 (2017)
7. Spear, M.J., Curling, S.F., Dimitriou, A., Ormondroyd, G.A.: Review of functional treatments for modified wood. Coatings. **11**(3) (2021)
8. Chen, C., Kuang, Y., Zhu, S., Burgert, I., Keplinger, T., Gong, A., Li, T., Berglund, L., Eichhorn, S.J., Hu, L.: Structure–property–function relationships of natural and engineered wood. Nat. Rev. Mater. **5**(9), 642–666 (2020)

9. Jebrane, M., Pockrandt, M., Cuccui, I., Allegretti, O., Uetimane, E., Terziev, N.: Comparative study of two softwood species industrially modified by Thermowood® and Thermo-vacuum process. Bioresources. **13**(1), 715–728 (2017)
10. Kubovský, I., Kačíková, D., Kačík, F.: Structural changes of oak wood main components caused by thermal modification. Polymers. **12**(2) (2020)
11. Yildiz, S., Gümüşkaya, E.: The effects of thermal modification on crystalline structure of cellulose in soft and hardwood. Build. Environ. **42**(1), 62–67 (2007)
12. Gaff, M., Kačík, F., Gašparík, M.: Impact of thermal modification on the chemical changes and impact bending strength of European oak and Norway spruce wood. Compos. Struct. **216**, 80–88 (2019)
13. Korkut, S.: The effects of heat treatment on some technological properties in Uludağ fir (Abies bornmuellerinana Mattf.) wood. Build. Environ. **43**(4), 422–428 (2008)
14. Korkut, S., Hiziroglu, S.: Effect of heat treatment on mechanical properties of hazelnut wood (Corylus colurna L.). Mater. Des. **30**(5), 1853–1858 (2009)
15. Yang, T.H., Lee, C.H., Lee, C.J., Cheng, Y.W.: Effects of different thermal modification media on physical and mechanical properties of moso bamboo. Constr. Build. Mater. **119**, 251–259 (2016)
16. Evangelista, W.V., Costa, E.D.: Evaluation of physical-anatomical properties of two woods used in the production of flooring. J. Agro-Environ. Sci. **15**(2), 132–138 (2017)
17. Longui, E.L., Yojo, T., Lombardi, D.R., Alves, E.S.: The potential of Ipê (Handroanthus spp.) and Maçaranduba (Manilkara spp.) woods in the manufacture of bows for string instruments. IAWA J. **31**(2), 149–160 (2010)
18. Paes, J.B., Morais, V.D.M., Lima, C.R., dos Santos, G.J.C.: Resistência natural de nove madeiras do semiárido brasileiro a fungos xilófagos em simulares de campo. Revista Árvore. **33**(3), 511–520 (2009)
19. Teixeira, D.E., Cunha, L.E., Wimmer, P., Andrade, A.: Resistência à abrasão, dureza Janka e a correlação com outras propriedades em 14 espécies de madeiras tropicais brasileiras com potencial para utilização em pisos de madeira. Revista Ciência da Madeira. **10**(2), 135–141 (2019)
20. Boonstra, M.J., Van Acker, J., Tjeerdsma, B.F., Kegel, E.V.: Strength properties of thermally modified softwoods and its relation to polymeric structural wood constituents. Ann. For. Sci. **64**(7), 679–690 (2007)
21. Brito, J.O., Silva, F.G., Leão, M.M., Almeida, G.: Chemical composition changes in eucalyptus and pinus woods submitted to heat treatment. Bioresour. Technol. **99**(18), 8545–8548 (2008)
22. Herrera-Díaz, R., Sepúlveda-Villarroel, V., Torres-Mella, J., Salvo-Sepúlveda, L., Llano-Ponte, R., Salinas-Lira, C., Peredo, M.A., Ananías, R.A.: Influence of the wood quality and treatment temperature on the physical and mechanical properties of thermally modified radiata pine. Eur. J. Wood Wood Prod. **77**(4), 661–671 (2019)

Temperature and Moisture Content Effects on Wood Compressive Properties

Hussein Daher[1,2(✉)], Sabine Caré[1], Loïc Payet[2], and Gilles Forêt[1]

[1] Navier Laboratory, Gustave Eiffel University, ENPC, CNRS (UMR 8205), Marne-la-Vallee, France
hussein.daher@cstb.fr

[2] CSTB, Champs-sur-Marne, France

Abstract. Wood may absorb or release water depending on the surrounding air conditions (relative humidity and temperature), impacting its mechanical properties, particularly stiffness and strength at room temperature. In the context of timber design, scant information is available on the effects of temperature, such as during exposure to fire. The current challenge involves understanding the residual properties of non-carbonized wood structural elements that are subjected to temperature up to 200 °C, to ensure occupant safety.

Building standards indicate that the compressive strength of wood at 100 °C is 25% lower than at 20 °C, yet the role of wood moisture content is not explicitly addressed. Experiments conducted by CSTB in 2020 revealed that, within the 20–100 °C temperature range, moisture content significantly influences strength. This paper aims to better investigate how temperature and moisture content affect the mechanical properties of wood. The experiments entailed examining the response of glued laminated timber (GL24H) to diverse levels of moisture content and temperature. Consequently, several heat protocols were examined to assess the impact of temperature and moisture content on timber design. Specifically, P_1 protocol served as the reference baseline at 20 °C for subsequent tests, while P_{1bis} et P_2 protocols were conducted at four temperatures (60, 100, 150 and 200 °C) with varying heating durations in an oven. Furthermore, the discussion of all results takes into account sample sizes.

Keywords: Drying temperatures · Fire · Moisture content · Mechanical properties · Timber design

1 Introduction

In recent times, there has been a global increase in the construction of wooden structures, in order to promote environmentally friendly construction practices and the utilization of forest resources. As a result, it has become crucial to assess the fire resistance of wooden elements exposed to fires, in order to ensure the credibility of wooden buildings in case of fire and to design appropriate structures.

One of the current challenges is to better understand the properties of non-carbonized wooden structural elements that may be subjected to temperatures ranging from 100 °C

G. Goli et al. (Eds.): ECWM 2024, SPM 86, pp. 101–108, 2025.
https://doi.org/10.1007/978-3-031-99418-0_14

to 200 °C. While temperature is an important factor affecting the mechanical properties of wooden elements exposed to fire heat, moisture content also plays a significant role. Research suggests that moisture content in the residual section (non-carbonized section) can vary due to complex processes such as water vaporization, transfer, and condensation [1–3].

According to the approach proposed in Eurocode 5 [4], the compressive strength of wood at 100 °C is 25% lower than that at 20 °C, but it does not specify the wood moisture content. However, preliminary tests conducted in 2020 at CSTB [5] revealed that beyond the temperature range of 20–100 °C, the compressive strength of wood is mainly influenced by its moisture content. At 100 °C, the compressive strength for a dry sample becomes nearly identical to the initial strength at 20 °C, constituting approximately 60% of the initial strength for a specimen with a moisture content of around 10%.

The behavior of wood in terms of its thermal, hygroscopic, and mechanical properties is complex and depends on various factors such as the type of wood species (solid or reconstituted wood), moisture content, temperature, and mechanical stresses. This study aims to investigate the changes in the mechanical properties of wood, specifically compressive strength, based on variations in temperature and humidity. The findings of this study will contribute to a better understanding of wood's behavior under different environmental conditions.

2 Materials and Methods

The tests were conducted on samples of "Glued Laminated Timber" (GL24H), with densities ranging from 0.38 to 0.46 g/cm^3 at room conditions.

2.1 Samples

The effect of five temperature levels, in the range from 20 to 200 °C, on the compressive strength parallel to grain was studied using three different sections (S_1, S_2, S_3), including 135 specimens with average density between 0.41 and 0.45 g/cm^3 and average moisture content between 11.5 and 12% at 20 °C (Table 1).

Table 1. Average density and moisture content of specimen groups at 20 °C

Section	Dimensions (mm)	Specimen	Density at 20 °C (kg/m^3)	Moisture content at 20 °C (%)
S_1	50 × 50 × 100	45	411	12
S_2	97 × 92 × 200	45	429	12
S_3	187 × 198 × 400	45	452	11.5

The samples were taken from glued-laminated beams of GL24H class in accordance with the NF EN 14080 standard, manufactured from spruce wood. These samples consisted of several spruce lamellas glued together, exhibiting diverse ring orientations. Then, these samples were stored in a climatic chamber at 20 °C and 65% relative humidity.

2.2 Experimental Protocols

Three protocols were implemented to conduct tests on glued-laminated timber (GL24H) to analyze its behavior under different conditions. These conditions included various temperatures (20 °C, 60 °C, 100 °C, 150 °C and 200 °C), moisture contents, and dimensions (S_1, S_2 and S_3). The initial protocol, P_1, involved compression tests at room temperature (20 °C) with a moisture content of approximately 12%. These tests served as a baseline for interpreting the results obtained from the other protocols. To simulate temperatures beyond room temperature (60 °C, 100 °C, 150 °C and 200 °C), samples underwent varying durations in an oven to achieve specific humidity or temperature conditions. Mechanical tests were conducted immediately after the samples were removed from the oven.

The integration of thermocouples into samples facilitated the verification of temperature consistency with that in the oven. The P_{1bis} protocol consisted of having about 10% mass loss for each kind of samples (S_1, S_2 and S_3). On the other hand, P_2 protocol aimed to maintain a uniform temperature throughout the entire height of the sample. Relative mass loss, drying time in oven and final moisture content are given in Table 2. For each protocol, drying time increases with the size of the samples and decreases with temperature. These experiments aim to have similar conditions in terms of heating temperature and moisture content for all sample sizes.

Table 2. Relative mass loss, drying time, and moisture content for the 3 sections for P_{1bis} and P_2 protocols

		60 °C		100 °C		150 °C		200 °C	
		P_{1bis}	P_2	P_{1bis}	P_2	P_{1bis}	P_2	P_{1bis}	P_2
Relative mass loss (%)	S_1	10.1	9.3	10.5	11.3	10.4	10.3	10.2	12.3
	S_2	9.6	5.3	10.3	10.3	10.5	13.8	10.3	12.3
	S_3	9.7	7.1	9.5	10.4	10.5	13.0	10.0	15.2
Drying time (h)	S_1	27	24	10	8	2	2	1	1.5
	S_2	144	28	48	29	19	10	3	6.5
	S_3	336	168	51	89	24	46	7	19
Moisture Content MC (%)	S_1	0	0	0	0	0	0	0	0
	S_2	1.6	6.1	0.5	0.5	0.3	0	1.1	0
	S_3	0.4	3.3	1.2	0.2	0.2	0	1.1	0

3 Results and Discussions

In this section, the data from the thermocouples, the effects of heating temperature on color change, moisture content, and chemical degradation, as well as the results of compressive strength parallel to grain of glued-laminated timber, are presented. All these results will be analyzed using statistical tests.

3.1 Thermocouple Data

In order to assess the thermal gradient within the samples, three thermocouples were inserted into each specimen, two at the sides and one at the center, at mid-thickness, as shown in Fig. 1.

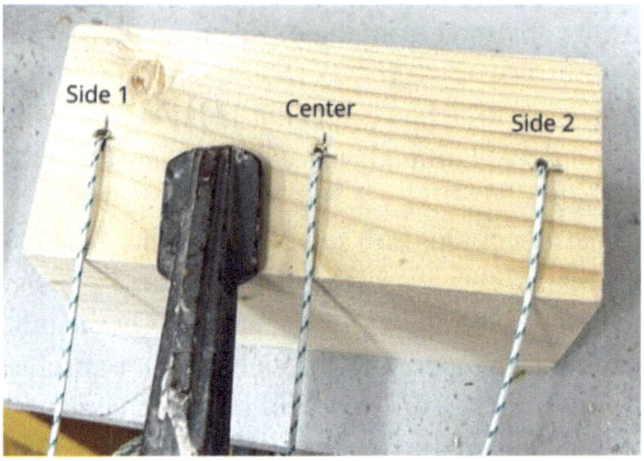

Fig. 1. Positioning of thermocouples at various heights within the samples

Figure 2 presents the thermocouple data for section S_3 according to P_2 protocol. The results indicate that the temperature remains uniform throughout the height after 40 h of heat treatment. Similar results have been obtained in the other cases (samples S_1, S_2 and S_3 and for each respective temperature). This uniformity is also evident in the case of P_{1bis} protocol (see Table 2).

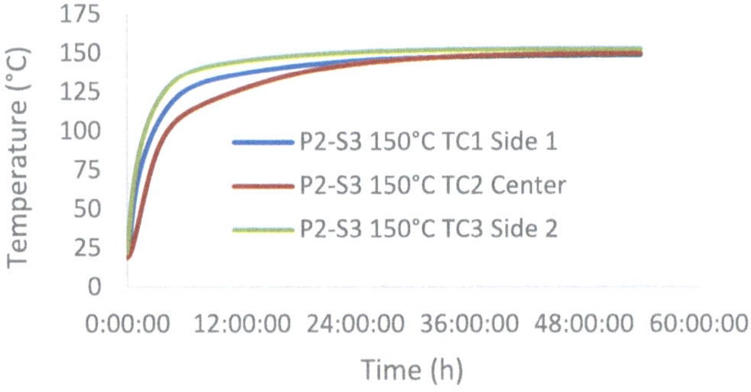

Fig. 2. Thermocouple data for S_3 at 150 °C in the oven for P_2 protocol

Figure 3 shows the thermocouple data for sample S_2 at 100 °C (results are similar for the other experiments). It is evident that the maximum strength has been reached with a uniform temperature throughout the sample height, thus confirming compliance with the conditions of the P_2 protocol. It is particularly noteworthy that, from the beginning of the mechanical test until reaching the maximum strength, no thermal gradient was observed inside sample, thereby reinforcing the obtained results.

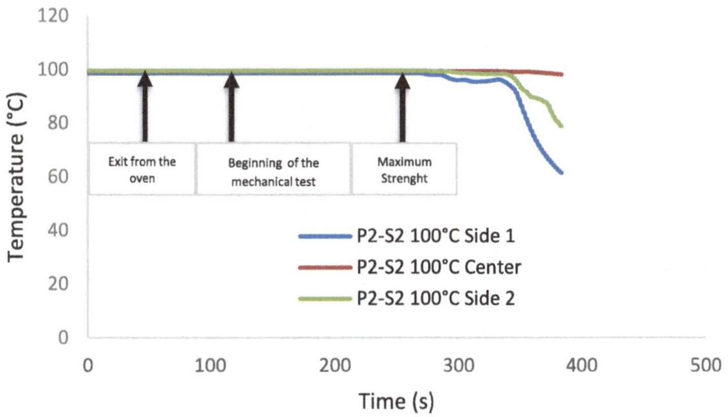

Fig. 3. Thermocouple data for S_2 at 100°C during the mechanical test for P_2 protocol

3.2 Color Changes of Specimens, Water Content Evolution, and Chemical Modification Due to Heating Temperature

The Fig. 4 presents the changes of wood color within the heating temperature range of 20–200 °C. These changes were observed in the external surfaces of the samples (due to thermal oxidation). The color changes became more noticeable at higher temperatures, particularly above 150 °C. Additionally, signs of wood combustion and checking were observed at the ends of the specimens. In these regions, the release of gas resulting from the thermal degradation of wood chemical components occurred more easily due to the grain orientation [6]. At lower temperatures, the color changes were less pronounced because the extent of chemical changes depends on the temperature level [7]. Within the temperature range of 40 °C–90 °C, the physical changes primarily involved the removal of the water and the emission and volatile extractives [8].

Fig. 4. Color changes of the specimens with heating temperatures. The colors between 20 °C and 60 °C are not significative because of the lighting conditions. Pictures have been taken after compressive tests that induced cracks.

These results suggest internal chemical modifications due to heating, although they have not been quantified. Additionally, temperature induces a decrease in moisture content (see Table 2). Considering the duration of heating, it can be assumed that the chemical modifications and moisture content are relatively homogeneous within the samples.

3.3 Compression Strength Parallel to Grain

Figures 5 and 6 respectively present the average compressive strength for the 3 types of samples S_1, S_2 and S_3 according to P_{1bis} and P_2 protocols. These results show that the compressive strength initially increases with temperature (up to 60 °C for P_{1bis} protocol and between 60 °C and 100 °C for P2 protocol), followed by a significant decrease. At room temperature, all 3 types of samples exhibit the same average strength of 34.5 MPa ± 1 MPa. The standard deviation for results obtained at higher temperatures and for each sample is less than 5 MPa. Considering the standard deviations, the evolution of compressive strength with heating temperature can be regarded as significant. This evolution will be discussed in the next section using statistical tests.

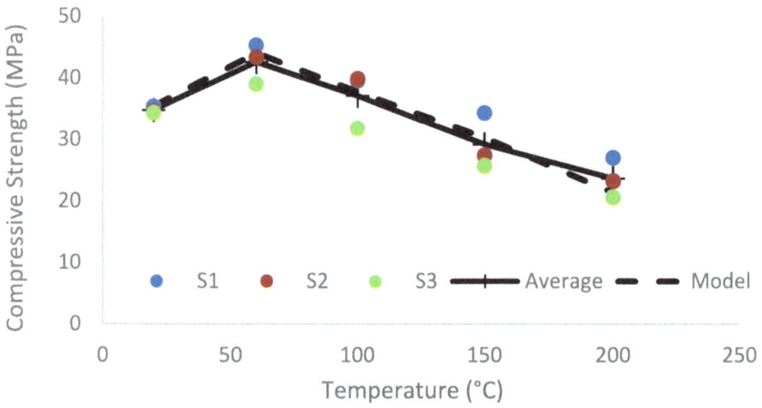

Fig. 5. Compressive Strength of the 3 types of samples S_1, S_2 and S_3 for P_{1bis} protocol. The solid line represents the average compressive strength for the three sections at each heating temperature, while the dashed line represents the average data predicted by the proposed model.

At 60 °C, the strength increase is approximately 25% compared to that at room temperature for S_1 (MC = 0%), 25% for S_2 (MC = 1.6%), and 14% for S_3 (MC = 0.4%). This increase may be explained by the decreases of moisture content (Table 1) without effect of thermal degradation (in coherence with no change in wood color, Fig. 4). At 200 °C, the strength reduction is approximately 33% for S_1, 35% for S_2 and 40% for S_3 compared to their strength at room temperature. These results may be attributed to the glass transition of polymers and the thermal degradation of wood. Otherwise, the compressive strength rises with a decrease in the moisture content of the samples.

3.4 Discussion

To examine the effect of the heating temperature on the compressive strength evolution, statistical tests were conducted. A multiple linear regression analysis was performed to assess the impact of moisture content and temperature on the compressive strength of glued laminated timber. The model was significant, $F(2, 27) = 48.1$, $p < .001$, with an R^2 of 0.781, indicating that 78.1% of the variation in compressive strength is explained by the model.

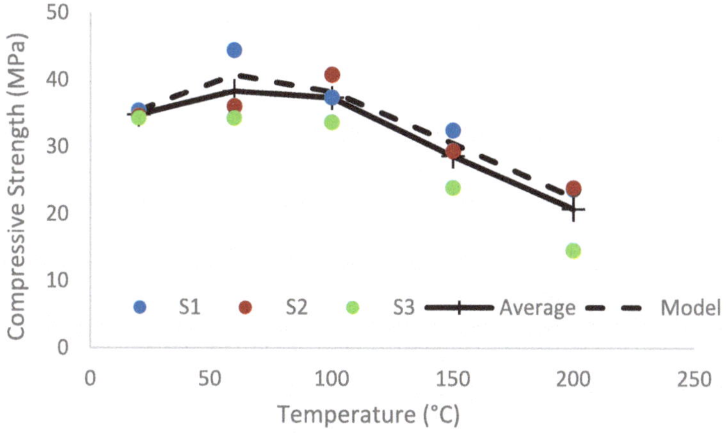

Fig. 6. Compressive Strength of the 3 types of samples S_1, S_2 and S_3 for P_2 protocol. The solid line represents the average compressive strength for the three sections at each heating temperature, while the dashed line represents the average data predicted by the proposed model.

The results revealed a significant effect of moisture content on compressive strength, $\beta = -1.406$, $p < .001$, suggesting that for every 1% increase in moisture content, compressive strength decreases by 1.406 MPa. Temperature also had a significant effect on compressive strength, $\beta = -0.155$, $p < .001$, indicating that for every 1 °C increase in temperature, compressive strength decreases by 0.155 MPa.

To further explore the sample size effect and the relationship between compressive strength, temperature, and moisture content, a model fitting compressive strength (σ) as a function of temperature (T) and moisture content (MC) was proposed. The model, as depicted in Figs. 5 and 6, follows the formula:

$$\sigma = A * MC + B * T + C\#(1)$$

The coefficients A, B, and C were determined using sample S_1, based on the P_{1bis} protocol at 20 °C, 60 °C, and 100 °C, and with the P_2 protocol at the same temperatures. To generate the model curve (Figs. 5 and 6), the average of coefficients A, B, and C for both the P_{1bis} and P_2 protocols was employed. For subsequent experiments (S_1 at 150 °C and 200 °C), as well for S_2 and S_3 at 5 temperatures, the previously obtained coefficients were utilized.

While the model performed well for samples S_1 and S_2, it showed reduced accuracy for S_3. This discrepancy is attributed to scaling effects, degradation of S_3 specimens (significant cracks were observed in S_3 specimens, especially at temperatures above 100 °C), and the presence of multiple nodes in all S_3 specimens.

4 Conclusion

The impact of heating temperatures up to 200 °C on the compressive strength parallel to the grain of glued-laminated timber (GL24H) was evaluated to assess its effect on chemical degradation and moisture content in wood samples of varying sizes. Based on these findings, building standards should be adjusted to account for the following observations:

- Up to approximately 100 °C, the compressive strength is primarily affected by the reduction in moisture content. Beyond 100 °C, the strength significantly decreases due to the effects of temperature. This suggests that the impact of heating extends beyond visible changes, such as color alteration, to internal chemical degradation of wood polymers at elevated temperatures.
- A similar trend was observed for S_1 and S_2 samples. However, S_3 samples consistently exhibited reduced performance across all heating temperatures, particularly with the P_2 protocol. This can be attributed to scale effects, degradation during heating (with significant cracks observed at temperatures exceeding 100 °C), and the presence of multiple nodes in all S_3 specimens.

References

1. Gerhards, C.C.: Effect of moisture content and temperature on the mechanical properties of wood. Wood Fiber. **1**, 4–36 (1982)
2. White, R.H., Shaffer, E.L.: Transient moisture content gradient in fire-exposed wood slab. Wood Fiber Sci. **13**(1), 17–38 (1981)
3. Wood Handbook: Wood as an Engineering Material. The Laboratory N°72, Department of Agriculture, United States (1987)
4. Eurocode 5: Conception et calcul des structures en bois. NF EN 1995-1-1,2 (2005)
5. Manthey, M.: Eude de la résistance en compression en fonction de la température du matériau bois-épicéa. Rapport d'essais n° EEM/EA2R 2026087828, CSTB, France (2020)
6. Marinquez, M.J., Moraes, P.D.: Influence of the temperature on the compression strength parallel to the grain of paricá. Constr. Build. Mater. **24**(1), 99–104 (2010)
7. LeVan, S.L.: Thermal degradation. In: Schniewind, A.P. (ed.) Concise Encyclopedia of wood and wood-based Material, pp. 271–273. Pergamon Press (1989)
8. Manninem, A.M., Parsanen, P., Holopainen, J.: Comparing the VOC emissions between air-dried and heat-treated scots pine wood. Atmos. Environ. **36**, 1763–1768 (2002)

Wood Modification Methods and Fire Resistance of Façades and Cladding

Joris Van Acker[1](✉), Marcy Durimel[1], Liselotte De Ligne[1], Bogdan Parakhonskiy[1,2], Andre G. Skirtach[2], and Jan Van den Bulcke[1]

[1] Laboratory of Wood Technology (UGent-Woodlab), Department of Environment, Faculty of Bioscience Engineering, Ghent University, Ghent, Belgium
{Joris.VanAcker,Marcy.Durimel,Liselotte.DeLigne, Bogdan.Parakhonskiy,Jan.VandenBulcke}@UGent.be

[2] Research Group Nano-Biotechnology, Department of Biotechnology, Faculty of Bioscience Engineering, Ghent University, Ghent, Belgium
Andre.Skirtach@UGent.be

Abstract. The last decade a considerable increase in façades/cladding applications have been linked to modified wood. It is becoming very common to use cladding products based on thermal modification using species such as spruce, poplar, ayous, and limba. Often these treatments alter the equilibrium moisture content and chemical composition and products are optimized in view of dimensional stability and decay resistance. Fire safety is of high importance for building with wood. Not only the structural building elements, like cross-laminated timber or light timber-frame structures are relevant to be assessed, also the building envelope needs consideration. A range of performance testing is available to asses fire resistance but this paper focuses on the cone calorimeter method. A variety of commercial products have been assessed and benchmarked with non-modified wood including some common (tropical) hardwoods used for cladding.

Regarding the impact of (thermal) wood modification on the fire resistance of fa-cades/cladding, it's essential to understand that while certain aspects of wood performance can be improved, it might have an impact on its fire resistance. To improve the fire resistance of wood facades/cladding, additional fire-retardant treatments or coatings can be applied. These treatments are designed to inhibit or delay the ignition and spread of fire.

The overall performance of wooden façades/cladding as part of the building envelope could benefit from combining wood protection options to enhance resistance against decay and to improve fire safety. Hence solutions for fire retardant treatments with adequate weather resistance could benefit from wood preservation and wood modification treatment technologies.

Keywords: Cladding · Façades · Fire resistance · Fire safety · Thermal wood modification · Calcium carbonate

G. Goli et al. (Eds.): ECWM 2024, SPM 86, pp. 109–123, 2025.
https://doi.org/10.1007/978-3-031-99418-0_15

1 Introduction

1.1 Fire Safety

The combustibility of timber is one of the main reasons that many building regulations limit the use of timber as a building material. Fire safety is an important contribution to feeling safe, and an important criterion for the choice of building materials. Historically, the combustibility of wood has been a disadvantage for using timber as a construction material. The main precondition for an increased use of timber in buildings is providing adequate fire safety [1]. The fire performance of timber cladding is a subject of particular interest when used in multi-storey buildings. Fire can spread on external walls due to an adjacent building being on fire, a fire directly outside of the building, or due to fire in a room with an opening through the cladding [2]. The recent fire events in buildings involving combustible cladding systems all over the world have raised concerns regarding the fire risk that these systems can pose. The fire behaviour of facades has come to be, more than ever, a topic of prime importance [3, 4]. Hakkarainen and Oksanen [5] performed research to lay the foundation for the reform of structural fire safety regulations for wooden facades. The basic principle of the acceptance criteria is, within a defined fire exposure, to ensure the safe evacuation of a multi-storey house from the compartments close to the origin of the fire. Two different fire scenarios can be defined to examine the behaviour of wooden facades: an external fire and a flashover of a compartment fire. It remains crucial to adhere to local building codes and regulations related to fire safety when choosing materials for facades and cladding, especially in areas prone to wildfires or when dealing with multi-storey buildings.

There has been a significant improvement in façade design to ensure excellent building performance in terms of energy efficiency and requirements on aesthetic appeal. These changes fundamentally alter the behaviour of modern façades in fire and pose a risk to building safety and economic loss in the event of a fire. Among the other factors affecting the fire performance of façade elements are the geometry, cavities, and external factors such as wind and spaces between buildings. Geometry greatly impacts the rapid spread of fire. The cavities between the external wall and the façade element, and within the façade element itself also impact the fire hazard [6]. Although designing wooden façades to improve fire safety is very relevant in relation to the concerns expressed above, the material used for wooden façades or cladding is a complementary factor that should not be neglected. Besides variability in types of wood products used, the low thermal conductivity and slow charring rate of wood products can even be used to protect underlying products from being heated and ignited and such fire protective behaviour of wood coverings has been verified by different methodologies [7]. Present-day engineers seek claddings with improved fire performance to satisfy state/government mandates against fire hazards. It is challenging to cater to the increased cladding demand using conventional materials while striving to comply with circular economy principles. As such, researchers and materials engineers have been paying more attention to identifying alternative cladding products [8].

1.2 Material Properties

The most durable species used for timber cladding in Europe are considered to be western red cedar (*Thuja plicata*), European larch (*Larix decidua*), Siberian larch (*Larix sibirica*), Douglas fir (*Pseudotsuga menziesii*) and European oak (*Quercus robur*). Other species, such as fir (*Abies* spp.), spruce (*Picea* spp.) and pine (*Pinus* spp.), are usually considered to be insufficiently durable for use unless they are treated in some way to protect or preserve them. Different modified claddings in use are based on thermally modified pine siding, Accoya siding (acetylation), Kebony siding (furfurylation), and charred spruce siding [2]. Wood modification has been defined by Hill [9] as a process that "involves the action of a chemical, biological or physical agent upon the material, resulting in a desired property enhancement during the service life of the modified wood." This is frequently considered to be a separate technology from the established wood preservative treatments using biocides.

Several systems are well-developed and commercially available, including: acetylation, thermal modification and resin or polymer modification [10]. There are several ways to chemically modify the wood cell wall polymers and the most abundant single site for reactivity is the hydroxyl groups. Chemical modification requires the penetration of the modifying chemical into the microstructure of wood. Wood species, which are difficult to impregnate, are therefore not suitable for chemical modification, at least at greater dimensions. The main form of chemical modification has been acetylation. Also relevant in this context is the furfurylation process which is based on the impregnation of furfuryl alcohol into the wood structure [11].

Thermally modified wood is particularly suitable for cladding as it provides an appropriate service life for easy replacement of elements that do not require high mechanical strength [12]. Thermal wood treating processing involves temperatures at 100–300 °C [11] and can be seen as thermal treatments (<150 °C), thermal modification and leading to thermal degradation (>240 °C) [13]. Two distinctly different purposes are mentioned: (a) softening the wood in steam or water to release internal stresses and make the wood easier to further process, or (b) controlled degradation of the wood involving temperatures between 150 and 260 °C with the purpose of improving shape stability and decay resistance. Resin or polymer modification relates to impregnated wood with phenol-formaldehyde (PF) resins and more recently with melamine-formaldehyde (MF) resins [12].

Charring of the wood surface also provides a degree of protection against fire [14, 15]. Although the flame charred surface is porous and rather friable, it resists weathering well because of compositional changes, measured by spectroscopical methods, that lead to a more ordered, graphitized carbon structure that is hardly reactive towards photodegradation [16]. Bartlett and co-authors [17] used a detailed meta-analysis of the fire resistance literature to show that the rate of burning as characterised by charring rate averaged over the full test duration and is observed to vary with material properties, in particular density and moisture content which induce a maximum 18% variability over the ranges expected in design.

Charring rate is demonstrated to be strongly dependent on density. In general, it has been found that charring rate under exposure to the standard temperature-time curve (to which the majority of recent work has been undertaken) can vary from around

0.8 mm/min for light, dry softwoods, decreasing to 0.4–0.5 mm/min for dense, moist hardwoods, with 0.6 mm/min being a reasonable average value. Decreasing density occurs due to an increase in void volume, reducing thermal conductivity and thus producing localised heating and heat accumulation, resulting in increased flame spread rates [17]. Njankouo and co-authors [18] tested spruce ($\rho = 478$ kg/m^3), fir ($\rho = 421$ kg/m^3), oak ($\rho = 557$ kg/m^3), azobe ($\rho = 1038$ kg/m^3), afzelia ($\rho = 860$ kg/m^3), balau ($\rho = 962$ kg/m^3), bilinga ($\rho = 692$ kg/m^3), meranti ($\rho = 522$ kg/m^3), merbau ($\rho = 779$ kg/m^3), and wengé ($\rho = 923$ kg/m^3) in a vertical furnace, with samples inserted in a frame with 20 holes measuring 280 $\times$ 170 mm and subjected to one-sided heating. The charring rate was found to decrease from around 0.60 mm/min at 500 kg/m^3 to around 0.40 mm/min at 1000 kg/m^3.

The presence of moisture is widely acknowledged as retarding pyrolysis due to a heat sink effect - the greater the moisture content, the greater the energy required to evaporate the water, and thus the less energy available for pyrolysis [17]. Permeability affects the charring rate as well. This is largely due to grain direction, as permeability along the grain is around four orders of magnitude higher than that across the grain [17]. Another significant factor affecting pyrolysis rate is the wood species. Whilst the species will affect factors discussed earlier such as density, moisture content, and permeability, there will also be additional factors specific to a species, such as chemical composition (primarily lignin content) and anatomy which influence the rate of pyrolysis [17]. The chemical constituents of wood behave independently during pyrolysis. At lower pyrolysis temperatures, lignin is the major contributor of char residue in the pyrolysis of wood. Hence, wood species with high lignin contents are more thermally stable. Aside from the composition of wood, external heat flux has a great influence on the pyrolysis process. Ignition occurs more rapidly in hardwoods than in softwoods. The ignition temperatures of wood constituents follow this rule: hemicellulose < cellulose < lignin. The moisture content of wood extends the time of ignition and increases the ignition temperature [19].

2 Materials and Methods

2.1 Reaction to Fire Test

To screen and compare materials on their reaction to fire the single-flame source test was selected according to EN ISO 11925-2. This test method measures the ignitability of building products when exposed to a small flame. The main purpose of the test is for classification according to Euroclass B, C, D, E, Bfl, Cfl, Dfl, Efl according to the European System. To be able to sell a construction product in Europe, the product must be tested, classified and if possible, CE-marked in accordance with the Construction Product Regulation. To meet the technical fire requirements, there are common test methods and a common classification system, EN 13501-1, also called Euroclass. This classification system is used by all EU countries, including Norway and Iceland.

The test specimen is mounted vertically and subjected to edge and/or surface exposure from a small gas flame. During the test, time of ignition, burning droplets and whether the flames reach the top marking of the test specimen within a prescribed time period, are registered. Although the method is designed to assess ignitability, this is merely addressed by measuring the spread of a small flame up the vertical surface of a specimen following the application of a small (match-sized) flame to either the surface or edge of a specimen for either 15 or 30 s. The determination of the production of flaming droplets/particles depends on whether or not the filter paper placed beneath the specimen ignites. For this research, a small flame was applied to the surface of each specimen for 30 s. At that stage flame height was recorded, the after-flame time was registered and finally the burn length was measured (Fig. 1).

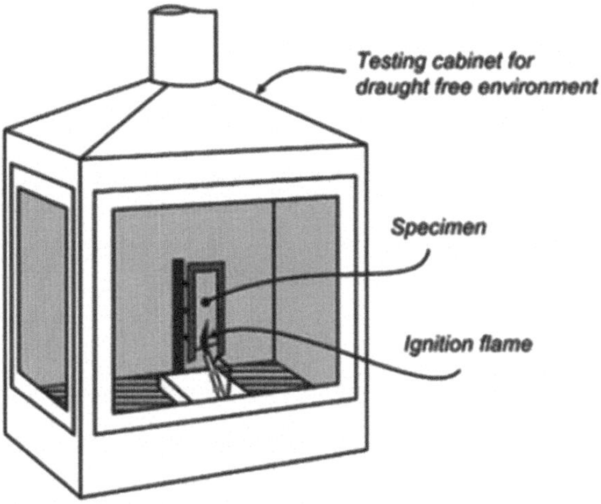

Fig. 1. Small flame test according to EN ISO 11925-2 [20].

2.2 Materials for Testing

A range of wood products are tested focusing on cladding (Table 1). Mainly commercially available cladding products based on wood modification were selected. Starting with softwoods both spruce and pine were included as untreated, thermally modified and preservative treated products. Also, chemical modifications like acetylation and furfurylation are included with radiata pine both as untreated and thermally modified material as reference. For general comparison also WRC (Western Red Cedar) cladding was tested and both larch and Oregon pine (Douglas fir) were added.

Next to softwood also thermally modified hardwoods are commonly used for cladding and façades. Here wood species like limba (fraké), ash, poplar and ayous are the main products present in this research. TMT (Thermally Modified Timber) of beech was also included besides beech as such. Furthermore, a range of wood species was added to cover density variability with those species that are more often used for cladding: European oak, chestnut, padouk, afrormosia and ipé. Less common thermally modified material of African celtis, movingui and bamboo was also included mainly related to density.

Specimen size according to EN ISO 1195-2 was altered regarding the length to ensure matching material for the different moisture content levels. Hence, the length

Table 1. Wood materials, number of test sets.

Wood spp.	Main botanical species	Type	CTRL[a]	TMT[b]	CMW[c]	IMP[d]
Pine	*Pinus sylvestris*	SW	1	3	–	1
Spruce	*Picea abies*		1	2	–	1
Radiata pine	*Pinus radiata*		1	1	2	–
WRC	*Thuya plicata*		2	–	–	–
Larch	*Larix decidua*		1	–	–	–
Oregon pine	*Pseudotsuga menziesii*		1	–	–	–
Ash	*Fraxinus excelsior*	HW	–	6	–	–
Limba (fraké)	*Terminalia superba*		1	5	–	–
Poplar	*Populus x euramericana*		1	2	–	–
Oak	*Quercus petraea & robur*		2	–	–	–
Padouk d'Afr.	*Pterocarpus soyauxii*		2	–	–	–
Ayous	*Triplochiton scleroxylon*		2	–	–	–
African celtis	*Celtis zenkeri*		–	1	–	–
Beech	*Fagus sylvatica*		1	1	–	–
Red oak	*Quercus rubra*		–	1	–	–
Chestnut	*Castanea sativa*		1	–	–	–
Afrormosia	*Pericopsis elata*		1	–	–	–
Ipê	*Handroanthus serratifolia*		1	–	–	–
Movingui	*Distemonanthus benthamianus*		–	1	–	–
Bamboo	*Phyllostachys edulis*	BB	–	1	–	–

Legend: [a]Control – not treated, [b]Thermally Modified Timber, [c]Chemically modified wood, [d]Vacuum pressure Cu impregnated; *SW* softwoods, *HW* hardwoods, *BB* Bamboo

was reduced to 100 mm, keeping width at 88 mm and thickness at 10 mm. For each test set (Table 1) 3 replicates were extracted from 3 different planks. For each test set 2 matching specimen sets were used to provide different conditioning up to equilibrium moisture content (EMC) at 50% RH and at 90% RH. Conditioning according to the standard EN 13238 should comply with the parameters (23 ± 2) °C and (50 ± 5) % relative humidity. All test pieces were defect free, planed and numbered on the back-side.

3 Results and Discussion

3.1 Density – Resin Treatments

As depicted in the Figs. 2, 3 and 4, density is an important factor in relation to reaction to fire. Test samples at over 800 kg/m^3 density show a mean flame height approaching a minimum value of approximately 30 mm (Fig. 2). At low density levels (<400 kg/m^3) the mean flame height is close to the maximum value of 100 mm. The burn length after testing confirms the relationship with density (Fig. 3). Only the highest densities under test, that are less relevant for cladding, show a scar close to 40 mm. The results for the bamboo product and ipé are underpinning this.

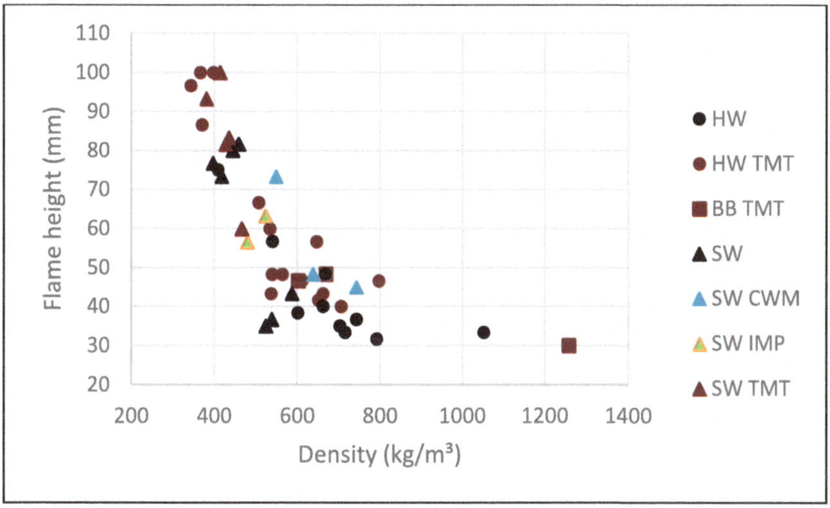

Fig. 2. Mean flame height in relation to wood density at 50% RH, legend according to Table 1.

The observation on flame height and burn length do not reveal any major difference between softwoods and hardwoods as density seems to be the overruling parameter. None of the wood modifications or treatments seem to alter results specifically. Only changes in density seem to be of major impact.

The after-flame time or continued burning is limited to lower density values, mainly below 500 kg/m^3 (Fig. 4). This parameter is seemingly more critical in identifying 2 major groups, one with and one without continued burning after the burner has been extinguished. Continued burning is mainly observed for softwood and thermally modified wood. This is related with observations that ignition is considered to be earlier for hardwoods.

The density value of 390 kg/m^3 has been used to indicate whether a fire class D can be attributed to a wood material. The results obtained here show that all non-modified

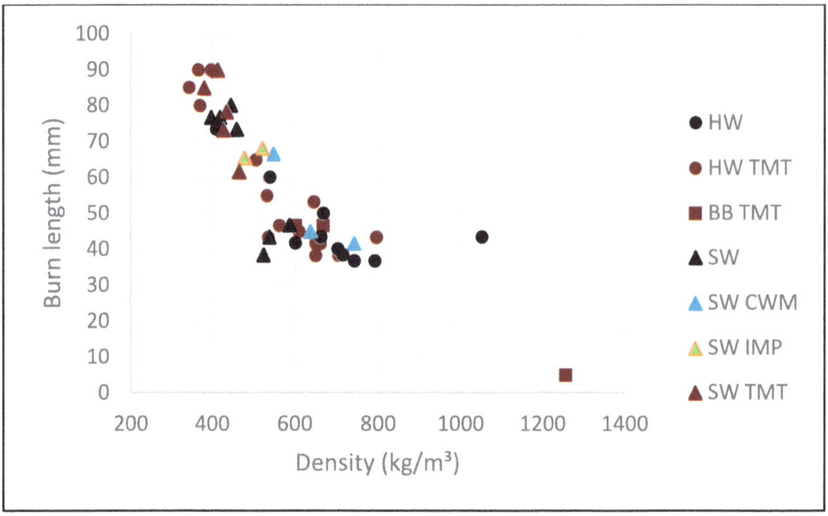

Fig. 3. Mean burn length in relation to wood density at 50% RH, legend according to Table 1.

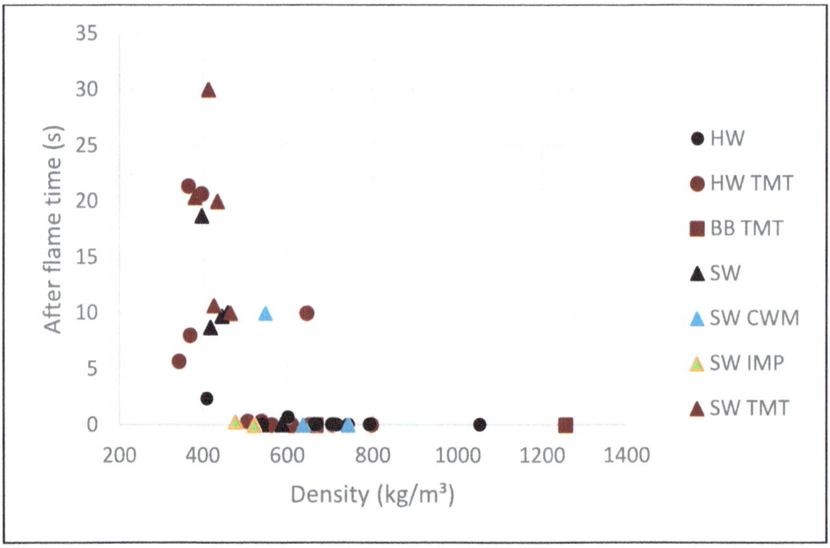

Fig. 4. Mean after-flame time in relation to wood density at 50% RH, legend according to Table 1.

wood could be considered as such. However, this cut-off value seems not to be really critical in identifying a separate class. Furthermore, there is no extra parameter that shows why thermal modification as such differentiates additionally to density. One could probably shift the criterion also to higher density values, e.g. 450 kg/m^3 covering nearly all materials showing significant after-flame time (Fig. 4). The results provide evidence that higher density can provide adequate fire performance when comparing wood species.

In relation to wood treatments mainly resin treatments are an option for improved fire performance and some results in literature underpin this.

Wood modification treatments often have an impact on density. In the following paragraphs some are listed as part of the discussion related to density. The incorporation of low molecular weight silicones into bulk aspen substrates and crosslinking them in situ improved the fire resistance [21]. Other treatments like the incorporation of DMDHEU seem to promote the combustion of oak wood due to the flammability of the condensed DMDHEU polymer [22]. Modification with 1,3-dimethylol-4,5-dihydroxyethyleneurea (DMDHEU) can dimensionally stabilize the wood and improve its durability, but does not change its flammability significantly. However, incorporating a fire retardant like combining methylolated guanylurea phosphate (MGUP) and boric acid (BA) to treat oak wood (*Quercus rubra* L.) caused a considerable decrease in the pyrolysis temperature, heat release, and smoke production [23].

Modification of wood with PF and MF changed the dynamics of thermo-oxidative decomposition and combustion patterns of the wood and as such retarded the decomposition process of the wood. The treated wood was less likely to ignite, and when it did burn, it yielded more char than the untreated wood during the cone calorimetric test [24]. Modification of wood with phenol-formaldehyde resin (PF) can improve many of its inherent properties, but does not upgrade its fire resistance [25]. Additional incorporation of fire retardants is also an option. However, Chinese fir treated with phenol-formaldehyde resin modified with boric acid (BPF) resin did improve its fire performance [26].

Sepiolite has a positive effect on fire properties of fir wood and could further be researched as a future fire-retardant in wood. The mineral nature of sepiolite with low reactivity to oxidize made it act as an incombustible layer towards the penetration of piloted fire to the substrate. Furthermore, the thermal conductivity coefficient of sepiolite is rather high in comparison to wood; therefore, the heat is more easily transferred and distributed throughout the substrate, preventing accumulation of heat at the spot nearest to the piloted fire, and ultimately delaying the process of catching fire [27].

Inspired by natural matrix-mediated biomineralisation, an artificial calcification approach for wood has been considered. This predominately targets the hardly accessible nanoporous cell wall structure rather than the micron-sized void system of the cell lumina. Porous $CaCO_3$ particles [28] can be deposited with this method deep inside the wood structure. Mineralisation of the wood cell wall architecture with $CaCO_3$ offers a green alternative to conventional fire-retardant systems [29]. Fu and co-authors [30] delignified balsa wood to form a hierarchically structured and nanoporous scaffold mainly composed of cellulose nanofibrils. This nanocellulosic wood scaffold was impregnated with colloidal montmorillonite clay to form a nanostructured wood hybrid with a high flame-retardancy. Densification as such eliminates the spaces between the cell walls, leading to a highly compact laminated structure that can block oxygen from infiltrating the material. In addition, an insulating wood char layer self-formed during the burning process obstructs the transport of heat and oxygen diffusion. These synergistic effects contribute to the material's excellent fire-retardant and self-extinguished properties, including a 2.08-fold enhancement in ignition time and 34.6% decrease in maximum heat release rate [31].

While the acetylation of beech leads to a slightly increased fire hazard, the thermal treatment of wood and crosslinking of cellulose microfibrils with dimethyloldihydroxyethyleneurea shows a limited flame retarding effect. Switching to woods with a higher lignin content, and thus higher char yield, however, results in a more pronounced improvement in flame retardancy performance [32]. This can also be linked to the better performance of some of the tropical wood species tested here.

3.2 Equilibrium Moisture Content – Thermal Wood Modification

The moisture content as such does not provide any critical protection level but should be regarded as an additional factor to the density (Fig. 5).

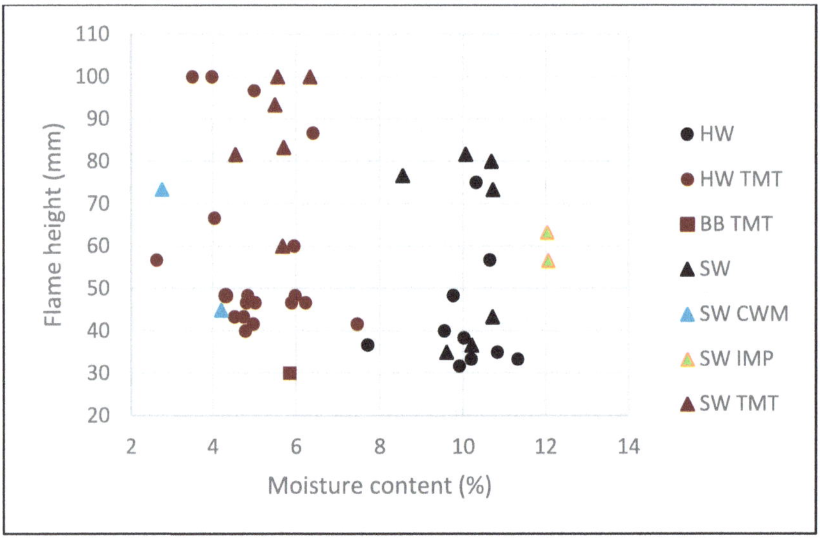

Fig. 5: Mean flame height in relation to MC conditioned at 50% RH prior to testing, legend according to Table 1.

The observations for chemically modified wood and wood impregnated with Cu-based wood preservatives underpin this. The impact of the different modification and treatment processes on the equilibrium moisture content (EMC) is however very pertinent (Fig. 6).

When comparing flame height results for each product at both EMC levels corresponding to 50 and 90% RH (Fig. 6) a shift towards lower values can be observed for higher moisture contents at 90% RH (Fig. 7). However, this seems merely valid for over 50 mm at 50% RH. Mean flame heights are 10–20 mm lower at EMC corresponding to 90% RH.

Although thermal modification does reduce the hygroscopicity of wood, water uptake properties of TMT are species dependent. For example, thermal treatment of pine sapwood was found to increase water sorption, whereas thermally treated spruce sapwood

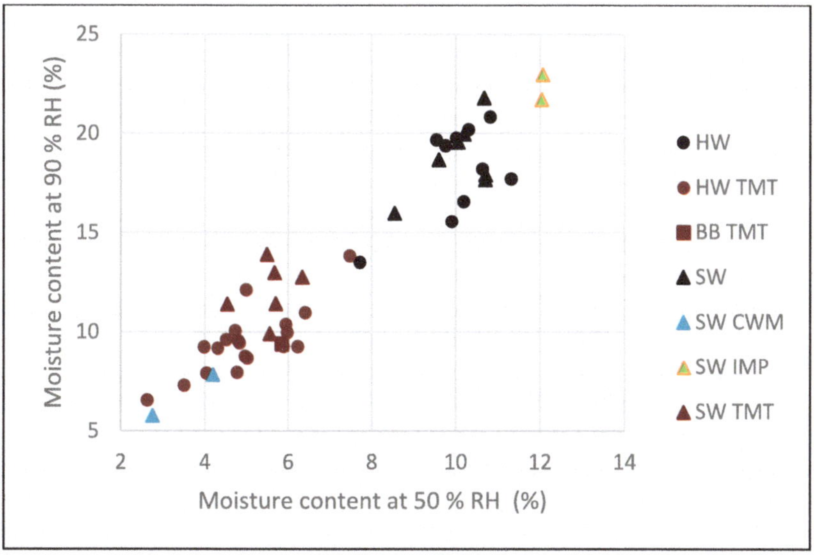

Fig. 6. Equilibrium moisture content (EMC) conditioned at 50 and 90% RH prior to testing, legend according to Table 1.

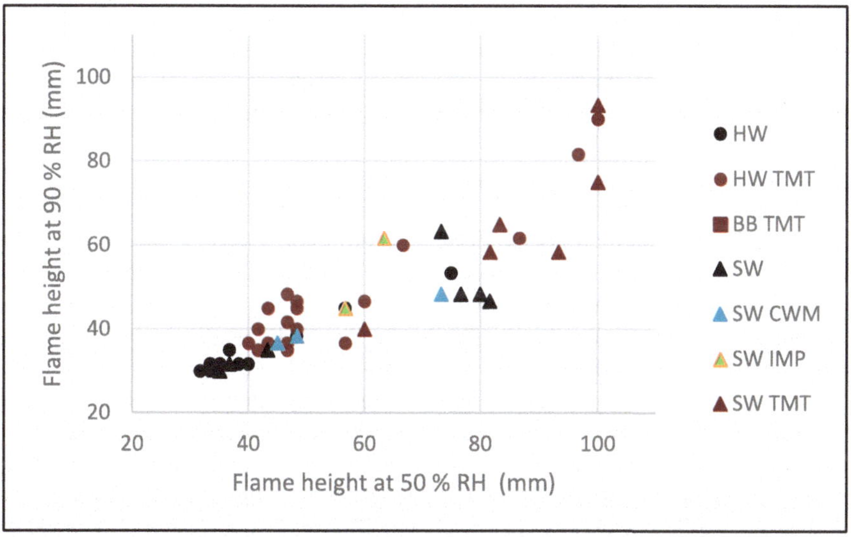

Fig. 7. Mean flame height after conditioning at 50 and 90% RH prior to testing, legend according to Table 1.

and heartwood showed much lower water uptake in floating tests [33]. Although moisture content in relation to dry (50% RH) and moist environments (90% RH) did not provide critical difference, it is possible that soaking or rain events can have a more significant impact.

Hydrothermal treatment of wood, even under relatively conservative conditions (i.e. 170 °C for 2 h) negatively affected its fire performance [34]. Thermally modified wood exhibited shorter ignition time and higher heat release rate than pristine wood. This is especially relevant considering that more and more thermally treated timber appears in tall buildings (i.e. over two stories), presenting significant fire hazard. Taking into account all results presented for TMT products, some caution is needed for using these in relation to fire hazard. Hence, architects are urged to consider the flammability of thermally modified wood seriously when planning for indoor or outdoor building applications.

Flammability characteristics of thermally modified oak (*Quercus robur*) wood did not show a clear effect [35]. Also, in our experiments there were no changes in parameters for oak due to thermal modification. Thermo-modified native black poplar (*Populus nigra* L.) wood was qualified as useful for insulation material [36] but in our experiments poplar wood showed very similar results as Scots pine and spruce either as pristine material or thermally modified.

Although using a somewhat deviating test setup, thermally modified spruce wood showed lower mass loss with open-flame burning. Čekovská and co-authors [37] showed that after the direct flame was removed there was no after flame. Our experiments revealed that all thermally modified softwoods showed significant levels of after flame time.

Reviewing potential chemical modification methods that may improve the reaction to fire of the wood material showed that the pure wood modification is not entirely suitable for significantly improving the reaction to fire of wood. None of the chemically modified materials tested significantly altered the (rather positive) test results of radiata pine. Therefore, it can be concluded that combined strategies of wood modification and the use of fire-retardant chemicals are needed to provide higher protection [38]. This aligns with the strategy to combined fire retardant and wood preservative treatments for outdoor wood applications [39].

In literature, some do discuss the impact of acetylation on reaction to fire. Mohebby and co-authors [40] acetylated beech layers prior to making plywood to be exposed to a burning flame from their edges for 60 s according to ISO 11925-3. Ignition time was increased due to raised weight percent gains and glowing was decreased instead. The acetylated plywood was burning with short blue flames; while the flame was long and yellowish in nonacetylated one. This study revealed that the acetylation retards slightly the fire in plywood. However, it does not help the resistance of wood against fire. Morozovs and Bukšāns [41] showed that the acetylation of ash diminished the performance of the wood on reaction to fire compared with untreated wood. It was supposed that acetylation decreased critical heat flux at extinguishment by acetic acid split off in pyrolysis conditions. Burning of acetylated wood produced less smoke than natural wood. We did not test any chemically modified hardwoods, but it is assumed that density and to some extent moisture content are still the most critical parameters.

4 Conclusion

In this work, close to 50 wood materials were tested (pristine or treated, mainly thermally modified) to check relevant parameters in relation to reaction to fire when used as cladding or façade material. The small flame test according to EN ISO 11925-2 was used as a generally applied simple approach using matched test material with 3 replicates for two levels of EMC. The conditioning at 50% RH is considered the worst case for exterior applications. The conditioning at 90% RH was also covered but the presence of liquid water (rain), which would be relevant for outdoor, was not tested. As such, moisture content was not found to be a limiting factor and lower EMC values for modified material did not reveal that this is a main factor. However, wood density was correlating very well with the reaction to fire parameters. For cladding or façades, some of the higher density materials tested were not suitable and hence, it can be relevant to avoid primarily low-density materials and, if necessary, add fire retardant treatments when considered of extra value.

Acknowledgement. The technical staff of UGent-Woodlab contributed with sample preparation: Stijn Willen and in performing the fire testing: Toon Gheyle. Timber trading and treating companies (alphabetical order) provided some of the material under test: Carpentier, Hanssens, Jadiwood, VCWood, Van Steenberge, LDCWood.

This paper is related to research activities in an ongoing project funded by the Research Foundation—Flanders (FWO) with the reference G043322N, "EnhanceWood: Nano-chemical treatment of wood for enhanced fungal and fire resistance".

References

1. Östman, B., Brandon, D., Frantzich, H.: Fire safety engineering in timber buildings. Fire Saf. J. **91**, 11–20 (2017)
2. Hill, C., Kymäläinen, M., Rautkari, L.: Review of the use of solid wood as an external cladding material in the built environment. J. Mater. Sci. **57**(20), 9031–9076 (2022)
3. Drean, V., Girardin, B., Chiva, R., Guillaume, E., Fateh, T.: Numerical investigation of the thermal exposure of façade during BS 8414 test series: influence of wind and fire source. Fire. Technol. **59**(1), 217–246 (2023)
4. Yuen, A.C.Y., Chen, T.B.Y., Li, A., De Cachinho Cordeiro, I.M., Liu, L., Liu, H., Lo, A.L.P., Chan, Q.N., Yeoh, G.H.: Evaluating the fire risk associated with cladding panels: an overview of fire incidents, policies, and future perspective in fire standards. Fire Mater. **45**(5), 663–689 (2021)
5. Hakkarainen, T., Oksanen, T.: Fire safety assessment of wooden facades. Fire Mater. **26**(1), 7–27 (2002)
6. Nguyen, K.T., Weerasinghe, P., Mendis, P., Ngo, T.: Performance of modern building façades in fire: a comprehensive review. Electron. J. Struct. Eng. **16**, 69–87 (2016)
7. Östman, B., Boström, L.: Fire protection ability of wood coverings. Fire. Technol. **51**, 1475–1493 (2015)
8. Thevega, T., Jayasinghe, J.A.S.C., Robert, D., Bandara, C.S., Kandare, E., Setunge, S.: Fire compliance of construction materials for building claddings: a critical review. Constr. Build. Mater. **361**, 129582 (2022)
9. Hill, C.A.S.: Wood Modification—Chemical, Thermal and Other Processes Wiley Series in Renewable Resources, p. 260. John Wiley & Sons, Chichester (2006)
10. Spear, M.J., Curling, S.F., Dimitriou, A., Ormondroyd, G.A.: Review of functional treatments for modified wood. Coatings. **11**(3), 327 (2021)
11. Jones, D., Sandberg, D., Goli, G., Todaro, L.: Wood Modification in Europe: A State-of-the-Art about Processes, Products and Applications, p. 123. Firenze University Press (2019)

12. Herrera, R., Arrese, A., de Hoyos-Martinez, P.L., Labidi, J., Llano-Ponte, R.: Evolution of thermally modified wood properties exposed to natural and artificial weathering and its potential as an element for façades systems. Constr. Build. Mater. **172**, 233–242 (2018)
13. Hill, C., Altgen, M., Rautkari, L.: Thermal modification of wood—a review: chemical changes and hygroscopicity. J. Mater. Sci. **56**, 6581–6614 (2021)
14. Hasburgh, L.E., Zelinka, S.L., Bishell, A.B., Kirker, G.T.: Durability and fire performance of charred wood siding (shou sugi ban). Forests. **12**(9), 1262 (2021)
15. Buksans, E., Laiveniece, L., Lubinskis, V.: Solid wood surface modification by charring and its impact on reaction to fire performance. Eng. Rural Dev. **20**(26), 899–905 (2021)
16. Kymäläinen, M., Lourençon, T.V., Lillqvist, K.: Natural weathering of soft-and hardwoods modified by contact and flame charring methods. Eur. J. Wood Wood Prod. **80**(6), 1309–1320 (2022)
17. Bartlett, A.I., Hadden, R.M., Bisby, L.A.: A review of factors affecting the burning behaviour of wood for application to tall timber construction. Fire. Technol. **55**, 1–49 (2019)
18. Njankouo, J.M., Dotreppe, J.C., Franssen, J.M.: Experimental study of the charring rate of tropical hardwoods. Fire Mater. **28**(1), 15–24 (2004)
19. Mensah, R.A., Jiang, L., Renner, J.S., Xu, Q.: Characterisation of the fire behaviour of wood: from pyrolysis to fire retardant mechanisms. J. Therm. Anal. Calorim. **148**(4), 1407–1422 (2023)
20. Janssens, M., Östman, B.: Reaction to fire performance. In: Fire Safe Use of Wood in Buildings, pp. 153–192. CRC Press (2022)
21. Eastman, S.A., Lesser, A.J., McCarthy, T.J.: Supercritical CO 2-assisted, silicone-modified wood for enhanced fire resistance. J. Mater. Sci. **44**, 1275–1282 (2009)
22. Xie, Y., Liu, N., Wang, Q., Xiao, Z., Wang, F., Zhang, Y., Militz, H.: Combustion behavior of oak wood (*Quercus mongolica* L.) modified by 1, 3-dimethylol-4, 5-dihydroxyethyleneurea (DMDHEU). Holzforschung. **68**(8), 881–887 (2014)
23. Jiang, T., Feng, X., Wang, Q., Xiao, Z., Wang, F., Xie, Y.: Fire performance of oak wood modified with N-methylol resin and methylolated guanylurea phosphate/boric acid-based fire retardant. Constr. Build. Mater. **72**, 1–6 (2014)
24. Xie, Y., Xu, J., Militz, H., Wang, F., Wang, Q., Mai, C., Xiao, Z.: Thermo-oxidative decomposition and combustion behavior of scots pine (*Pinus sylvestris* L.) sapwood modified with phenol-and melamine-formaldehyde resins. Wood Sci. Technol. **50**, 1125–1143 (2016)
25. Wu, M., Emmerich, L., Kurkowiak, K., Militz, H.: Fire resistance of pine wood treated with phenol-formaldehyde resin and phosphate-based flame retardant. Wood Mater. Sci. Eng. **18**, 1933–1939 (2023)
26. Yue, K., Chen, Z., Lu, W., Liu, W., Li, M., Shao, Y., Tang, L., Wan, L.: Evaluating the mechanical and fire-resistance properties of modified fast-growing Chinese fir timber with boric-phenol-formaldehyde resin. Constr. Build. Mater. **154**, 956–962 (2017)
27. Taghiyari, H.R., Tajvidi, M., Soltani, A., Esmailpour, A., Khodadoosti, G., Jafarzadeh, H., Militz, H., Papadopoulos, A.N.: Improving fire retardancy of unheated and heat-treated fir wood by nano-sepiolite. Eur. J. Wood Wood Prod. **79**, 841–849 (2021)
28. Huang, Y., Cao, L., Parakhonskiy, B.V., Skirtach, A.G.: Hard, soft, and hard-and-soft drug delivery carriers based on $CaCO_3$ and alginate biomaterials: synthesis, properties, pharmaceutical applications. Pharmaceutics. **14**(5), 909 (2022)
29. Merk, V., Chanana, M., Keplinger, T., Gaan, S., Burgert, I.: Hybrid wood materials with improved fire retardance by bio-inspired mineralisation on the nano-and submicron level. Green Chem. **17**(3), 1423–1428 (2015)
30. Fu, Q., Medina, L., Li, Y., Carosio, F., Hajian, A., Berglund, L.A.: Nanostructured wood hybrids for fire-retardancy prepared by clay impregnation into the cell wall. ACS Appl. Mater. Interfaces. **9**(41), 36154–36163 (2017)

31. Gan, W., Chen, C., Wang, Z., Song, J., Kuang, Y., He, S., Gan, W., Chen, C., Wang, Z., Song, J., Kuang, Y., He, S., Mi, R., Sunderland, P.B., Hu, L.: Dense, self-formed char layer enables a fire-retardant wood structural material. Adv. Funct. Mater. **29**(14), 1807444 (2019)
32. Rabe, S., Klack, P., Bahr, H., Schartel, B.: Assessing the fire behavior of woods modified by N-methylol crosslinking, thermal treatment, and acetylation. Fire Mater. **44**(4), 530–539 (2020)
33. Metsä-Kortelainen, S., Viitanen, H.: Wettability of sapwood and heartwood of thermally modified Norway spruce and scots pine. Eur. J. Wood Wood Prod. **70**(1–3), 135–139 (2011)
34. Song, K., Ganguly, I., Eastin, I., Dichiara, A.: High temperature and fire behavior of hydrothermally modified wood impregnated with carbon nanomaterials. J. Hazard. Mater. **384**, 121283 (2020)
35. Gašparík, M., Osvaldová, L.M., Čekovská, H., Potůček, D.: Flammability characteristics of thermally modified oak wood treated with a fire retardant. Bioresources. **12**(4), 8451–8467 (2017)
36. Todaro, L., Liuzzi, S., Pantaleo, A.M., Lo Giudice, V., Moretti, N., Stefanizzi, P.: Thermo-modified native black poplar (*Populus nigra* L.) wood as an insulation material. iForest-Biogeosci. For. **14**(3), 268–273 (2021)
37. Čekovská, H., Gaff, M., Osvald, A., Kačík, F., Kubš, J., Kaplan, L.: Fire resistance of thermally modified spruce wood. Bioresources. **12**(1), 947–959 (2017)
38. Popescu, C.M., Pfriem, A.: Treatments and modification to improve the reaction to fire of wood and wood based products—an overview. Fire Mater. **44**(1), 100–111 (2020)
39. Marney, D.C.O., Russell, L.J.: Combined fire retardant and wood preservative treatments for outdoor wood applications–a review of the literature. Fire. Technol. **44**(1), 1–14 (2008)
40. Mohebby, B., Talaii, A., Najafi, S.K.: Influence of acetylation on fire resistance of beech plywood. Mater. Lett. **61**(2), 359–362 (2007)
41. Morozovs, A., Bukšāns, E.: Fire performance characteristics of acetylated ash (*Fraxinus excelsior* L.) wood. Wood Mater. Sci. Eng. **4**(1–2), 76–79 (2009)

Standards

42. EN 13501–1: Fire classification of construction products and building elements - Classification using data from reaction to fire tests (2019)
43. EN 13238. Reaction to fire tests for building products—Conditioning procedures and general rules for selection of substrates (2010)
44. ISO 11925-2. Reaction to fire tests - Ignitability of products subjected to direct impingement of flame. Part 2: Single-flame source test (2020)

Comparison of Major Wood Hygro-Thermal Modification Technologies Paves the Way for a Generalized Mass Loss Kinetic Model

Bertrand Marcon[1](✉), Paola Mazzanti[2], and Giacomo Goli[2]

[1] Laboratoire des Matériaux et Procédés (LaBoMaP), Arts et Metiers Institute of Technology (ENSAM), Cluny, France
bertrand.marcon@ensam.eu

[2] Dipartimento di Scienze e Tecnologie Agrarie, Alimentari, Ambientali e Forestali (DAGRI), Università di Firenze, Firenze FI, Italy

Abstract. Bibliographic data about dry mass loss coming from different thermal and hygrothermal modification processes versus time were collected in this work and analyzed together. The data sets were collected from 2 experimental campaigns involving different modification technologies: (A) spruce wood hygro-thermally modified under superheated steam conditions performed at relative humidity ranging from 35 to 92% at temperatures ranging from 110 to 170 °C; and (B) poplar wood samples thermally modified with Thermo-vacuum® technology performed at temperatures ranging from 150 to 240 °C. For both processes, conversation rates master curves at 150 °C were identified on the experimental points using the time-temperature and the time-temperature-humidity superposition method.

The two master curves were then compared and a generalized kinetic model able to predict the mass losses when modifying wood at different temperatures and relative humidity implemented. A kinetic model is expressed with a master curve composed of 2 kinetic stages. That model can predict the mass variation occurring during any hygrothermal modification whatever the temperature applied and the environment medium relative humidity.

Keywords: Poplar · Hygro-thermal Modification · Mass Loss Kinetic · Modelling · Time-Temperature Equivalency

1 Introduction

The idea to define a model for the description of physical and mechanical properties variation during thermal (T) or hydro-thermal (HT) modification of wood comes from very far. This idea has been pursued by researchers applying kinetic models based on Arrhenius' law and time temperature equivalency (TTE) principle. The method has been successfully applied to different wood modification processes by different authors. Among them, Goli et al. [1] have shown the effect of different air exchange rates on the dry mass loss kinetic during thermal modification of poplar wood. Marcon et al. [2] have shown the effect of thermal modification on several physical and mechanical properties of poplar wood.

G. Goli et al. (Eds.): ECWM 2024, SPM 86, pp. 124–129, 2025.
https://doi.org/10.1007/978-3-031-99418-0_16

Other authors have worked under steam environment such as Esteves et al. [3] who have shown how the thermal modification under superheated steam has different kinetics processing with different relative humidity levels. Endo et al. [4] behind kinetic experiments at different relative humidity levels with superheated steam approached the idea of a time-humidity superposition principle for hydro-thermal modification with superheated steam.

In most of the cases the main physical descriptor of the state of degradation of the sample is recognized to be the dry mass loss as discussed by Candelier [5]. Authors have recorded a simple kinetic process for thermal modification (T) and a double kinetic process for hygrothermal modification ones (HT) such as Marcon et al. [6]. The double kinetic for HT process was explained with a completely different degradation of cell-wall polymers and in particular with a very rapid degradation of hemicelluloses as documented by ATR FT-IR analysis presented in the same experimental work. This behavior, was highlighted thanks to the very large mass losses attained in the study, mass losses never documented until there in research works for degradation kinetics. Most of the kinetic studies stop before 20% of dry mass loss (DML) because after 10% ML wood has lost most of its mechanical performances.

In this article, a model to describe dry mass loss kinetics according to modification time, temperature and relative humidity is developed based on data-sets coming from two different experimental works already published.

2 Material and Methods

Two different hygro-thermal wood modification processes are analyzed separately in a first approach and together in a second approach in order to find a common mass loss kinetic model. Firstly, a data set from treatment performed in an open-air ventilated oven and secondly an experimental data set from saturated steam.

The modification procedures as well as the samples used are detailed below.

2.1 Dry Modification in Ventilated Oven (T)

The data set as well as the relative master curve building up were presented by [2, 6] and include a set of experiments performed on poplar wood (*Populus alba* L.) separately by Goli in Italy and by Marcon in Japan. The specimens were modified under dry air conditions at atmospheric pressure in a conventional ventilated oven up to a dry mass loss of 76%. The modification temperatures ranged from 90 to 262 °C. Samples were conditioned at dry state before starting the modification process. Both the experiments were successfully modelled with a unique master-curve. The specimen sizes were 30 (L) by 8 (R) by 8 (T) mm^3 for the Goli's sample in Italy, and 120 (L) by 20 (R) by 2 (T) mm^3 for the Marcon's samples in Japan; and their over whole average density (ρ) of 340 ± 2 $kg.m^{-3}$ at standard environmental conditions, 20 °C of temperature (T) and 65%of relative humidity (RH). The annual ring width was 10 ± 2 mm and the estimated age of the tree is 30 years old.

2.2 Saturated Steam in Autoclave (HT)

The data set as well as the relative master curve was presented by [6] and include a set of experiments performed on poplar wood (*Populus alba* L.). The specimens were modified under saturated steam atmosphere in autoclave up to a dry mass loss of 65%. Both oven and saturated steam samples were sawn from boards extracted from the same tree. The specimens were oven dried at 103 ± 2 °C before the thermal modification in order to know the oven dry mass. The samples were than equilibrated at standard environmental conditions (20 °C of T and 65% of RH) and modified in autoclave at the desired temperature and relative saturated steam pressure for the desired time. The modification temperatures ranged from 110 °C to 170 °C. Samples were conditioned at standard environmental conditions (20 °C of temperature and 65% of relative humidity before starting the modification process. The same modelling approach was successful.

2.3 Mass Loss Kinetic Models for both Modification Processes

For the sake of brevity, the model strategy is not described again since fully explained in the previous published research works in details [1, 2, 6] but to resume it is inspired by rheological considerations (exponential law) to identify a kinetic master curve including a maximum mass loss $ML_{(t \rightarrow \infty)}$, a time constant τ, and a slope parameter p (giving the basic mass variation other treatment duration at a chosen reference temperature). After the master curve identified on the experimental data set at the reference temperature, the Arrhenius' law, involving the concept of energy of activation (Ea usually expressed in kJ.mol^{-1}) to compute a time/temperature equivalency (TTE), enable to shift the master curve at any other heat modification temperature. This general part was developed in [6] and readers can referee to previous work for a better understanding and see the master identified master curves in Fig. 1.

The ventilated oven set exhibits a single macroscopic behavior even though a lot of different chemical reactions are acting during the modification as summarized in [7]: oxidation, depolymerization of hemicelluloses and lignin, dehydration, crystallinity modifications. On the other hand, the saturated steam mass loss kinetic shows a different pattern with two stages trends having different kinetics.

As stated in [6], during the first stage of mass loss a very high relative humidity in the medium accelerates substantially the kinetic, while In the second stage the two processes (dry and saturated steam) have a similar shape.

The basic similarities between the two processes and the fact that a real effect of the humidity can be clearly observed on the first stage suggests that a more generalized system of equations could feature both process only by the addition of the relative humidity of the medium dependency.

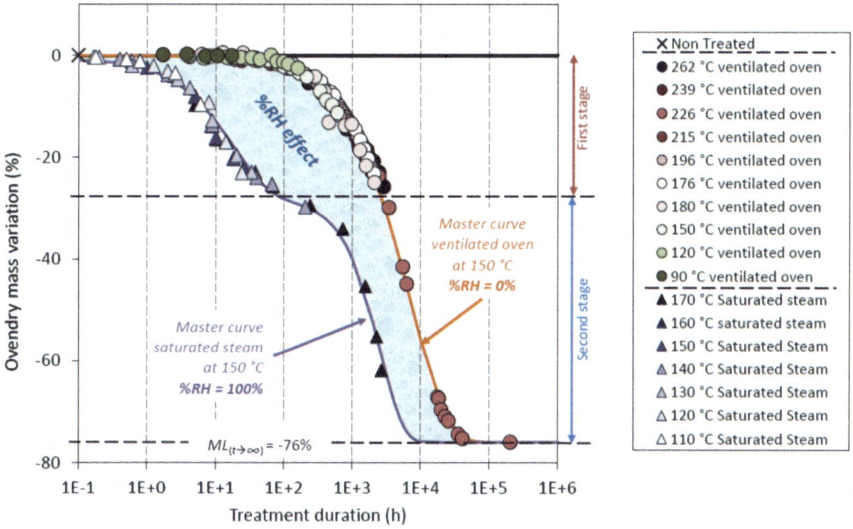

Fig. 1. Dry mass loss kinetics master curves identified at the reference temperature of 150 °C for the open-air ventilated oven (0% RH) and the saturated steam modification processes (100% RH); the time shifting between the two master curves emphases the relative humidity effect on the mass loss kinetic.

3 Results and Discussions

3.1 Generalization of the Mass Loss Kinetic Model for both Modification Processes

As given in [9], the time constant τ of the open-air ventilated oven (under %RH = 0%) first stage is $\tau^1_{\text{oven}} = 7000$ h when it becomes only $\tau^1_{\text{SatSteam}} = 18$ h for the saturated steam one (under %RH = 100%); on the other hand, the second stage of the saturated steam is characterized by $\tau^2_{\text{SatSteam}} = 2600$ h (under %RH = 100%). As mentioned before, the two stages are not influenced in the same way by the relative humidity of the medium involving that two different dependencies toward this environmental peculiar condition have to be considered.

As for the temperature dependency, where a reference temperature (θ_{ref}) had been chosen arbitrarily at 150 °C, it had been chosen here a reference relative humidity of 0% (corresponding to a moisture content of 0%). Without having any information about the effect of the relative humidity over the mass loss kinetic out of the dry and saturated steam, a linear trend is considered to model the mass loss behavior in correspondence with the relative humidity of the treatment media. The Fig. 2 plots the identified linear trends relative humidity dependencies of both time constants τ^1 and τ^2 of the two stages respectively.

Taking into account the relative humidity dependencies, it is now possible to propose a more generalize approach to mass loss kinetic compared to the one proposed in [9]. This new approach is proposed in Table 1.

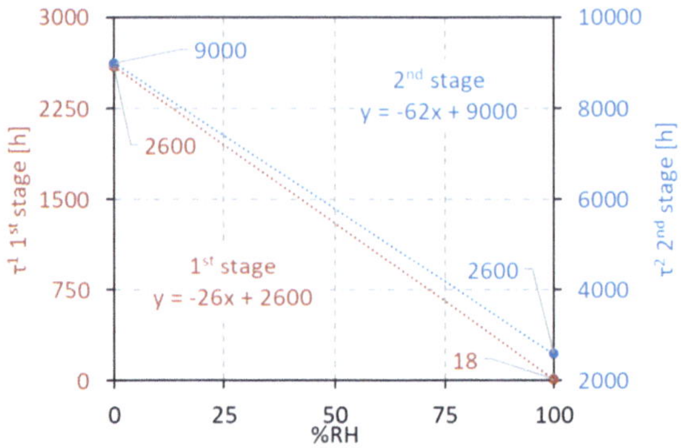

Fig. 2. Assumption of the relative humidity linear effect on the characteristic times based on the experimental data considered in the present study to generalize the mass loss kinetic model.

Table 1. Parameters of the single mass loss kinetic model for both processes considered together taking into account the treatment duration effect, the temperature effect and the relative humidity of the medium effect.

	1st stage			2nd stage		
	$ML^1_{(t\to\infty)}$ (%)	$\tau^1_{(RH_{ref})}$ (h)	p^1	$ML^2_{(t\to\infty)}$ (%)	$\tau^2_{(RH_{ref})}$ (h)	p^2
	−28	2600	0.8	−48	9000	1.2
Temperature effect ($\theta_{ref} = 150$ °C)	$Ea = 115$ kJ/mol and $A_{(T)}/A_{(Tref)} = 0.0064 \times T$					
Relative humidity effect ($RH_{ref} =$ 0%)	$\tau^1_{(RH)} = -26 \times R + \tau^1_{(RH_{ref})}$			$\tau^2_{(RH)} = -62 \times RH + \tau^2_{(RH_{ref})}$		

Of course, the linearity assumption in terms of relative humidity effects on the dry mass loss kinetics should be verified.

4 Conclusion

Even though no industries would apply such extensive mass loss to any wood at least in the aim to reach a great durability without compromising drastically the mechanical strength of the final product (most industrial modifications are conducted below or close to 10% of mass loss), it is interesting to understand what's occurring to the material with these modelling techniques. The mass loss is considered to be a great estimator of the modification extent and has an accurate relationship to a large amount of material features such as equilibrium moisture content, color, stiffness and strength. The Arrhenius' law has already shown in previous work a strong capacity to predict the evolution of different properties of wood during modification with different methods. The possibility to include the relative humidity in the model is a great opportunity to develop a fully generalized kinetic model to be applied in hygro-thermal modification of wood.

References

1. Goli, G., Marcon, B., Fioravanti, M.: Poplar wood heat treatment: effect of air ventilation rate and initial moisture content on reaction kinetics, physical and mechanical properties. Wood Sci. Technol. **48**, 1303–1316 (2014). https://doi.org/10.1007/s00226-014-0677-5
2. Marcon, B., Goli, G., Matsuo-Ueda, M., Denaud, L., Umemura, K., Gril, J., Kawai, S.: Kinetic analysis of poplar wood properties by thermal modification in conventional oven. IForest Biogeosci. For. **11**, 131–139 (2018). https://doi.org/10.3832/ifor2422-010
3. Esteves, B., Marques, A., Domingos, I., Pereira, H.: Influence of steam heating on the properties of pine (Pinus pinaster) and eucalypt (Eucalyptus globulus) wood. Wood Sci. Technol. **41**, 193–207 (2007)
4. Endo, K., Obataya, E., Zeniya, N., Matsuo, M.: Effects of heating humidity on the physical properties of hydrothermally treated spruce wood. Wood Sci. Technol. **50**, 1161–1179 (2016). https://doi.org/10.1007/s00226-016-0822-4
5. Candelier, K., Hannouz, S., Elaieb, M., Collet, R., Dumarçay, S., Pétrissans, A., Gérardin, P., Pétrissans, M.: Utilization of temperature kinetics as a method to predict treatment intensity and corresponding treated wood quality: durability and mechanical properties of thermally modified wood. Maderas Cienc. Tecnol. **17**, 253–262 (2015). https://doi.org/10.4067/S0718-221X2015005000024
6. Marcon, B., Tondi, G., Procino, L., Goli, G.: Thermal modification kinetics and chemistry of poplar wood in dry and saturated steam media. Holzforschung, 000010151520200166 (2021). https://doi.org/10.1515/hf-2020-0166
7. Esteves, B.M., Pereira, H.M.: Wood modification by heat treatment: a review. Bioresources. **4**, 370–404 (2009)

Detection of the Aromatic Profile of Thermally Modified Wood Species

Valentina Lo Giudice[1], Angelo Rita[2], and Luigi Todaro[1](✉)

[1] School of Agricultural, Forestry, Food and Environmental Science, University of Basilicata, Potenza, Italy
{valentina.logiudice,luigi.todaro}@unibas.it

[2] Department of Agriculture, University of Naples Federico II, Portici, NA, Italy
angelo.rita@unina.it

Abstract. Thermo-modified wood material for indoor use produces chemical compounds that significantly affect indoor environmental issues and human health. The thermal modification method influences the chemical properties of wood. Our study aimed to evaluate the effect of thermal modification on the aromatic profile of seven wood species namely *Cedrus deodara* Roxb, *Pinus laricio* L, *Alnus cordata* L, *Populus nigra* L, *Fagus sylvatica* L (thermo-modified at a temperature of 200 °C), *Quercus Cerris* L, and *Castanea sativa* Mill (thermo-modified at a temperature of 170 °C). An olfactory machine so-called PEN 3 AIRSENSE electronic nose (e-nose) consisting of 10 doped semi-conductive metal-oxide gas sensors (MOS) was used for this purpose. Wood planks were thermally modified and cut into small wood blocks for autoclave sterilization to obtain the final extracts in water solution for olfactory detection. Principal component analysis (PCA) and linear discriminant analysis (LDA) provided an evaluation of the ability of the e-nose sensor array to associate a response pattern with the corresponding class of compounds. The loadings analysis was used to identify the sensors responsible for discrimination in the current pattern file. Results showed that the W5S sensor, with its characteristic for react to nitrogen oxides (NO_2), explained most of the variance attributed to the odor of both thermo-modified and unmodified woods.

Keywords: Chemical Compounds · Electronic Nose · Heat-treated Wood · Sensors

1 Introduction

Thermal modification has been developed to improve certain wood properties for several applications (Wehsener et al. 2016). This process modifies the chemical structure of wood, resulting in changes in some wood properties, such as resistance to moisture absorption and dimensional stability. The changes in the chemical structure of wood lead to the release of volatile organic compounds (VOCs) during and after the modification process which are responsible for release of various odors. Thermal modification can lead to the disappearance of some chemical components, transform existing ones, or create new ones (Pohleven et al. 2019). A previous study revealed the presence of toxic

G. Goli et al. (Eds.): ECWM 2024, SPM 86, pp. 130–137, 2025.
https://doi.org/10.1007/978-3-031-99418-0_17

polyaromatic compounds during the thermal treatment of wood (Kamdem et al. 2000). The increasing use of thermo-modified wood in indoor applications requires an odor pattern investigation to identify chemicals that significantly affect indoor air quality. The electronic nose (e-nose) is a cost-effective method for odor analysis for various application fields and can detect the class of VOCs without identifying specific chemical compounds (Anwar et al. 2023). Cordeiro et al. (2012) used the e-nose to identify several wood species by their VOCs. Our study aimed to evaluate the effect of thermal modification on the aromatic profile of seven wood species namely, deodar cedar (*Cedrus deodara* Roxb), laricio pine (*Pinus laricio* L), alder (*Alnus cordata* (Loisel) Duby), black poplar (*Populus nigra* L), European beech (*Fagus sylvatica* L) (thermo-modified at a temperature of 200 °C), Turkey oak (*Quercus Cerris* L), and sweet chestnut (*Castanea sativa* Mill) (thermo-modified at a temperature of 170 °C).

2 Materials and Methods

2.1 Sample Preparation and Thermal Treatment

The wood species were harvested in the Apennine forest of Basilicata Region (Southern Italy). The logs were sawn in planks of 40 mm (tangential) × 200 mm (radial) × 2200 mm (longitudinal). Half of the planks were initially dried to 0% moisture content in a vacuum ranging from 185–200 mbar at a temperature of 90 °C for 4 h. The drying process started at a temperature level of 30 °C and increased to 90 °C at 5 °C per hour intervals. The thermal treatment was applied to the boards by gradually increasing to a final temperature of 200 °C, under pressure conditions of 220–320 mBar for 4 h in a thermo-vacuum cylinder (WDE-Maspell s.r.l., Terni, Italy). Turkey oak and chestnut were thermally modified at a temperature of 170 °C for 4 h under pressure conditions of 200 mBar (Todaro et al. 2015).

2.2 Autoclave Extraction

The unmodified and thermo-modified planks were reduced into small pieces and subjected to the water extraction method to obtain the extracts. Wood samples (25 g) were placed in an airtight glass jar with 125 mL of distilled water and placed in an autoclave (Vapor Matic 770 Asal srl, Cernusco, Italy) for 20 min, at a temperature of 120 °C and a pressure of 1 bar for extraction. Once the cooling temperature was reached (40 ± 1 °C), the wood samples were separated from the extracts. The wood extracts were dried in an oven for 7 days at a temperature of 37 °C to remove the remaining water (Mecca et al. 2021).

2.3 Electronic Nose, Data Acquisition and Analysis

The PEN3 AIRSENSE e-nose device (Fig. 1) equipped with 10 metal oxide sensors thermoregulated in the range of 150–500 °C, was used for aroma detection of wood extracts. Table 1 lists all used sensors, their main applications, and the specified reaction. Each wood extract (10 mL) was placed into a glass vial and subjected to the aroma analysis in triplicate.

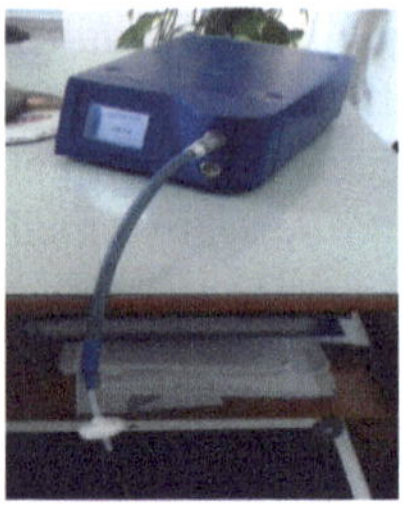

Fig. 1. Electronic nose (e-nose) AIRSENSE Analytics GmbH, Schwerin, Germany

Table 1. Sensors used and their main applications in PEN3 (Gómez et al. 2006).

Sensor number	Sensors	General description	Reference
1	W1C	Aromatic compounds	Toluene, 10 mg/kg
2	W5S	React on nitrogen oxides	NO_2, 1 mg/kg
3	W3C	Ammonia, aromatic compounds	Benzene, 10 mg/kg
4	W6S	Mainly hydrogen	H_2, 0.1 mg/kg
5	W5C	Alkanes, aromatic compounds	Propane, 1 mg/kg
6	W1S	Methane	CH_3, 100 mg/kg
7	W1W	Sulfur compounds, terpenes, limonene, pyrazine	H_2S, 1 mg/kg
8	W2S	Alcohol, partially aromatic compounds	CO, 100 mg/kg
9	W2W	Aromatic compounds, sulfur organic compounds	H_2S, 1 mg/kg
10	W3S	React on high concentrations >100 mg/kg, sometimes very Selective (methane)	CH_3, 10 CH_3, 100 mg/kg

A 20 g luer-lock needle connected to a 3 mm Teflon-tubing was used to perforate the vial seal and absorb the air accumulated inside it. The headspace gas was pumped over the sensors of the e-nose with a flow of 400 mL/min. During the measurement, the bomb pushes the volatiles through a closed loop that includes the measurement and concentration chambers. When sensors are exposed to volatiles, the computer records the electrical resistance changes experienced by the sensors. The measurement lasts 60 s. Data resulting from the sensor response was chosen as the average of the last 10 s. The signals of all sensors were analyzed using Principal component analysis (PCA) and linear discriminant analysis (LDA) by WinMuster (version 1.5.2.4 Jun 2003, copyright 1996–2002 WMA Airsense Analysentechnik GmbH 2003) software.

3 Results and Discussion

3.1 E-Nose Response to Wood Extracts Aroma

PCA and LDA analysis were applied for both unmodified and thermo-modified wood samples. Figures 2 and 3 are related to unmodified wood samples and show analysis results on a two-dimensional plane, principal component 1 (PC1) and principal component 2 (PC2), as well as the first and second linear discriminant LD1 and LD2, respectively.

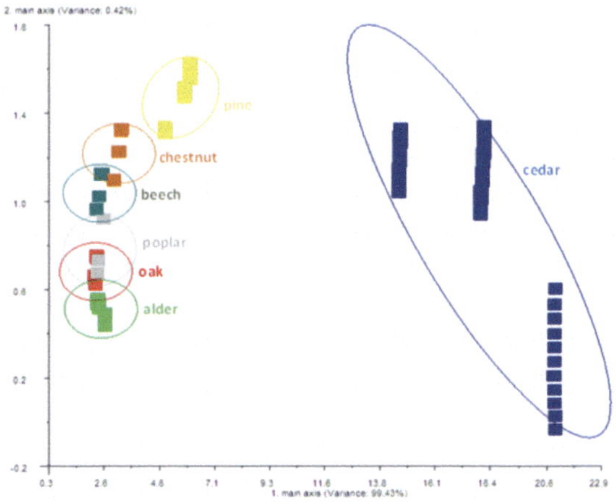

Fig. 2. PCA analysis for unmodified wood samples.

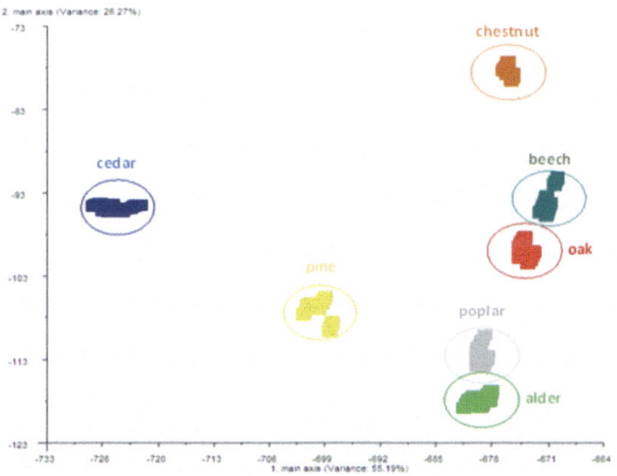

Fig. 3. LDA analysis for unmodified wood samples.

The score plot is tracked inside the ellipses that represent the variation around each picking data (aromatic characteristic) in the space. The processed data have exhibited a particular shift of the different picking data along the first principal component, PC1, explaining 99.43% of the total variance of 99.848%. The second principal component (PC2) explains 0.42% of the variation (Fig. 2). Cedar wood differentiated from the rest of the wood species in the case of the unmodified samples. In contrast to the clear separation that was achieved by cedar wood, other unmodified wood species (pine, chestnut, beech, poplar, oak, and alder) have overlapped each other. This observation could be related to the "strength" of the olfactory imprint of cedar wood as the odors of the other wood species were perceptibly weaker and similar to each other according to the sensor's response (Ye et al. 2023). Hardwoods seem to belong to fairly different classes overall. LDA function 1 (LD1) and function 2 (LD2) accounted for 55.19% and 26.27% of the variance respectively (see Fig. 3). A small overlap joint was achieved between beech and oak, as well as between poplar and alder, meaning that volatiles do not differ much among them. Loading analysis (Fig. 4) helped to identify the sensors responsible for discrimination in the current pattern file. Sensors with loading parameters near zero have a low contribution to the total response of the array, whereas high values indicate a discriminant sensor (Gómez et al. 2006). Some sensor groups have almost identical loading parameters (W1C, W3C, W5C) and they might be represented by only one sensor. Figure 4 shows that sensor W5S, with its characteristic of reacting to harmful substances such as nitrogen oxides (NO_2) (Ozgen et al. 2021), explained most of the variance attributed to the odor of unmodified wood.

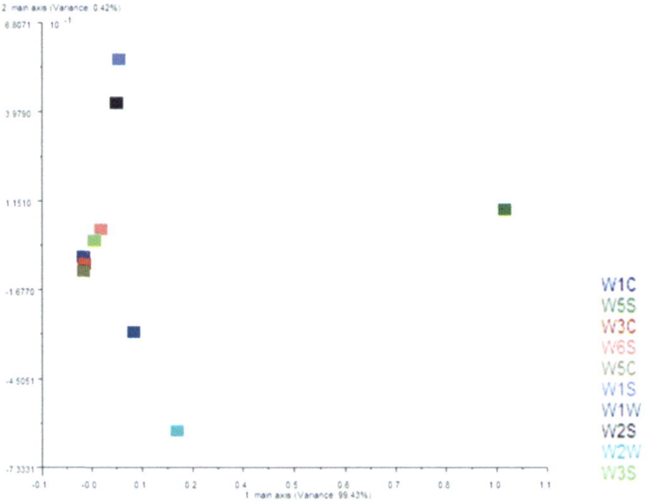

Fig. 4. Loading analysis related to PC1 and PC2 for unmodified wood samples total variance 99.848%.

In the case of thermo-modified wood samples, the first major component (PC1) and the second principal component (PC2), 99.77% and 0.16% respectively, explained a value of 99.928% of the total data (Fig. 5). Thermal modification has led chestnut wood to stand out from the rest of the wood species. As a consequence of the thermal modification, the wood species have changed their chemical composition and, consequently, their aromatic profile, leading in this case chestnut wood to distinguish itself from the other wood species (González-Peña et al. 2009). The LDA was able to clearly distinguish the aromatic profile of thermo-modified wood species. In this case, the plot showed 95.3% of the total variance. LDA function 1 (LD1) and function 2 (LD2) accounted for 57.39% and 37.91% of the variance respectively (Fig. 6). The plot of the loading factors associated with PC1 and PC2 for thermo-modified wood samples is shown in Fig. 7. In this case, not much has changed. One sensor could represent both the W1C sensor, which is characterized by its sensitivity level to toluene, and the W3C sensor, with a sensitivity level to benzene. Sensors located near the center of the diagram (0, 0) have a minor responsibility for the distribution pattern in the PCA plot. Therefore, nearly a subset of fewer sensors can be chosen to explain all variance. Figure 7 shows that sensor W5S explained most of the variance attributed to the odor of thermo-modified woods.

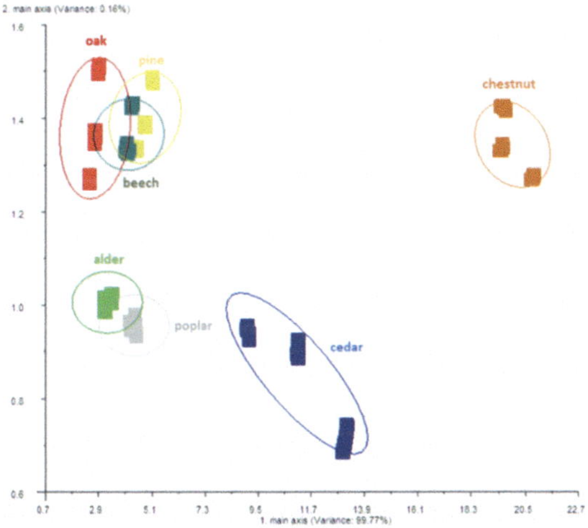

Fig. 5. PCA analysis for thermo-modified wood samples.

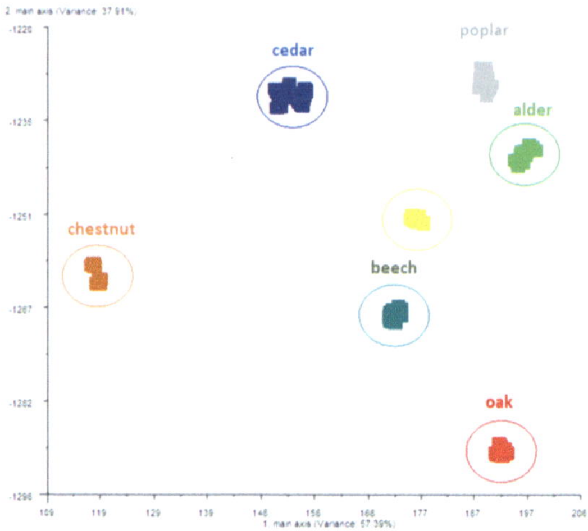

Fig. 6. LDA analysis for thermo-modified wood samples.

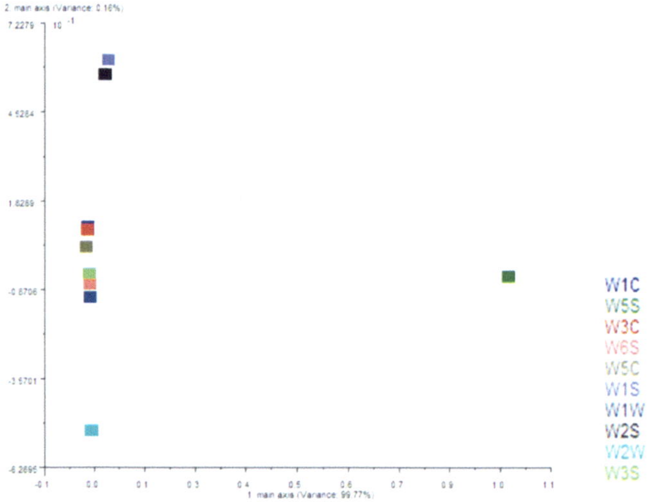

Fig. 7. Loading analysis related to PC1 and PC2 for thermo-modified wood samples total variance 99.928%.

4 Conclusions

An e-nose composed of ten sensors was used to identify the aromatic profile of seven unmodified and thermo-modified wood species: cedar, pine, alder, poplar, beech, oak, and chestnut. The volatile compounds were altered by the thermal modification. Cedar wood differentiated itself in the case of unmodified wood species, whereas chestnut in the case of thermo-modified ones. The e-nose was able to detect a distinct separation in all thermo-modified wood species using LDA analysis. The W5S sensor, which is sensitive to nitrogen oxides (NO_2) explained most of the variance associated with the odor of both unmodified and thermo-modified wood species, the other sensors produced similar responses showing a low discriminant power. This result could be used in further studies to optimize the number of sensors involved in the component aromatic analysis.

Acknowledgments. This study was carried out at Agritech National Research Center and received funding from the European Union Next-GenerationEU (PIANO NAZIONALE DI RIPRESA E RESILIENZA (PNRR)—MISSIONE 4 COMPONENTE 2, INVESTIMENTO 1.4—D.D. 1032 17/06/2022, CN00000022).

This manuscript reflects only the authors' views and opinions; neither the European Union nor the European Commission can be considered responsible for them.

References

Anwar, H., Anwar, T., Murtaza, S.: Review on food quality assessment using machine learning and electronic nose system. Biosens. Bioelectron. **14**, 100365 (2023)

Cordeiro, J.R., Martinez, M.I., Li, R.W., Cardoso, A.P., Nunes, L.C., Krug, F.J., Paixão, R.L.C., Nomura, S., Gruber, J.: Identification of four wood species by an electronic nose and by LIBS. Int. J. Electrochem. **2012**, 1–5 (2012)

Gómez, A.H., Wang, J., Hu, G., Pereira, A.G.: Electronic nose technique potential monitoring mandarin maturity. Sensors Actuators B Chem. **113**(1), 347–353 (2006)

González-Peña, M.M., Curling, S.F., Hale, M.D.: On the effect of heat on the chemical composition and dimensions of thermally-modified wood. Polym. Degrad. Stab. **94**(12), 2184–2193 (2009)

Kamdem, D.P., Pizzi, A., Triboulot, M.C.: Heat-treated timber: potentially toxic byproducts presence and extent of wood cell wall degradation. Eur. J. Wood Wood Prod. **58**(4), 253–257 (2000)

Mecca, M., Todaro, L., Lo Giudice, V., Lovaglio, T., D'Auria, M.: GC-MS and SPME techniques highlighted contrasting chemical behaviour in the water extractives of modified Castanea sativa mill. And Fagus sylvatica L. wood. Forests. **12**(8), 986 (2021)

Ozgen, S., Cernuschi, S., Caserini, S.: An overview of nitrogen oxides emissions from biomass combustion for domestic heat production. Renew. Sust. Energ. Rev. **135**, 110113 (2021)

Pohleven, J., Burnard, M.D., Kutnar, A.: Volatile organic compounds emitted from and thermally modified wood- a review. Wood Fiber Sci. **51**(3), 231–254 (2019)

Todaro, L., Rita, A., Moretti, N., Cuccui, I., Pellerano, A.: Assessment of thermo-treated bonded wood performance: comparisons among Norway spruce, common ash, and Turkey oak. Bioresources. **10**(1), 772–781 (2015)

Wehsener, J., Brischke, C., Haller, P., Hartig, J., Meyer-Veltrup, L.: Thermally and thermo-mechanically treated wood for outdoor applications–bending strength, structural integrity and set recovery. In: World Congress on Timber Engineering WCTE, pp. 22–25 (2016)

Ye, Z., Li, Y., Jin, R., Li, Q.: Towards accurate odor identification and effective feature learning with an AI-empowered electronic nose. IEEE Internet Things J. (2023)

Natural Weathering of Thermally Modified Wood Cladding Treated with Fire Retardants at Different Exposure Levels

Imke De Windt[1], Inge Wuijtens[1](✉), Kurt De Proft[2], and Lieven De Boever[1]

[1] Wood.be, Brussel, Belgium
inge@wood.be
[2] Buildwise, Limelette, Belgium

Abstract. Due to stricter regulations regarding fire safety, fire retardant treatments of wooden cladding has become more common for low and medium height buildings. Knowledge is available on how to perform these treatments to obtain a sufficient retention for a specified reaction to fire classification, both on modified and unmodified timber. However, little is known about the impact of added fire retardants to the appearance of these surfaces under natural weathering. On the other hand, the extent to which a non-fire retardant treated wooden surface visually changes has been discussed largely in literature.

A mock-up was constructed to encompass different exposure levels to test different wood species in 5 combinations. Simultaneously, a test set-up was monitored for the selected wood species without the fire retardants. The appearance was monitored by means of colour measurements and visual observations, during 1 year of outdoor exposure.

The findings reveal a spectrum of discolouration effects, including fading and colour shifts. Moreover, significant differences were observed between softwoods and hardwoods. These differences are linked to a difference in fire retardant. Furthermore, the rate and scale of the above described effects are significantly linked to the exposure level.

A such, the study highlights how these discoloration patterns contribute to the overall aesthetic evolution of wooden cladding applied to an existing building project in Belgium.

Keywords: Fire retardant treatment · Cladding · Natural weathering

1 Introduction

Since 2022, fire regulations have become more strict in Belgium. For cladding on the outside of low-rise residential buildings (<10 m height) with self-reliant occupants, a reaction to fire class D-s3,d1 is sufficient. However, in the case of not self-reliant occupants, more severe requirements are in use: a reaction to fire class C-s3,d1 is required. For medium height buildings (between 10 and 25 m height) the rules are more severe with a demand for B-s3,d1 or better [1]. In general, fire retardants are added to the wooden cladding elements to obtain Euroclass B or C.

G. Goli et al. (Eds.): ECWM 2024, SPM 86, pp. 138–147, 2025.
https://doi.org/10.1007/978-3-031-99418-0_18

In addition to this, thermally modified wood has become very common as a raw material for cladding due to its elevated dimensional stability and natural durability [2, 3]. However, not all thermally modified wood species can be used for a cladding system in end use conditions where class Euroclass D or better is required [4].

As a consequence, fire retardant treatments for wooden cladding have become more common.

Knowledge is available on how these treatments should be applied to achieve sufficient retention for a given reaction to fire classification on both modified and unmodified timber [5, 6]. On the other hand, the extent to which a non-fire retardant treated wood surface is visually altered has been extensively discussed in the literature [6–11]. However, little is known about the effect of added fire retardants on the appearance of these surfaces during natural weathering.

Understanding the effects of fire retardant treatment and environmental exposure on the visual and colour characteristics of thermally modified wood is crucial for optimising its performance and longevity in various applications.

The present work has been carried out from an industrial point of view. For a building project in Belgium, the architect envisaged the use of wooden cladding. For the cladding of this building Euroclass B-s3,d1 was required by law, which implied the need of fire retardant treatment of the wood.

The architect wanted to have an insight into the ageing behaviour and the phenomena to be expected over time, in order to communicate this to the end users of the building. Therefore, a mock-up was created to study the visual appearance of thermally modified wood treated with fire retardants, focusing on leaching effects, colour changes over time, the influence of treatment product and the degree of exposure.

2 Materials and Methods

2.1 Materials

Four wood species were examined: *Terminalia superba* (fraké), *Pinus Sylvestris* (pine), *Picea abies* (spruce) and *Triplochiton scleroxylon* (ayous). The thermal modification process was based on the same method, but TMT processes 1 and 2 were carried out in different production plants. Two phosphate based fire retardants were applied. Based on availability (AVCP system 1 certified), 5 combinations were tested in the mock-up. The test specimens were industrially produced: wooden cladding elements of 18 mm thickness and 131 mm width with tongue and groove (Table 1).

Simultaneously, a test set-up was monitored for the selected wood species without the fire retardants. For this test, small test specimens (150 × 75 mm) were used.

2.2 Methods

2.2.1 Mock-Up

A mock-up (Fig. 1) with full scale dimensions (2.4 m height) was set up to compare the ageing behaviour of wooden cladding, produced in different wood species and treated with different fire retardants.

Table 1. Combinations of wood species, thermal modification process and type of fire retardant treatment tested in the mock-up setup

Test combination	Wood species	Thermal modification process	Fire retardant treatment
1	Heat treated fraké	TMT 1	FR1
2	Heat treated pine	TMT 1	FR1
3	Heat treated fraké	TMT 2	FR2
4	Heat treated spruce	TMT 2	FR2
5	Heat treated ayous	TMT 2	FR2

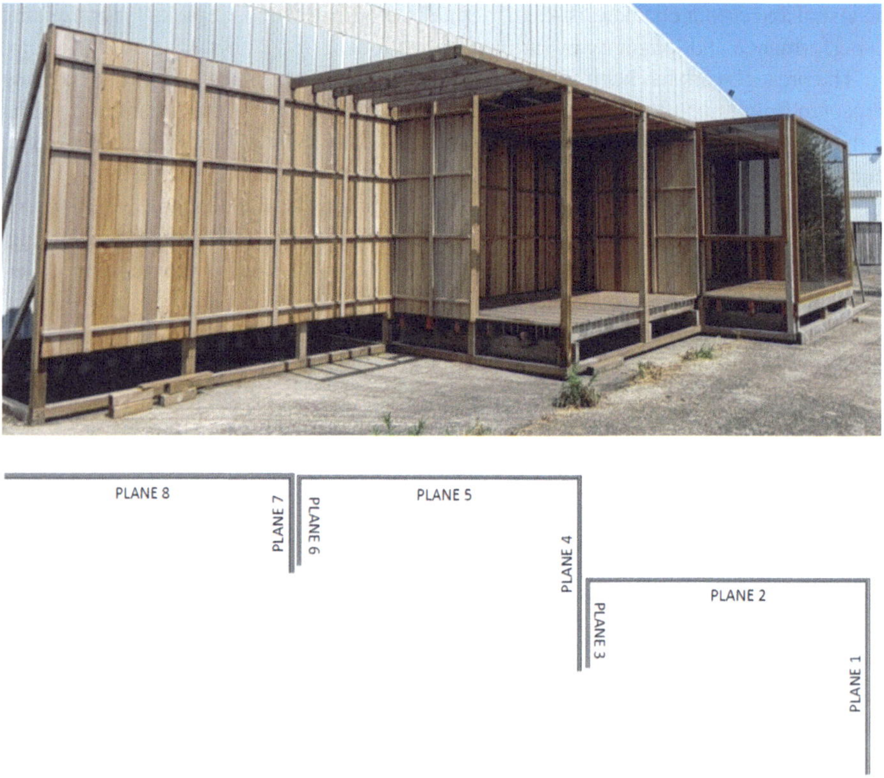

Fig. 1. Mock-up

The mock-up was constructed to encompass 3 different exterior exposure levels: behind window glass, protected by a ceiling without glass, and fully exposed. Three geographical orientations were included in the setup as well: south west (plane 1, 4 and 7), south east (plane 2, 5 and 8) and north east (plane 3 and 6).

This natural weathering test was set up with an industrial focus. Different parameters were combined to allow the architects of a building project in Antwerp to get an insight into the phenomena that may occur during ageing of fire retardant treated wooden cladding.

2.2.2 MultiFacade Exposure Rig (MFER)

Small specimens (150 ± 2 mm × 74 ± 1 mm × 18 ± 1 mm (L × B × H)) were exposed on a 'multifacade exposure rig' (MFER, Fig. 2). This setup offers the possibility of simultaneously weathering the test specimens in the different wind directions at angles of 90°, 45° and 180° (multiple exposure). Thus, a simulation is created for cladding, outdoor exposure according to EN 927–3 and decking respectively.

Fig. 2. Multifacade exposure rig (MFER).

2.2.3 Visual Assessments

Visual assessments, focusing on leaching effects, were done on the cladding surface of the mock-up. The aim of these observations was to identify specific phenomena or defects in the fire retardant treated wooden elements, both at the moment of installation and over time.

2.2.4 Colour Measurements

Colour measurements were carried out with a Hunterlab Miniscan EZ spectrophotometer, using a D65 standard illuminant and a standard observer of 10°. The diameter of the measuring aperture was 25 mm and measuring geometry of the device was 45°/0°.

On each cladding element 9 measurements were carried out. In each of the 3 zones that were created by the horizontal elements that were added on the vertical cladding 3 measurements were carried out with a distance of about 10 cm between them. To avoid edge effects, the measurements were done in the middle of the cladding elements.

On each of the MFER specimens, 5 colour measurements were done.

3 Results and Discussion

3.1 Visual Assessment

Leaching effects were observed on several panels (Fig. 3).

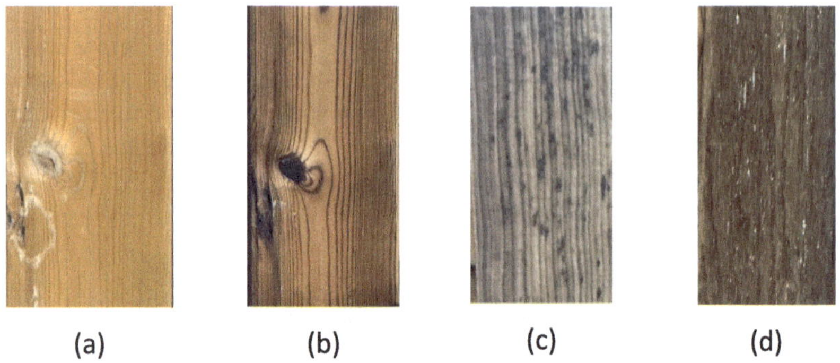

Fig. 3. Leaching effects: TMT spruce treated with FR 2 in plane 5 initially (a) and after 1 year (b), TMT fraké treated with FR1 after 6 months of exposure, plane 2 (c) and plane 8 (d)

White salt deposits were clearly visible on the FR2-treated panels at the time of installation, but were no longer visible after 1 year of exposure (plane 5, Fig. 3a, b). This is a known phenomenon [5].

Initially leaching was also observed on fraké TMT1FR1 elements, but in the form of small dark spots (plane 2, Fig. 3c). Probably, the additives added to the fire retardant to prevent it from leaching interacted with the wood and caused a leaching effect. After 6 months these dark stains turned white and could be removed from the wood by rubbing the wood surface (plane 8, Fig. 3d). On sheltered planes the dark stains were still visible.

3.2 Colour Measurements

3.2.1 Effect of Fire Retardant Treatment

Colour data from 1 year of natural weathering of thermally modified spruce, ayous and fraké treated with FR2 in plane 8 (SE orientation) were compared with their untreated references which were exposed to the south on a MFER in vertical position during 1 year (Table 2).

In general the untreated thermally modified hardwood samples showed a slightly higher L*, a* and b*-value at the beginning. This implies that these specimens were respectively slightly lighter, more red and more yellow in colour than those treated with the fire retardant. This was not the case for softwood where the opposite effect was observed.

The colour differences between the treated samples were initially related to the wood species. Ayous TMT2R2 and fraké TMT2FR2 were less red, less yellow and darker than spruce (respectively lower a*, b* and L*). These differences shifted during the 6 months of exposure. The colour of ayous TMT2FR2 and spruce TMT2FR2 became lighter and

Table 2: Average L*a*b* values and standard deviation of three treated and untreated thermally modified wood species during outdoor exposure (SE orientation. Plane 8; S orientation. vertical) and indication of the colour differences relative to the initial colour ($\Delta E^*_{initial}$).

Time [days]	TMT2				TMT2FR2			
	L*	a*	b*	$\Delta E^*_{initial}$	L*	a*	b*	$\Delta E^*_{initial}$
Spruce								
0	40.8 ± 2.2	10.2 ± 1.1	20.8 ± 2.5		44.2 ± 1.0	13.3 ± 1.1	24.7 ± 1.2	
180	57.3 ± 2.4	3.6 ± 0.3	10.2 ± 1.9	20.9 ± 2.7	50.3 ± 3.7	12.8 ± 0.5	27.4 ± 0.6	7.5 ± 2.8
365	58.1 ± 1.8	2.6 ± 0.4	7.5 ± 1.0	23.3 ± 1.3	59.6 ± 2.0	7.9 ± 1.0	18.9 ± 1.8	17.5 ± 2.3
Ayous								
0	42.8 ± 3.7	12.7 ± 0.7	24.1 ± 2.8		40.1 ± 1.0	9.7 ± 1.0	18.5 ± 1.4	
180	60.0 ± 2.4	3.6 ± 0.3	9.3 ± 3.5	24.6 ± 1.2	50.4 ± 4.0	11.1 ± 1.0	24.8 ± 1.5	12.9 ± 1.8
365	56.6 ± 3.2	2.1 ± 0.9	6.0 ± 1.6	25.3 ± 0.5	58.0 ± 3.2	7.6 ± 0.7	19.5 ± 1.7	18.1 ± 2.7
Fraké								
0	42.1 ± 1.2	11.2 ± 0.3	23.7 ± 0.3		41.8 ± 1.6	11.2 ± 0.5	20.7 ± 1.0	
180	56.1 ± 6.6	4.8 ± 0.4	10.8 ± 0.3	20.5 ± 4.0	53.3 ± 2.7	8.0 ± 1.0	20.4 ± 1.5	12.1 ± 1.8
365	56.7 ± 2.8	3.2 ± 0.6	8.3 ± 1.6	22.8 ± 1.1	54.8 ± 2.1	6.0 ± 0.5	14.6 ± 1.3	15.4 ± 2.5

reached the same average L*-level. A small difference in redness was observed with spruce a little more red than ayous. Yellowness showed the greatest difference, with spruce being a bit more yellow than ayous. After 1 year of weathering, both materials showed similar L*, a* and b*-values. Their colour had become lighter, less red and less yellow and was comparable to the colour of fraké TMT2FR2 after 6 months weathering. Fraké TMT2FR2 showed a colour shift towards less red (a* decreased) and more white (L* increased). No significant change in yellowness was noted. The change in yellowness was seen after 1 year of natural exposure, while lightness and redness showed little change.

The colour of the untreated samples appeared to be independent of the wood species. After 6 months of outdoor exposure, the untreated modified wood had become significantly lighter and less saturated. After 1 year there was only a small, insignificant decrease in yellowing and redness. The L* value didn't change significantly. This means that the discolouration of the untreated thermally modified samples was already stable after 6 months, as can be seen from the ΔE*, which didn't change significantly between 6 and 12 months.

3.2.2 Effect of Type of Fire Retardant Treatment

Colour measurements were compared between fraké TMT1FR1 and fraké TMT2FR2 in plane 7 during natural weathering at three time points: initial, approximately 180 days, and approximately 365 days (Table 3).

A clear difference in brightness was observed between fraké TMT1FR1 and fraké TMT2FR2 at the beginning of the weathering process, with fraké TMT1FR1 being significantly darker. However, this difference decreased with increasing exposure time

Table 3: Average L*a*b* values and standard deviation of treated TM fraké during outdoor exposure (SW exposure, plane 7) for two different fire retardant treatments (FR1 and FR2) and indication of the colour differences relative to the initial colour ($\Delta E^*_{initial}$)

Time [days]	TMT1FR1				TMT2FR2			
	L*	a*	b*	$\Delta E^*_{initial}$	L*	a*	b*	$\Delta E^*_{initial}$
Fraké								
0	30.9 ± 2.1	7.8 ± 1.2	12.1 ± 2.2		38.8 ± 2.9	9.7 ± 1.3	17.3 ± 1.7	
180	48.6 ± 2.9	9.0 ± 0.8	21.1 ± 1.0	20.0 ± 1.8	53.6 ± 1.8	6.0 ± 1.1	14.9 ± 2.7	15.5 ± 1.4
365	56.2 ± 3.3	5.9 ± 0.9	14.6 ± 1.8	25.5 ± 1.8	55.9 ± 1.2	4.5 ± 1.3	10.1 ± 1.7	19.5 ± 1.0

and was eliminated after 1 year. Both samples showed an increase in L* values, indicating increasing brightness.

The chromaticity of fraké TMT1FR1 was significantly lower than that of fraké TMT2FR2, showing lower levels of redness and yellowness. After 6 months of natural weathering, the redness of the samples TMT1FR1 had slightly increased, while the redness of the samples TMT2FR2 had slightly decreased. Similar changes were observed for the yellowness.

After 1 year of outdoor exposure facing southwest, there was no longer any distinction between the different treatments. All of the fraké specimens that were treated had similar L*, a*, and b* values. This was due to small increases in lightness combined with decreases in yellowness and redness.

In general, the highest variations in colour were observed for Frake TMT1FR1 ($\Delta E^*_{TMT1FR1} = 25.5$; $\Delta E^*_{TMT2FR2} = 19.5$).

3.2.3 Effect of Degree of Exposure

The differences in colour after 1 year exposure between the different exposure levels was studied for treated thermally modified softwood (pine TMT1FR1/spruce TMT2FR2) and hardwood (fraké TMT1FR1/TMT2FR2).

Softwood Table 4 clearly showed that the higher the exposure level, the higher the colour difference ΔE*.

After 1 year, TMT softwood treated with FR1 was slightly less red than TMT softwood treated with FR2. The discolouration was most pronounced in the samples treated with FR1 (ΔE* was higher).

TMT softwood treated with FR1 followed a similar discolouration pattern along the different exposure degrees compared to TMT softwood treated with FR2. The wood surfaces became lighter when sheltered but not shielded by glass. The fully exposed surfaces were even more light and showed less redness and yellowing.

The colour changes, ΔE*, were most distinct in the fully exposed areas, plane 7 and plane 8. Despite the fact that the boards in plane 4 were sheltered (not shielded by glass), a difference in lightness compared to the other sheltered boards can be observed for both softwood samples as this plane was oriented SW.

Table 4: Average L*a*b* values and standard deviation of treated TM softwood along different exposure levels and indication of the colour differences relative to the initial colour ($\Delta E^*_{initial}$).

Plane	TMT1FR1				TMT2FR2			
	L*	a*	b*	$\Delta E^*_{initial}$	L*	a*	b*	$\Delta E^*_{initial}$
Softwood								
1	48.1 ± 4.6	13.2 ± 1.4	27.3 ± 2.3	11.2 ± 2.1	46.5 ± 2.7	14.8 ± 1.6	29.2 ± 2.4	9.6 ± 3.3
2	50.4 ± 2.5	12.9 ± 0.6	28.1 ± 1.4	13.6 ± 1.6	47.5 ± 4.1	14.0 ± 1.2	28.7 ± 2.4	9.1 ± 2.5
3	50.4 ± 3.3	12.4 ± 0.6	27.3 ± 1.7	9.3 ± 1.4	47.0 ± 1.1	13.2 ± 0.8	27.0 ± 1.7	6.5 ± 1.9
4	52.4 ± 3.6	12.4 ± 1.6	27.2 ± 1.9	12.9 ± 1.8	53.6 ± 4.4	13.0 ± 2.5	27.8 ± 4.2	10.7 ± 3.4
5	51.7 ± 3.7	12.7 ± 1.2	28.5 ± 1.4	12.7 ± 2.0	47.7 ± 3.2	14.8 ± 1.7	29.9 ± 2.2	10.0 ± 1.6
6	48.1 ± 2.6	12.8 ± 0.4	27.2 ± 1.3	8.0 ± 1.7	50.9 ± 2.1	13.3 ± 1.1	28.9 ± 2.0	8.6 ± 1.3
7	58.6 ± 1.6	7.0 ± 1.6	16.5 ± 4.0	20.6 ± 2.4	56.4 ± 1.7	6.5 ± 2.2	12.8 ± 2.5	17.5 ± 1.7
8	55.0 ± 2.3	9.9 ± 1.4	23.7 ± 3.2	13.6 ± 6.6	59.6 ± 2.8	7.9 ± 1.6	18.9 ± 3.3	17.0 ± 2.7

The greatest differences in colour were seen in plane 7. This plane was facing SW, the wind direction with the highest weathering load.

Hardwood (Table 5) For fraké TMT2FR2 the colour parameters, L*a*b*, varied less with the degree of exposure than they did for fraké TMT1FR1. The difference in effect of treatment was best seen in the colour data of the sheltered specimens. For fraké TMT2FR2 samples sheltered and/or shielded by glass no significant changes in colour were noticed along different orientations, except for the SW oriented planes. When sheltered and/or behind glass fraké TMT1FR1 showed more pronounced colour differences between the orientation SW (plane 1, 4) and SE (plane 2, 5) with slightly lower yellowness and redness for the SW-exposed specimens.

Table 5: Average L*a*b* values and standard deviation of treated TM hardwood along different exposure levels and indication of the colour differences relative to the initial colour ($\Delta E^*_{initial}$).

Plane	TMT1FR1				TMT2FR2			
	L*	a*	b*	$\Delta E^*_{initial}$	L*	a*	b*	$\Delta E^*_{initial}$
Hardwood								
1	34.2 ± 5.4	8.7 ± 2.0	14.1 ± 5.2	6.1 ± 3.2	52.9 ± 2.5	11.2 ± 1.1	26.9 ± 1.7	12.0 ± 3.3
2	40.0 ± 3.8	11.6 ± 1.8	21.1 ± 4.0	10.3 ± 3.7	51.2 ± 2.4	10.5 ± 0.7	25.4 ± 0.9	14.1 ± 1.9
3	37.2 ± 4.6	9.3 ± 0.6	16.1 ± 2.7	6.0 ± 4.5	48.4 ± 2.1	10.8 ± 1.1	25.1 ± 1.9	10.7 ± 2.2
4	44.2 ± 8.3	9.1 ± 1.6	18.1 ± 2.0	15.2 ± 7.8	54.3 ± 3.7	9.6 ± 2.0	23.8 ± 4.0	14.6 ± 4.4
5	35.1 ± 4.1	10.7 ± 1.3	16.7 ± 3.6	6.6 ± 3.6	49.8 ± 2.8	10.8 ± 1.0	25.1 ± 1.5	13.7 ± 2.0
6	36.6 ± 28	8.7 ± 0.6	15.7 ± 2.2	8.6 ± 1.7	48.6 ± 3.0	10.3 ± 0.9	24.5 ± 1.7	12.8 ± 1.3
7	56.2 ± 3.3	5.9 ± 0.9	14.6 ± 1.8	25.5 ± 1.2	57.3 ± 5.8	5.0 ± 3.0	11.2 ± 6.3	20.0 ± 4.8
8	51.9 ± 3.8	8.2 ± 1.2	19.2 ± 2.0	18.1 ± 8.6	54.8 ± 2.8	6.0 ± 1.0	14.6 ± 2.6	12.4 ± 6.2

For the least exposed planes e.g. plane 3 (sheltered, behind glass) and 6 (sheltered), the L*a*b*-values did not differ significantly for both treatments.

For both treatments the exposure to sunlight caused significant colour changes (highest ΔE^*). Redness and yellowness decreased while lightness increased. At the highest exposure level, plane 7, fraké TMT2FR2 showed no significant colour difference with fraké TMT1FR1 exposed in plane 8.

Compared to softwood, the fire retardant treatment of thermally modified fraké has a more pronounced effect on the colour differences between the different exposure levels. In sheltered areas, treatment FR1 showed a wider range of chromaticity (a* and b*).

4 Conclusion

The study presented an in-depth analysis of the visual assessment and colour measurements of thermally modified wood treated with fire retardants. Visual observations revealed leaching effects on several panels, with initial salt deposits disappearing over time. Overall, a treatment with fire retardants delayed the discolouration during weathering, although the fading process did not appear to be stable yet after 1 year of weathering when compared to untreated thermally modified wood, and varied depending on the wood species used. While initial differences in colour between the different fire retardant treatments were observed, these variations diminished over time, resulting in similar colour characteristics after 1 year of exposure.

These findings underscore the complex interplay between fire retardant treatment, wood species, exposure level, and colour evolution, providing valuable insights for the practical application and long-term performance of fire retardant treated thermally modified wood in various settings. Further research in this area could deepen our knowledge on leaching of fire retardant treatments and its relation to visual observations.

Acknowledgements. This research was funded by VLAIO (Flemish Government, project HBC.2019.2510 'Recurwood') and by FOD Economy (Federal Belgium Government, project PN18B05 'Woodforlife' and PN23A69 'THERMO-FIT').

References

1. Koninklijk Besluit van 7 juli 1994 tot vaststelling van de basisnormen voor de preventie van brand en ontploffing waaraan de nieuwe gebouwen moeten voldoen. Brussel, FOD Binnenlandse Zaken, Belgisch Staatsblad van 26 april 1995 (+ erratum BS 19/3/1996) en zijn wijzigingen van 4/4/1996 (BS 20/4/1996), 18/12/1996 (BS 31/12/1996), 19/12/1997 (BS 30/12/1997), 4/4/2003 (BS 5/5/2003), 13/6/2007 (BS 18/7/2007) + erratum (BS 17/8/2007), 18/9/2008 (BS 16/10/2008), 1/3/2009 (BS 15/7/2009) + erratum (BS 4/2/2011), 12/7/2012 (21/9/2012) + erratum (BS 10/1/2014) en 7/12/2016 (BS 18/01/2017), en 20/05/2022 (BS 23/06/2022)
2. Hill, C., Altgen, M., Rautkari, L.: Thermal modification of wood—a review: chemical changes and hygroscopicity. J. Mater. Sci. **56**(11), 6581–6614 (2021)
3. Gérardin, P.: New alternatives for wood preservation based on thermal and chemical modification of wood—a review. Ann. For. Sci. **73**(3), 559–570 (2016)

4. European Commission: Commission Delegated Regulation (EU) 2024/1399 on the conditions for classification, without testing, of solid wood panelling and cladding with regard to their reaction to fire and amending Decision 2006/213/EC (2024)
5. Popescu, C.-M., Pfriem, A.: Treatments and modification to improve the reaction to fire of wood and wood based products—an overview. Fire Mater. **44**, 100–111 (2020)
6. Östman, B., Tsantaridis, L.: Durability of the reaction to fire performance of fire-retardant-treated wood products in exterior applications—a 10-year report. Int. Wood Prod. J. **8**(2), 1–7 (2017)
7. Rüther, P., Jelle, B.P.: Color changes of wood and wood-based materials due to natural and artificial weathering. Wood Mater. Sci. Eng. **8**(1), 13–25 (2013)
8. Oltean, L., Teischinger, A., Hansmann, C.: Wood surface discolouration due to simulated indoor sunlight exposure. Holz Roh-Werkst. **66**, 51–56 (2008)
9. Nuopponen, M., Wikberg, H., Vuorinen, T., Maunu, S., Jämsä, S., Viitaniemi, P.: Heat-treated softwood exposed to weathering. J. Appl. Polym. Sci. **91**(4), 2128–2134 (2004)
10. Tomak, E.D., Ustaomer, D., Ermeydan, M.A., Yildiz, S.: An investigation of surface properties of thermally modified wood during natural weathering for 48 months. Measurement. **127**, 187–197 (2018)
11. Godinho, D., Araujo, S., Quilhó, T., Diamantino, T., Gominho, J.: Thermally modified wood exposed to different weathering conditions: a review. Forests. **12**(10), 1400 (2021)

Modification with Chemicals

Studies on the Durability of the Reaction to Fire Performance of Melamine Formaldehyde Resin and Phosphorus Polyol Treated Wood

Muting Wu[1](✉), Lukas Emmerich[2], and Holger Militz[1]

[1] Wood Biology and Wood Products, Georg-August-Universität Göttingen, Büsgenweg 4, 37077 Gottingen, Germany
muting.wu@uni-goettingen.de, hmilitz@gwdg.de

[2] Landesbetrieb Wald und Holz NRW, Zentrum Holz, Team Holzwirtschaft, Carlsauestraße 91A, 59939 Olsberg, Germany
lukas.emmerich@wald-und-holz.nrw.de

Abstract. Wood modification using melamine formaldehyde (MF) resin has been recognized for its efficacy in enhancing the resistance against wood-destroying organisms and reducing moisture-induced swelling and shrinking processes. However, this treatment shows a low impact on the flammability and combustion behavior of wood. To address the aspect of fire protection, one approach could be a combination of MF resins with compatible flame retardant agents. In this study, Scots pine sapwood (*Pinus sylvestris* L) has been treated with an aqueous MF resin formulation, to which a phosphorus polyol was added. Both, the fixation of the flame retardant (FR) inside the wooden structure and the fire performance were studied on treated wood specimens. The fixation of the FR was assessed by weight percent gain with an accelerated ageing process according to EN 84 (2020). The reaction to fire of the treated material was assessed through the Bunsen burner test and mass loss calorimeter measurements (ISO 13927). Results indicated that MF and FR can penetrate the wood cell wall and be deposited inside the wooden structure. After an accelerated ageing process (EN 84), pure flame retardant treatment does not fix the flame retardant, while a certain fixation could be achieved by the incorporation in a polymeric network formed by cured MF resin. The combined treatment of MF + FR demonstrated a significant improvement in flammability and reduction in heat release. After the leaching process, the THR_{600s} values for MF + FR treated wood (9.9 MJ/$m^{2)}$ exhibited an 80% reduction compared to untreated wood (48.5 MJ/m^2).

Keywords: Chemical modification · EN84 · flame retardant · impregnation treatment

1 Introduction

Wood is a renewable biomass with many unique properties, such as a high weight-to-strength ratio, less energy-intensive to process and aesthetic values compared to concrete and steel [1]. However, its outdoor application is limited by inherent drawbacks,

G. Goli et al. (Eds.): ECWM 2024, SPM 86, pp. 151–157, 2025.
https://doi.org/10.1007/978-3-031-99418-0_19

including low dimensional stability [2], low resistance to biological attacks [3], and susceptibility to burning [4].

Wood impregnation with melamine formaldehyde (MF) resin could enhance the resistance against wood-destroying organisms, while simultaneously mitigating the swelling and shrinking by moisture [5]. Despite its effectiveness in these areas, the MF resin treatments have shown limitations in improving fire resistance [6]. To address this issue, a promising strategy involves combining MF resins with compatible flame retardants [7].

Previous research demonstrated that introducing 10% phosphorus polyol into MF resin significantly enhances the fire resistance of wood [8]. However, the fixation of this combined treatment during water leaching was still unclear. Additionally, the flame retardancy after the accelerated aging process remains to be further explored.

This study focuses on the durability of the reaction to fire performance of melamine formaldehyde resin and phosphorus polyol treated wood. The fixation of MF resin and flame retardant was evaluated through weight percent gain (WPG) and cell wall bulking (CWB) during an accelerated aging process following EN 84 (2020) standards. Fire performance was assessed through flammability tests using the Bunsen burner test (Pries and Mai) [9] and heat release measurements with a mass loss calorimeter (ISO 13927).

2 Experimental

2.1 Materials

Scots pine (*Pinus sylvestris* L.) sapwood with an average oven-dry density of 0.47 g/cm^3 was acquired from Lower Saxony state forest (Germany). MF resin (MW840) was obtained from Prefere Resins (Germany) with pH (20 °C) 9–11. Additionally, FR (Exolit OP 550 (Clariant, Switzerland), a polymeric ester of ethylene glycol and phosphoric acid with ethyl and hydroxyalkyl groups, pH (20 °C) 5) was introduced into the impregnation solution.

2.2 Treatment

All specimens underwent an initial oven-drying process at 103 °C for 48 h to determine the dry density and then conditioned at 65% relative humidity (RH) and 20 °C for 2 weeks. Aqueous solutions of flame retardant and MF resins were prepared as described in Table 1. The conditioned specimens were impregnated with a vacuum (5 kPa, 1 h) followed by a pressure phase (1200 kPa, 2 h). Subsequently, the impregnated specimens were air-dried for 1 week and then subsequently underwent a dry-curing process at 40 °C (24 h), 60 °C (24 h), 80 °C (24 h), 103 °C (24 h) and a final curing stage at 120 °C for (24 h).

Table 1. Formulation of aqueous solutions: solid content of resin (MF) and flame retardant (FR) in the impregnation formulation.

Treatment code	MF	FR
MF	20%	
FR		10%
MF + FR	20%	10%

2.3 Experimental Methods

Fixation of Impregnation Compounds

To calculate the weight percent gain (WPG, in %) and cell wall bulking (CWB, in %), ten wood specimens per group of 25 × 25 × 10 (ax.) mm^3 were used. The WPG and CWB were determined based on the oven-dry weight and cross-sectional dimensions before ($M_{untreated}$, $A_{untreated}$) and after treatment ($M_{treated}$, $A_{treated}$) according to (1) and (2), respectively. The WPG_{EN84} and CWB_{EN84} were determined based on the oven-dry weight and cross-sectional dimensions before ($M_{untreated}$, $A_{untreated}$) and after ($M_{after\ leaching}$, $A_{after\ leaching}$) an accelerated ageing process of EN 84 (2020) according to (3) and (4), respectively.

$$WPG = (M_{treated} - M_{untreated})/M_{untreated} \tag{1}$$

$$CWB = (A_{treated} - A_{untreated})/A_{untreated}\Big) \tag{2}$$

$$WPG_{EN84} = \left(M_{after\ leaching} - M_{untreated}\right)/M_{untreated} \tag{3}$$

$$CWB_{EN84} = \left(A_{after\ leaching} - A_{untreated}\right)/A_{untreated}\Big) \tag{4}$$

Bunsen Burner Test

For each treatment, five wood specimens (13 × 4 × 125 mm) with an annual ring orientation of ca. 45° were used for the flammability test, following the methodology of Pries and Mai [9]. Prior to testing, the specimens underwent conditioning at 20 °C and 65% RH for 2 weeks. The wood specimens were positioned 80 mm away with 45° angles from the Bunsen burner flame. The specimen tip was then ignited with the Bunsen burner with the flame exposed lasting 30 seconds. The test continued until the specimen stopped burning. Mass loss was calculated based on the weight of the specimens before ($m_{mod,\ before\ burning}$) and after ($m_{mod,\ after\ burning}$) the flammability test. Additionally, the time it took for the specimens to extinguish after the Bunsen burner flame was removed was measured. The mass loss was calculated according to (5) and the time to extinguish was determined using (6).

$$Mass\ loss = \left(m_{\mathrm{mod},before\ burning} - m_{\mathrm{mod},after\ burning}\right)/m_{\mathrm{mod},before\ burning} \tag{5}$$

$$Time\ to\ extinguish = Time_{total\ burning\ of\ wood} - Time_{remove\ the\ Bunsen\ burner\ flame} \quad (6)$$

Mass Loss Calorimeter Test

Mass loss calorimeter (Fire Testing Technology Ltd., UK) was used to evaluate the heat of untreated and treated wood specimens (18 × 100 × 100 mm) with an annual ring orientation of ca. 45°. Prior to testing, the specimens were conditioned at 20 °C and 65% RH. The test was performed according to ISO 13927 (2015) under an external heat flux of 50 kW/m^2. Three replicates were tested for each group.

3 Results and Discussion

3.1 Fixation of Impregnation Compounds

As shown in Table 2, a positive weight percent gain (WPG) indicated, that both MF and FR were deposited inside the wooden structure. A permanent cell wall bulking (CWB) effect, which was measured after the treatments, confirmed that MF resin and the phosphorus polyol were able to penetrate into the wood cell wall, respectively. After an accelerated ageing process (EN 84, 2020), noticeable reductions in WPG were measured on FR and MF + FR treated specimens. In case of purely FR treated wood specimens, a WPG around 0% indicated FR compound had been removed during a cold-water leaching (EN 84, 2020), thus had not been fixed inside the wooden structure. Even though WPG reductions were measured on MF + FR treated wood specimens, the extent of the latter indicated that the FR compounds experienced a certain fixation by the incorporation in a polymeric network formed by cured MF resin. However, macroscopic WPG measurements before and after a cold-water leaching did not allow to differentiate between the removal of MF and FR compounds. Thus, it remains questionable to which extent WPG reductions in MF + FR treated specimens were based on non-fixed MF or FR compounds.

Table 2. Weight percent gain (WPG) and cell wall bulking (CWB) of MF, FR and MF + FR treated Scots pine sapwood specimens before and after an accelerated ageing according to EN 84 (2020). The prefix "EN84" in the sample ID indicates that the collective was leached (EN 84).

Sample	WPG [%]	CWB [%]	Sample	WPG [%]	CWB [%]
MF	33.5(±1.5)	4.1(±0.4)	EN84-MF	31.6(±1.4)	2.9(±0.3)
FR	10.2(±1.4)	2.1(±0.7)	EN84-FR	-1.5(±0.4	-2.2(±0.7)
MF+FR	52.2(±2.2)	4.2(±0.4)	EN84-MF+FR	37.7(±2.0)	0.9(±0.3)

3.2 Bunsen Burner Test

According to Table 3, MF-treated wood demonstrated a decrease in mass loss and shorter flaming time compared to untreated wood due to the presence of nitrogen compounds of MF. Both FR and MF + FR collectives exhibited a significant reduction in mass loss to around 12% and 9%, respectively, accompanied by rapid self-extinguishing ability. After leaching, the fire performance of the wood treated solely with FR showed significant deterioration, as evidenced by prolonged burning time and increased mass loss. Interestingly, despite the WPG of pure FR treated wood being around 0% after leaching, there was still an improvement in flame resistance compared to untreated wood, indicating that some flame retardant remained within the wood matrix and did not completely leach out. On the contrary, the EN84-MF + FR collectives demonstrated good self-extinguishing properties even after leaching. The results indicate the flame retardant was effectively fixed in the wood matrix by MF resin, contributing to low flammability.

Table 3. Results of the Bunsen burner test: time to extinguish (TTE) and mass loss (ML) of the wood specimens (13 × 4 × 125 mm).

Sample	TTE [s]	ML [%]	Sample	TTE [s]	ML [%]
R	57.0 (±9.0)	70.2 (±1.9)	EN84-R	54.2 (±8.0)	71.1 (±1.4)
MF	39.4 (±9.3)	19.0 (±2.3)	EN84-MF	47.3 (±11.0)	22.1 (±3.2)
FR	2.3 (±1.0)	11.7 (±2.4)	EN84-FR	29.8 (±9.7)	22.4(±6.0)
MF+FR	1.8 (±0.8)	8.9 (±1.9)	EN84-MF+FR	3.4 (±0.5)	11.6 (±1.7)

3.3 Mass Loss Calorimeter Test

Figure 1 presents the heat release rate (HRR) and total heat release (THR) curves measured during the test. MF-treated wood displayed a heat release pattern similar to that of untreated wood. Purely FR-treated wood specimens exhibited a considerable reduction of heat release. With the phosphorus and nitrogen (P-N) synergistic effect [10] of phosphorus polyol and MF resin, the combined treatment of MF + FR significantly reduces heat release. The THR_{600s} values for MF + FR (5.2 MJ/m^2) showed an 89.3% and 91% reduction compared to untreated wood (48.5 MJ/m^2) and pure MF treated wood (57.3 MJ/m^2), respectively. This improvement also outperformed the pure FR-treated wood. After leaching, due to the smaller specific surface area of the larger specimens, the leaching for mass loss colorimeter specimens was not as complete as in small specimens subjected to the fixation and Bunsen burner test. After leaching, the larger specific surface area of the specimens in the mass loss calorimeter revealed that leaching was not as complete as in the small specimens subjected to the Bunsen burner test. The WPG for pure FR treated wood dropped from 17.7% before leaching to 3.7%, while MF + FR treated dropped from 53.2% to 46.8%. The lower weight loss in WPG further confirms that MF can fix the more flame retardant inside the wood. Due to the leaching of flame retardant, both the time to heat release peak and heat release rate were increased for the FR treated wood [11]. In contrast, for MF + FR treated wood, this enhancement remained strong even after the leaching process (9.9 MJ/m^2), surpassing the performance of other collectives.

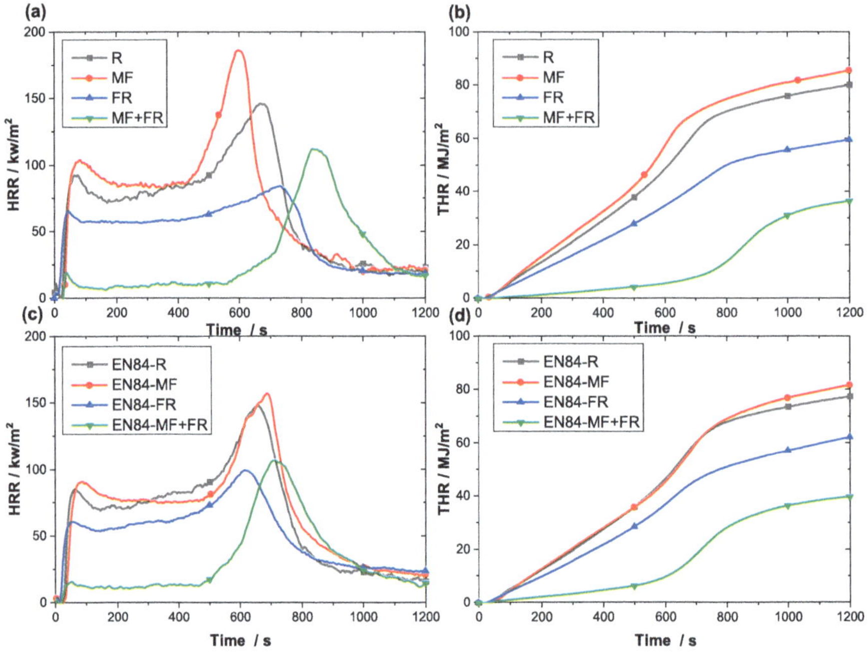

Fig. 1. Heat release rate (HRR) and total heat release (THR) curves of the wood specimens before (a, b) and after (c, d) the cold-water leaching test according to the EN 84 standard.

4 Conclusion

In conclusion, our study demonstrates that wood modification using melamine-formaldehyde resin and phosphorus polyol exhibits a certain fixation within the wood matrix after the accelerated ageing of EN 84 standard. This integrated treatment holds promise for expanding its application in outdoor environments, contributing to enhanced fire safety.

References

1. Ramage, M.H., Burridge, H., Busse-Wicher, M., Fereday, G., Reynolds, T., Shah, D.U., Wu, G., Yu, L., Fleming, P., Densley-Tingley, D., et al.: The wood from the trees: the use of timber in construction. Renew. Sust. Energ. Rev. **68**, 333–359 (2017)
2. Kocaefe, D., Huang, X., Kocaefe, Y.: Dimensional stabilization of wood. Curr. For. Rep. **1**(3), 151–161 (2015)

3. Schmidt, O.: Wood and Tree Fungi. Springer (2006)
4. Östman, B., Brandon, D., Frantzich, H.: Fire safety engineering in timber buildings. Fire Saf. J. **91**, 11–20 (2017)
5. Wood Modification in Europe: A state-of-the-art about processes, products and applications. In: Jones, D., Sandberg, D., Goli, G., Todaro, L. (eds.) Proceedings e report, vol. 124, 1st edn. Firenze University Press, Florence (2020)
6. Xie, Y., Xu, J., Militz, H., Wang, F., Wang, Q., Mai, C., Xiao, Z.: Thermo-oxidative decomposition and combustion behavior of scots pine (Pinus Sylvestris L.) sapwood modified with phenol- and melamine-formaldehyde resins. Wood Sci. Technol. **50**(6), 1125–1143 (2016)
7. Lin, C.-F., Karlsson, O., Mantanis, G.I., Sandberg, D.: Fire performance and leach resistance of pine wood impregnated with Guanyl-urea phosphate/boric acid and a melamine-formaldehyde resin. Eur. J. Wood Prod. **78**(1), 107–111 (2020)
8. Wu, M., Emmerich, L., Kurkowiak, K., Militz, H.: Combined treatment of wood with thermosetting resins and phosphorous flame retardants. Eur. J. Wood Prod. **82**, 167–174 (2023)
9. Pries, M., Mai, C.: Fire resistance of wood treated with a cationic silica sol. Eur. J. Wood Prod. **71**(2), 237–244 (2013)
10. Alongi, J., Han, Z., Bourbigot, S.: Intumescence: tradition versus novelty. A comprehensive review. Prog. Polym. Sci. **51**, 28–73 (2015)
11. Kurkowiak, K., Wu, M., Emmerich, L., Militz, H.: Fire-retardant properties of wood modified with sorbitol, citric acid and a phosphorous-based system. Holzforschung. **77**(1), 38–44 (2023)

Improving the Energy Storage Properties of Wood by Using Lauric Acid

Ahmet Can(✉)

Faculty of forestry, Bartin University, Bartin, Turkey
acan@bartin.edu.tr

Abstract. While global energy consumption varies based on countries' economic prosperity, a significant portion is attributed to approximately 30–40% buildings in the world. The predominant share of this energy usage is allocated for the purposes of heating and cooling. On a yearly basis, our nation's energy demands are escalating by 8–9%. To cope with this mounting energy requirement, there is a steady upsurge in the reliance on fossil fuels. The advancement of energy storage systems holds equivalent importance to the exploration of novel energy sources. Energy storage serves as the mechanism that adeptly captures and transforms energy into the desired usable form. This framework stands as the optimal approach to amplify the integration of renewable energy sources. Phase-change materials (PCMs) are extensively favored for thermal energy storage due to their elevated energy density and uniform temperature behavior during heat exchange. In this study, three different wood species (*Populus euroamericana* I214, *Pinus sylvestris*, *Fagus orientalis*) were impregnated with lauric acid under vacuum. The results show that the lauric acid modified wood thermal stability at the working temperature. After modification, approximately 14.87%, 12.45%, and 15.41% weight gain was obtained in poplar, beech, and pine wood, respectively. Latent heat storage of 50.8 J/g, 48.1 J/g, 52.0 J/g, and phase change temperatures of 48.29 °C, 45.22 °C and 50.46 °C (medium-temperature zone (buildings fields)) was measured for poplar, beech and pine wood, respectively.

Keywords: Phase change materials · Lauric acid · DSC · Wood

1 Introduction

The idea of lessening pollution to the environment and enhancing human health protection is driving the demand for biodegradable materials these days. In this context, it is equally important to discuss the usage of wood. Compared to substitute materials, the utilization of wooden materials is growing daily [1]. Wooden materials are renewable resources with excellent mechanical and physical qualities [2]. Owing to its significant benefits, engineered wood is extensively employed in numerous industries. The primary uses of wooden materials are in furniture design, interior architecture, and structural applications. When phase-change materials transition between solid and liquid states, they have the capacity to store and release significant amounts of energy.

G. Goli et al. (Eds.): ECWM 2024, SPM 86, pp. 158–164, 2025.
https://doi.org/10.1007/978-3-031-99418-0_20

Phase-change compounds, acting as wood impregnators, can assist to control the moisture content and temperature of wood, enhancing its resistance to rot and increasing its durability and dimensional stability. When choosing a correct phase-change materials (PCM) for construction applications, all physical, chemical, thermal, and dynamic qualities should be taken into account. Other aspects, such as security, usability, cost, compatibility, and dependability, must also be taken into account. When choosing an appropriate PCM, operating temperature and latent heat of fusion are crucial factors to consider. Significant latent heat of fusion and suitability for the operational temperature range are characteristics of ideal PCM candidate materials. To ascertain which PCM is appropriate for a particular building application, it is crucial to thoroughly assess and compare various PCMs based on these criteria and other pertinent considerations [3–11]. The potential application of bio-based phase change materials (BPCM) in wall panels and other construction materials has been studied. Phase change materials are thought to be an efficient way to lower energy usage when included into conventional building structures [12]. It has been demonstrated in earlier studies that wall panels with a PCM content of up to 30% can be produced with success in order to offer thermal energy storage capabilities [13]. The incorporation of PCM into building materials has been made possible by recent developments in materials research and microencapsulation technology [14]. The use of cement-based materials modified with PCM in building construction has been the subject of experimental and numerical investigations. According to research, incorporating PCM into cement-based components can significantly lower the energy consumption of buildings [15]. All things considered, utilizing bio-based phase change materials in construction materials might greatly lower energy usage and improve thermal comfort in buildings. Compounds are generally represented by the formula $CH_3(CH_2)_2NCOOH$, and they can be found in both plant and animal sources. The fatty acid groups that are most frequently used include lauric, myristic, caprylic, capric, and palmitic. The study's conclusions suggest that fatty acids (stearic, palmic, lauric, myristic, caprylic, and capric), which may be derived from both plant and animal sources, are a suitable material for use as PCM [3, 7]. Among the special and remarkable properties of fatty acids are congruent melting, strong thermal and chemical stabilities, non-toxicity, biodegradability, and a melting temperature range that, when mixed or esterified, is appropriate for employment in construction [16–18]. The results of the study show that because of their exceptional and special qualities, fatty acids make a great material to use as PCM. When combined or esterified, they are non-toxic and biodegradable. They also exhibit excellent chemical and thermal stabilities, a congruent melting point, and a melting temperature range appropriate for building applications [3, 7, 10].

2 Materials and Methods

2.1 Materials

The materials used in our study include three different wood species (*Populus euroamericana* I214, *Pinus sylvestris, Fagus orientalis*), and lauric acid (LA) as PCM. Wood samples were selected from timber boards with a density of 0.42 g/cm^3, 0.62 g/cm^3, 0.66 g/cm^3, respectivelly. Lauric acid (LA), ($C_{12}H_{24}O_2$, melting point: 43.2 °C) is the material for PCM.

2.2 Impregnation of Wood

The wood samples underwent a vacuum-pressure treatment in an autoclave (ALP CL 32S, Tokyo/Japan) to impregnate them with PCM. Before being impregnated, the wood samples were dried in a climate chamber to a moisture content of 12%. Wood samples were impregnated with oil/water formulations for 40 min at 60 °C under 600 mm-Hg vacuum and for 10 h at 60 °C under atmospheric pressure. Following the impregnation process, samples were removed from the system and any remaining solutions were cleaned off the wood surface. Using the weights prior to and following the impregnation, a formula (Eq. 1) was used to determine the formulation's weight percent gain (WPG).

$$\text{WPG}, \% = (W_2 - W_1)/W_1{}^* \ 100 \tag{1}$$

Where,

W_2: Sample weight after impregnation (g).
W_1: Sample weight before impregnation (g).

2.3 Characterization of Samples

The thermal properties of the LA, and LA-impregnated wood were measured using a differential scanning calorimeter (DSC, Perkin Elmer Pyris™1 DSC, temperature accuracy: ±0.2 °C, enthalpy accuracy: ±5%, Waltham, USA) under a sustained stream of argon, with the cooling and heating rate of 5 °C/min. The test temperature range was set from 0 to 50 °C. The graphics of the melting and solidifying temperature were drawn with the Origin 2019b program.

3 Result and Discussion

The results of the impregnation method showed that the WPG of the specimens was consistently high for each specimen (Table 1). Following the impregnation procedure, samples of beech, pine, and poplar showed 12.45%, 15.41%, and 14.87% WPG, respectively. Poplar samples have higher WPG, which is most likely caused by their wider cell walls than those of other species.

Table 1. Results of the impregnation process and mass loss after a leakage test for the wood specimens.

	Density (g/cm^3)	WPG (%)	Density after impregnation (g/cm^3)	Rate of leakage (%)
Beech	0.73	12.45	0.98	1.8
Scots pine	0.68	15.41	0.76	2.3
Poplar	0.45	14.87	0.63	2.8

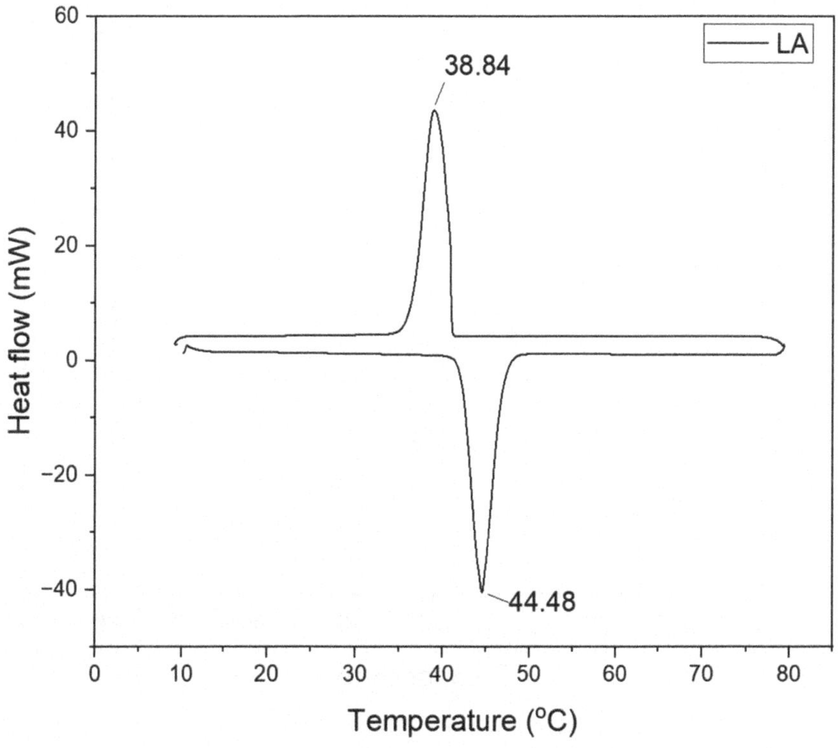

Fig. 1. DSC curves of the lauric acid

According to studies by Nazari et al. [19] and Hartig et al. [20], the viscosity of the PCM and the impregnation parameters have a significant impact on the variables determining PCM absorption into beech. Although the BPCM readily permeates the wood during impregnation, once it becomes liquid, it also has a tendency to leak out of the wood. A straightforward leakage test that lasted 24 h at 50 °C revealed that the PCM was continually leaching at the conclusion of the test 1.8% for beech, 2.3% for Scots pine, and 2.8% for poplar.

Figures 1 and . 2 shows the DSC results indicating the melting temperature of the lauric acid at 44.48 °C. The melting temperature is observed at 41.41 °C for beech, 41.79 °C for pine, and 42.17 °C for poplar wood impregnated with LA. Notably, the impregnated wood specimen's end set temperature is lower than that of pure lauric acid. The influence of capillary forces between the BPCM molecules and the wood's cell walls can be responsible for the observed temperature shift of lauric acid incorporated into wood, as determined by DSC analysis. This may affect both the interfacial interactions and the overall performance of the composite material [21]. The LA, LA-impregnated pine, and LA-impregnated poplar wood had respective melting latent heat temperatures of 207.8 J/g, 4.6 J/g, and 5.6 J/g. Samples of Scots pine wood treated with an 80% concentration of capric acid showed a retention value of 267 kg/m^3, according to Amini et al. [22]. A latent heat of 70.5 J/g was observed at this retention level.

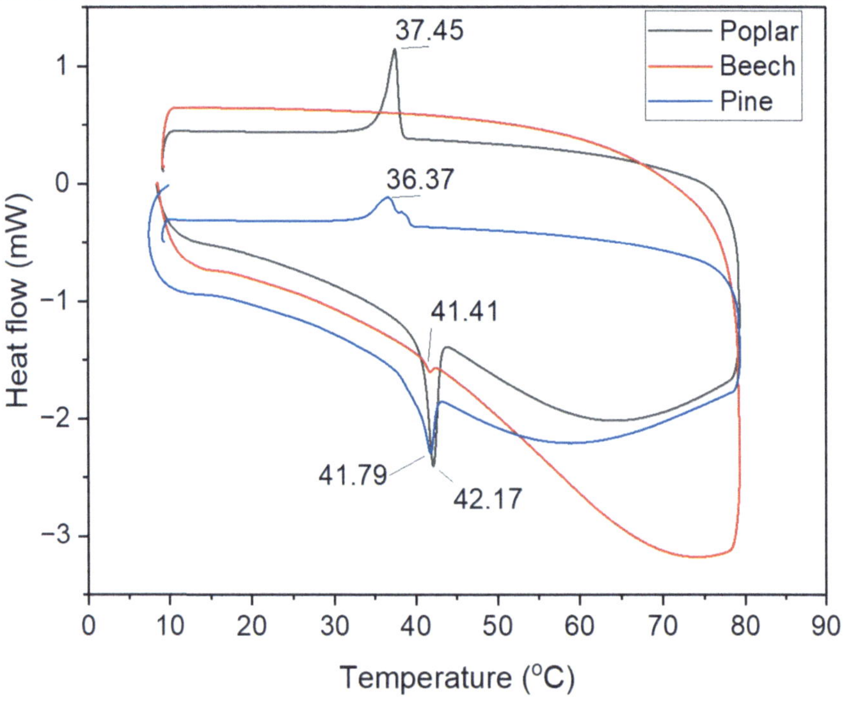

Fig. 2. DSC curves of the PCM impregnated woods

4 Conclusion

In this study, three wood species—*Populus euroamericana* I214, *Pinus sylvestris*, and *Fagus orientalis*—were impregnated with LA under vacuum to enhance their thermal energy storage capabilities. The modification resulted in significant weight gains of 14.87%, 12.45%, and 15.41% for poplar, beech, and pine, respectively. Additionally, the treated woods exhibited substantial latent heat storage capacities of 50.8 J/g, 48.1 J/g, and 52.0 J/g, with phase change temperatures well-suited for medium-temperature applications in buildings (48.29 °C for poplar, 45.22 °C for beech, and 50.46 °C for pine). These findings indicate that LA impregnation improves the thermal stability and energy storage potential of wood, making it a viable material for enhancing energy efficiency in building applications. The ability of these modified woods to store and release heat efficiently offers promising opportunities for reducing energy consumption in heating and cooling systems, contributing to more sustainable building practices.

Acknowledgements. This work was supported by the Conference Grant from TÜBİTAK 2214-A. I am grateful for the invaluable support provided by TÜBİTAK, which significantly contributed to the realization of this research.

References

1. Öztürk, O.: Alternative Methods for Improving Physical and Mechanical Properties of Fiberboard Composites with Polymers, Master Thesis, p. 89. Hacettepe University Institute of Science, Ankara (2003)
2. Salca, E.A., Hiziroglu, S.: Evaluation of hardness and surface quality of different wood species as function of heat treatment. Mater. Des. **62**, 416–423 (2014)
3. Pasupathy, A., Velraj, R., Seeniraj, R.V.: Phase change material-based building architecture for thermal management in residential and commercial establishments. Renew. Sust. Energ. Rev. **12**, 39–64 (2008)
4. Agyenim, F., Hewitt, N., Eames, P., Smyth, M.: A review of materials, heat transfer and phase change problem formulation for latent heat thermal energy storage systems (LHTESS). Renew. Sust. Energ. Rev. **14**, 615–628 (2010)
5. Baetens, R., Jelle, B.P., Gustavsen, A.: Phase change materials for building applications: a state-of-the-art review. Energy Build. **42**, 1361–1368 (2010)
6. Pielichowska, K., Pielichowski, K.: Phase change materials for thermal energy storage. Prog. Mater. Sci. **65**, 67–123 (2014)
7. Kalnæs, S.E., Jelle, B.P.: Phase change materials and products for building applications: a state-of-the-art review and future research opportunities. Energy Build. **94**, 150–176 (2015)
8. Sharma, R.K., Ganesan, P., Tyagi, V.V., Metselaar, H.S.C., Sandaran, S.C.: Developments in organic solid–liquid phase change materials and their applications in thermal energy storage. Energy Convers. Manag. **95**, 193–228 (2015)
9. Konuklu, Y., Ostry, M., Paksoy, H.O., Charvat, P.: Review on using microencapsulated phase change materials (PCM) in building applications. Energy Build. **106**, 134–155 (2015)
10. Souayfane, F., Fardoun, F., Biwole, P.H.: Phase change materials (PCM) for cooling applications in buildings: a review. Energy Build. **129**, 396–431 (2016)
11. Nazir, H., Batool, M., Bolivar Osorio, F.J., Isaza-Ruiz, M., Xu, X., Vignarooban, K., Phe-lan Inamuddin, P., Kannan, A.M.: Recent developments in phase change materials for energy storage applications: a review. Int. J. Heat Mass Transf. **129**, 491–523 (2019)

12. Zhou, D., Shire, G.S.F., Tian, Y.: Parametric analysis of influencing factors in phase change material wallboard (PCMW). Appl. Energy. **119**, 33–42 (2014)
13. Stovall, T.K., Tomlinson, J.J.: What are the potential benefits of including latent storage in common wallboard? J. Solar Energy Eng. **117**, 318–325 (1995)
14. Voelker, C., Kornadt, O., Ostry, M.: Temperature reduction due to the application of phase change materials. Energy Build. **40**, 937–944 (2008)
15. Essid, N., Eddhahak-Ouni, A., Neji, J.: Experimental and numerical thermal properties investigation of cement-based materials modified with PCM for building construction use. Int. J. Archit. Eng. Technol. **26**, 04020018 (2020)
16. Jeong, S.G., Lee, J.H., Seo, J., Kim, S.: Thermal performance evaluation of bio-based shape stabilized PCM with boron nitride for energy saving. Int. J. Heat Mass Transf. **71**, 245–250 (2014)
17. Yu, S., Jeong, S.G., Chung, O., Kim, S.: Bio-based PCM/carbon nanomaterials composites with enhanced thermal conductivity. Sol. Energy Mater. Sol. Cells. **120**, 549–554 (2014)
18. Gulfam, R., Zhang, P., Meng, Z.: Advanced thermal systems driven by paraffin-based phase change materials-a review. Appl. Energy. **238**, 582–611 (2019)
19. Nazari, M., Jebrane, M., Terziev, N.: Solid wood impregnated with a bio-based phase change material for low temperature energy storage in building application. J. Therm. Anal. Calorimet. **147**, 10677–10692 (2022)
20. Hartig, J.U., Hilkert, F., Wehsener, J., Haller, P.: Impregnation of wood with a paraffinic phase change material for increasing heat capacity. Wood Mater. Sci. Eng. **18**, 19–28 (2023)
21. Can, A., Lee, S.H., Antov, P., Abd Ghani, M.A.: Phase-change-material-impregnated wood for potential energy-saving building materials. Forests. **14**, 514 (2023)
22. Amini, M.H.M., Temiz, A., Hekimoglu, G., Demirel, G.K., Sarı, A.: Properties of Scots pine wood impregnated with capric acid for potential energy saving building material. Holzforschung. **76**, 744–753 (2022)

Effect of Aspen Face Veneer Thickness on the Fire Performance of Post-Manufacture Fire-Retardant Treated Birch Plywood

Percy Alao[1,2(✉)], Anti Rohumaa[1,3], Karl Harold Dembovski[1], Jussi Ruponen[4], Heikko Kallakas[1], and Jaan Kers[1]

[1] Material and Environmental Technology, Tallinn University of Technology, Tallinn, Estonia
{percy.alao,kademb,heikko.kallakas,jaan.kers}@taltech.ee, anti.rohumaa@xamk.fi

[2] Division of Wood Science and Technology, Department of Forest Biomaterials and Technology, Swedish University of Agricultural Sciences, Uppsala, Sweden

[3] Fiber Laboratory, South Eastern Finland University of Applied Sciences, Savonlinna, Finland

[4] Department of Bioproducts and Biosystems, Aalto University, Espoo, Finland
jussi.ruponen@aalto.fi

Abstract. The growing interest in renewable energy affects the demand for plywood in transportation and construction. However, fire safety demands frequently restrict its application as subfloors, roof sheathing, and wall panels. Moreover, using underutilized timber species like aspen is essential to maximizing the efficiency of wood resources. Hence, this study investigates the impact of fire retardant (FR)-treated aspen face veneer (AFV) thickness on the fire performance of combi-plywood (ply) composed of 1.0 mm, 1.5 mm, and 2.0 mm AFVs and a 12 mm birch core. The FR consists of a novel fire retardant (FR) formulated from bisphosphonate acid and an alkanol amine. Results indicate that thicker AFV reduces plywood density, with a 5.7% decrease observed using 2.0 mm AFV compared to 1.0 mm. The FR treatment enhanced fire performance, particularly for 2.0 mm AFV. While both 1.5 mm and 2.0 mm AFV exhibited comparable total heat release (THR) and maximum average rate of heat release (MARHE), the 2.0 mm AFV demonstrated lower temperature transmission compared to thinner veneers. Notably, FR treatment reversed the fundamental protection time increase found with increasing AFV thickness in untreated panels, indicating a complicated interplay between FR and AFV on fire behavior. These findings suggest the promising potential of further developing FR treatments for plywood.

Keywords: Aspen · Fire retardant · Fire test · Heat release · Plywood · Veneer

1 Introduction

Plywood is a valuable and versatile material [1] with properties superior to that of wood, but the type of wood species, adhesives, application, and bonded veneer layers affect the performance and durability of the plywood. Softwoods are abundant and reasonably

G. Goli et al. (Eds.): ECWM 2024, SPM 86, pp. 165–172, 2025.
https://doi.org/10.1007/978-3-031-99418-0_21

priced, which is why fir (*Abies*), pine (*Pinus*), and spruce (*Picea*) are the most used wood species for plywood production in Scandinavia, Central and Northern Europe. But, due to relative affordability, plentiful supply, robustness, longevity, pale color, and appealing grain, birch (*Betula pendula* Roth) is the primary wood species used for plywood in Estonia and the Baltic region [2]. Nonetheless, the drive for sustainable wood consumption and the desire to valorize natural resources have spurred interest in exploiting these underutilized hardwood species. One study [3] already presents the possibility to use alder and aspen veneers as plywood core materials by replacing birch veneers under various lay-up configurations. Their findings demonstrated that these underutilized wood species can be successfully employed in the veneer-based products industry if appropriate lay-up designs are implemented. Additional studies have further explored the potential of incorporating underutilized wood species like alder and aspen into plywood production [2, 4]. Besides, aspen (*Populus tremula*) is lighter, has a lower heat combustion rate, is cheaper, is more pliable, less dense and grows faster than birch.

The flammability of plywood restricts its applicability to a wide range of applications. For instance, the mandatory requirement for fire protection in construction often hinders its use as a building material. To address this, the application of fire retardant (FR) treatments is essential to meet safety standards in the construction industry [5]. The FRs reportedly used to treat wood consists of a combination of organic and inorganic salts. Common inorganic salts employed include monoammonium phosphate (MAP), diammonium phosphate (DAP), ammonium sulfate, ammonium polyphosphate, borax, and boric acid [6]. These salts are inexpensive and can be impregnated into wood using pressure impregnation techniques. However, it has been reported that these FRs lack environmental acceptability and may not effectively prevent fires, inhibiting plywood from achieving the B-s1,d0 fire classification even at low retention levels [7]. Recent advancements in the FR products have brought forth the introduction of new FRs based on ionic liquids, which are more environmentally friendly and effective [8]. This presents a need for further studies, especially with the adoption of the new wood species.

Hence, the current study examines a post-manufacture fire-retardant (FR) treatment approach to investigate the influence of aspen face veneer (AFV) thickness as a fire retardant treated layer over birch core plywood. In this study, we set our first hypotheses, that increasing the face veneer thickness will increase the fire performance. Several studies [1, 9–11] have only examined treatment of wood or plywood via penetration, FR type, FR retention or methods of application. Other studies [5, 12, 13] have examined the impact of wood species. Nevertheless, for plywood, there is a paucity of evidence regarding the interaction between FR application, particularly for novel FR formulations, and face veneer thickness. Furthermore, incorporating aspen face veneer with birch core plywood, referred to as combination plywood (combi-plywood), enables more utilization of undervalued wood species.

The FR considered in this study (Palonot F1–10 (p10)), comprises an aqueous solution of bisphosphonate acid, an alkanol amine, and optionally an alkaline agent [7]. The study outcomes provide valuable insights into the potential for enhancing the FR treatment process of plywood in the industry.

2 Materials and Method

2.1 Materials

12 mm thick birch plywood panels were obtained from Estonian Plywood AS (ESTPLY, Viruvere, Estonia), produced from 1.5 mm thick birch veneers and phenolic resin. The plywood grade is WG/WG [14], representing plywood suitable for building needs when aesthetics is unimportant. According to the ESTPLY data sheet, the birch plywood for the core has a density that ranges from 640 to 700 kg/m^3. In addition, aspen veneers of various thickness (1 mm, 1.5 mm and 2 mm) were peeled in the Laboratory of Wood Technology at Tallinn University of Technology, Estonia. The novel fire retardant from Palonot Oy, Palonot F1–10 (p10) was used in the FR treatment of the AFV.

2.2 Methods

2.2.1 Veneer Peeling and Bonding to Core Birch Plywood

The aspen veneers were peeled using a rotary peeling lathe (Model 3HV66; Raute Oyj, Finland) at a speed of 60 m/min and knife angle of 19°. Prior to peeling, the aspen logs were soaked in a water bath at 40 °C for 48 h. The veneer thickness considered were 1.0 mm, 1.5 mm, and 2.0 mm. The veneers were glued using phenol-formaldehyde (PF) Prefere (14 J025) resins (Finland Oy, Hamina, Finland). The adhesive mix comprises 68% phenolic resin, 14% hardener, and 18% water. The spread rate of the adhesive was 160 g/m^2. The dimension of the veneers and core birch plywood was 865 mm × 410 mm. Two aspen sheets were glued on top of the core plywood to mitigate the impact of the underlying birch veneers on the initial combustion behavior of the plywood. The layup of the PF-coated aspen face veneers (AFV) and birch plywood core were cold pressed for 5 min and then hot pressed for 11 min at a temperature of 130 °C and pressing pressure of 1.8 MPa. Pressing parameters were constant, regardless of the AFV thickness. Such a plywood comprising birch core and AFV can be regarded as a combination plywood (combi-ply).

2.2.2 Treatment of the Plywood with Palonot F1–10 (p10)

The panels (360 mm × 750 mm) were sanded lightly using a 60-grit size sanding paper. The p10 was then applied by roller coating with an average spread rate of 302.5 g/m^2. The panels were wet stacked, underweight in between extruded polystyrene panels to prevent too quick cooling for simulation of industrial process.

2.2.3 Reaction to Fire Test

The reaction to fire test was assessed using cone calorimeter testing in accordance with ISO 5660-1. Both piloted and autoignition conditions were investigated. For the unguided ignition, 0.25 mm type K thermocouple (Pentronic AB, Västervik, Sweden) was used to record temperature readings. The test was conducted with a heat flux of 50 kW/m^2 and distance of 25 mm. All samples were conditioned (50% relative humidity and 23 °C) and density (EN 323) and thickness were measured before the test. The unguided ignition was used to ascertain temperature transfer recorded via the thermocouple at the interface with the timber. The piloted ignition test was conducted without a thermocouple to obtain data on time to ignition (TTI), peak/heat release rate (pHRR), total smoke production (TSP), maximum average rate of heat emission (MARHE), total heat release (THR_{600}) for the test duration (600 s), and mass loss.

3 Results

3.1 Measured Physical Parameters

Table 1 shows that the density of the specimens decreases with increase in the AFV thickness. The panel thickness for both the control non-FR (C) and FR (p10)-treated specimens was consistent, with no major difference observed after treatment. The 2.0 mm AFV, however, exhibited a 0.3 mm increase in thickness, which is considered minor in this context.

Table 1. Density and thickness of examined panels (C for control, p10 for FR-treated panels).

AFV thickness	Panel thickness	Density	
(mm)		(kg/m^3)	
		C	p10
1.0	15.1	695.9	705.2
1.5	17.2	658.3	678.5
2.0	19.2	656.1	645.6

The densities of birch plywood with 1.5 mm veneers and aspen plywood with 2.6 mm veneers were previously reported to be approximately 708 and 550 kg/m^3, respectively [2]. Besides, the core birch plywood is certified a density of 640 to 700 kg/m^3. The density in the current study falls within these values. The decrease in density with thicker face veneer is typical and has also been previously reported [2]. This decrease in density is likely attributable to an increase in volume without a substantial gain in mass, suggesting the presence of more air gaps within the plywood structure. This is not unexpected, as pressing regimes, particularly duration, are typically controlled based on veneer thickness. As previously mentioned, the pressing parameters for all the AFV were the same. Although the control and treated samples originated from the same log, they were not part of the same panel batch. Therefore, a direct correlation between the pre- and post-treatment densities was not established. Partly because, the density of the treated samples was not controlled prior to treatment, since the panel size used during treatment was 360 mm × 750 mm. These factors might have influenced the observed reduction in density for the FR-treated plywood with 2.0 mm AFV and the astronomical increase for the 1.5 mm AFV variant. Therefore, the density of the panel is primarily reported for this study, as a defining characteristic, as it influences the fire performance of a material.

3.2 Piloted Reaction to Ignition

Regarding pHRR, Wang et al. 2018 [9] reported a pHRR of 217 kW/m^2 for aspen plywood with polyurethane resin while birch plywood with 1.5 mm thick veneers, with a density of 727 kg/m^3 seemingly produced a pHRR of 301.7 kW/m^2 [15]. The pHRR obtained in the current study (Table 2) falls within these values. Regardless of the p10 treatment, the density of a material influences the reaction to fire properties [16]. Despite this, the pHRR for panels with 1.0 mm AFV was the lowest, while the batch with 1.5 mm AFV produced a higher outcome than 2.0 mm AFV plies. The pHRR result, in this case, does not indicate a trend with the AFV thickness; however, in all other measured fire-related parameters, an improvement in effect is consistently observed with increasing AFV thickness. Fire performance was significantly better with p10 treatments applied than with control samples. A distinct trend was also attained with AFV thickness. These findings show that the thickness of the AFV has some impact on the fire behavior of plywood. Thicker AFV increases the effectiveness of the FR treatment, which in turn improves fire performance. This is because the FR does not penetrate the bond line, as already observed in a similar study [15]. A thinner face veneer results in a proportionally thinner FR layer, even when using the same FR quantity.

Table 2. Observed fire performance of control (C) and FR-treated (p10) panels.

AFV (mm)	pHRR (kW/m^2)		MARHE (kW/m^2)		THR_{600} (MJ/m^2)		TSP (m^2)		Mass loss (%)	
	C	p10	C	p10	C	p10	C	p10	C	p10
1.0	250.8	226.5	160.3	141	97.8	85.9	6.2	3.7	71.0	60.7
1.5	273.9	177.4	161.1	123.3	98.3	74.9	5.3	3.3	65.6	53.6
2.0	267.7	169.6	154.3	122.6	94.1	74.7	4.3	2.8	56.9	48.3

3.3 Unguided Reaction to Ignition

Figure 1a displays two sets of temperature curves: top curve shows the exposed panel surface, bottom curve shows the underside resting on a timber block. At the beginning of exposure (up to 260 s) the p10-treated samples display a delay in ignition and slower temperature rise compared to untreated variants, which is likely because the FR treatment inhibits the chemical reactions that causes ignition. The linearization of temperature rise, which begins at about 110 s and over 650 °C for the unmodified samples, is significantly delayed for the p10-treated variants, especially in the batch with 2.0 mm AFV, which suggests that the p10 treatment with face veneer thickness affects the post-ignition combustion behavior of the plywood. No meaningful changes in temperature transfer were observed between all the samples treated with p10. Nonetheless, examining the reaction to fire patterns in isolation in Fig. 1b shows that the unexposed side temperature rise of the p10 specimens decreases with increasing AFV thickness, indicating that the thicker AFV slows down the thermal breakdown to the plywood's bottom. This consequence is primarily attributable to the plywood's increased thickness due to its thicker face veneers.

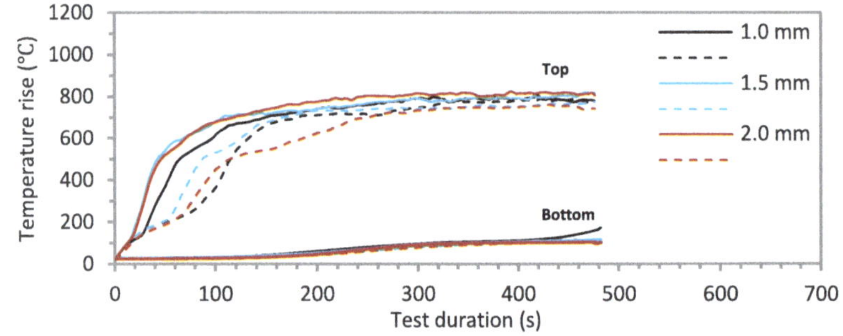

(a)

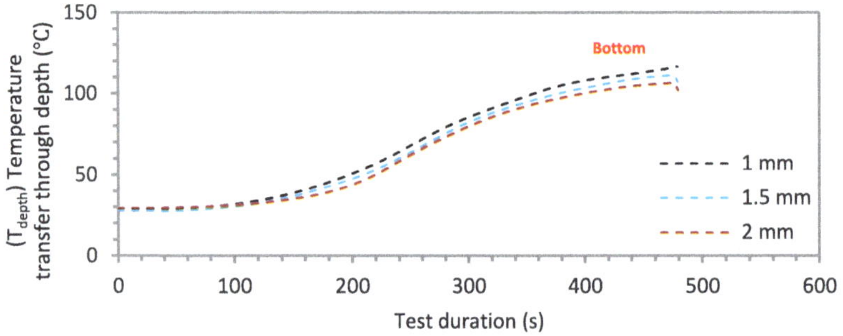

(b)

Fig. 1. The temperature-time curve showing: (a) surface and bottom temperature; (b) temperature transfer through depth of untreated (–) and p10-treated (– –) AFV-birch plywood.

Figure 2a shows a noticeable increase in ignition temperature (t_{ig}) for p10-treated samples compared to the untreated controls. The increase was 73%, 44%, and 32.5% for the 1.0 mm, 1.5 mm, and 2.0 mm thicknesses, respectively. Additionally, the p10 treatment resulted in a significant delay in Time-to-Ignition (TTI) compared to the control samples. This outcome systematically indicates beneficial results for thinner AFV, since both t_{ig} and TTI were higher for the 1.0 mm AFV. Despite this, the control panel with 1.0 mm AFV (Fig. 2b) shows the higher char depth, exceeding the 1.5 mm and 2.0 mm AFVs by 9% and 10%, respectively. While there was no discernible difference in charring following p10 treatment. It is established that the reduction in char depth for the p10 sample is due to the presence of FR decreasing the burning of the plywood [17]. Theoretically, because the p10 penetrates the thicker face veneer (AFV) more deeply, it could offer superior fire protection. Nevertheless, the p10 concentration in the top layer is impacted by the thicker AFV at the same application rate, potentially negating the benefits of thicker AFVs. Consequently, the char depth remained consistent throughout AFV thicknesses. The char depth of a substance indicates how much of it has been consumed by fire. Additional studies are needed to ascertain the exact reason for early ignition in the thicker AFVs despite the better outcome in all other aspects of the reaction to fire.

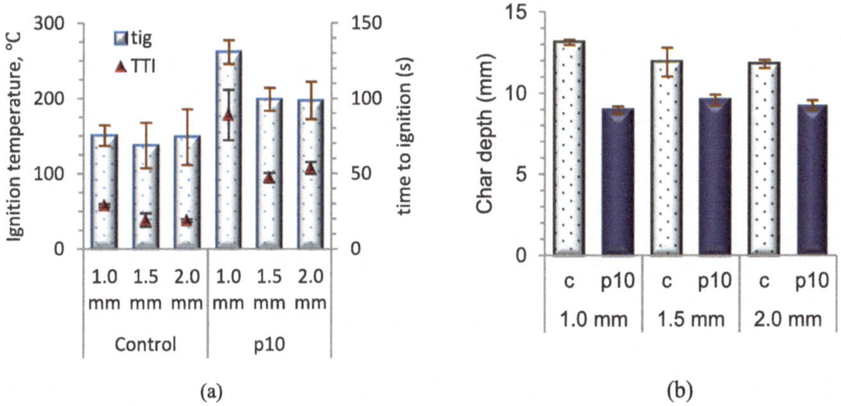

Fig. 2. Basic ignition outcomes measured under piloted ignition (a) ignition temperature and TTI; (b) char depth.

4 Conclusion

In this study combination (combi)-plywood (ply) from core birch plywood (12 mm) and aspen of various thickness (1.0, 1.5-, and 2.0-mm) as face veneers was examined with fire retardant (FR) treatment.

The combi-ply characteristics of the studied face veneer thicknesses show that whereas the FR treatment resulted in a modest increase in density due to FR retention, panel density decreases with thicker aspen face veneer (AFV). The ISO 5660-1 tests, especially with piloted ignition, uncovered the influence of AFV thickness on the combustion characteristics of the combi-ply, simultaneously introducing a novel application for the underappreciated aspen wood.

The study shows that fire resistance parameters, such as T_{depth}, pHRR, MARHE and mass loss, improve with increasing AFV thickness. However, further fire testing is required to fully confirm these results.

Acknowledgement. The Estonian Environmental Investment Centre grant (MNKE22048) and Estonian Research Council grant (PRG2213) are hereby acknowledged for supporting the research. The State Forest Management Centre Estonia is also acknowledged for helping to obtain the wood material for this research.

References

1. Kawalerczyk, J., Dziurka, D., Mirski, R., Grześkowiak, W.: The effect of veneer impregnation with a mixture of potassium carbonate and urea on the properties of manufactured plywood. Drewno. **62**(203), 107–116 (2019)

2. Akkurt, T., Kallakas, H., Rohumaa, A., Hunt, C.G., Kers, J.: Impact of Aspen and Black Alder substitution in Birch plywood. Forests. **13**(2), 142 (2022)
3. Kallakas, H., Rohumaa, A., Vahermets, H., Kers, J.: Effect of different hardwood species and lay-up schemes on the mechanical properties of plywood. Forests. **11**(6), 1–13 (2020)
4. Rohumaa, A., Kallakas, H., Mäetalu, M., Savest, N., Kers, J.: The effect of surface properties on bond strength of Birch, Black Alder, Grey Alder and Aspen veneers. Int. J. Adhes. Adhes. **110**, 102945 (2021)
5. Demir, A., Aydin, I., Ozturk, H.: Effect of fire retardant chemicals on formaldehyde emission of plywood. In: 25th International Scientific Conference New Materials and Technologies in the Function of Wooden Products: Proceedings, pp. 63–66 (2014)
6. Russell, L.J., Marney, D.C.O., Humphrey, D.G., Hunt, L.J., Dowling, A.C., Cookson, V.P.: Combining Fire Retardant and Preservative Systems for Timber Products in Exposed Applications—State of the Art Review. The Forest and Wood Products Research and Development Corporation (2004)
7. Ruponen, J., Kukkonen, J.: Ionic liquid (ILS) based fire retardant for wood-based panels. In: World Conf. Timber Eng. 2021, WCTE 2021, pp. 2–5 (2021)
8. Jia, Y.-W., et al.: Synergy effect between quaternary phosphonium ionic liquid and ammonium polyphosphate toward flame retardant PLA with improved toughness. Compos. Part B Eng. **197**(10–11), 108192 (2020)
9. Wang, W., et al.: A novel application of silicone-based flame-retardant adhesive in plywood. Construct. Build Mater. **189**, 448–459 (2018)
10. Mali, P., Sonawane, N.S., Patil, V., Lokhande, G., Mawale, R., Pawar, N.: Morphology of wood degradation and flame retardants wood coating technology: an overview. Int. Wood Prod. J. **13**(11), 1–20 (2021)
11. Ali, S., Aslina Hussain, S., Tohir, M.Z.M.: Science & Technology Fire Test and effects of fire retardant on the natural ability of timber: a review. Pertanika J. Sci. Technol. **27**(2), 867–895 (2019)
12. La, H., Zhilin, C., Feng, F., Mizi, F.: Investigation of factory fire retardant treatment of Eucalyptus plywood. For. Prod. J. **65**(7–8), 150227105121004 (2016)
13. Michael Albert, C., Chiang Liew, K.: Recent development and challenges in enhancing fire performance on wood and wood-based composites: a 10-year review from 2012 to 2021. J Bioresour Bioprod. **9**, 27–42 (2024)
14. Finnish Forest Industries Federation: Hand Book of Finnish Plywood. Kirjapaino Markprint Oy, Lahti, Finland (2002) ISBN 952-9506-63-5
15. Alao, P.F., Dembovski, K.H., Rohumaa, A., Ruponen, J., Kers, J.: The effect of Birch (Betula pendula Roth) Face Veneer thickness on the reaction to fire properties of fire-retardant treated plywood. Construct. Build Mater. **426** (2024)
16. R. H. White: Reprint from Standard Technical Publication 826 Use of Coatings to Improve Fire Resistance of Wood (1984)
17. Ilyas, R.A., et al.: Polymer composites filled with metal derivatives: a review of flame retardants. Polymers (Basel). **13**(11), 1701 (2021)

Exploring the Solid Wood Modification with Preserved Hierarchical Structure Via Non-Cellulosic Substances Removal

Yi Hien Chin[1,2,3](✉), Pascal Biwole[1,4], Joseph Gril[1,5], Christophe Vial[1], Rostand Moutou Pitti[1,6], Salah-Eddine Ouldboukhitine[1], Nicolas Labonne[2], and Yoshiki Horikawa[3]

[1] Université Clermont Auvergne, Clermont Auvergne INP, CNRS, Institut Pascal, Clermont-Ferrand, France
yi_hien.chin@uca.fr
[2] Dagard Company, Boussac, France
[3] Institute of Agriculture, Tokyo University of Agriculture and Technology, Fuchu, Tokyo, Japan
[4] MINES Paris, PERSEE - Center for Processes, Renewable Energies and Energy Systems, PSL Research University, Paris, France
[5] Université Clermont Auvergne, INRAE, PIAF, Clermont-Ferrand, France
[6] CENAREST, IRT, Libreville, Gabon

Abstract. Wood is a bio-based material that has always been used for various purposes due to its excellent mechanical and thermal properties. Innovations have been made in wood modification technology to create new functionalities while improving its current advantages. Wood is composed of three main chemical elements: cellulose, hemicellulose, and lignin. This study focuses on the modification of wood by the removal of non-cellulosic substances to create a cellulose scaffold. The creation of pores after treatments contributes to lower density and therefore lower thermal conductivity. Besides, a new composite can be invented by impregnation of other materials between the cellulose fibers. The top-down modification approach increases the production scalability but the drying of the modified wood remains challenging for the industrial professionals. This work starts by reviewing the protocol of various wood delignification methods that preserve the natural hierarchical structure of solid wood. Then, it focuses on experimental trials on French poplar wood (*Populus* spp.) using soda pulping, alkaline sulfite pulping, organosolv pulping, hydrogen peroxide bleaching and sodium chlorite bleaching. To look for potential drying solutions, a comparative study was done on different drying techniques. It was found that dielectric heating has the highest drying rate across all moisture contents whereas convective blow drying exhibits the lowest sectional shrinkage. The findings encourage further exploration of delignified wood's properties leading to large-scale modified wood manufacturing in the future.

Keywords: Bio-based material · Delignification · Drying technologies · Nanotechnology · Poplar wood · Wood modification

G. Goli et al. (Eds.): ECWM 2024, SPM 86, pp. 173–182, 2025.
https://doi.org/10.1007/978-3-031-99418-0_22

1 Introduction

To reduce CO_2 emissions towards sustainable development, especially in the building industry, it is necessary to substitute non-renewable materials with low embodied energy and bio-based products. Delignified wood was reported to have low thermal conductivity, making it a potential biodegradable thermal insulator [1]. Despite numerous wood delignification methods that had been proposed for various functionalizations, there was not a thorough comparative analysis of these techniques, a single method was usually applied to each research study. French poplar wood (*Populus* spp.) had been identified as an abundant and fast-growing resource. The aims of this work are to study the degree of lignin removal by three delignification techniques and to suggest potential drying techniques for large-scale production. This work starts by reviewing the current state of knowledge in bulk wood delignification. Soda pulping, alkaline sulfite pulping and organosolv pulping were chosen to be combined with or without bleaching steps to modify poplar wood. The NaOH modified wood samples were subjected to ambient-air drying, convective blow drying, oven heating and dielectric heating. The information reported allows a better understanding of wood modification and drying mechanisms for the design of appropriate manufacturing procedures.

2 State-of-the-Art

Wood tissue is a hierarchically composite material with complex organizations from macro to micro scales as presented in Fig. 1. Among the total mass of wood, cellulose makes up 40 to 50%, hemicelluloses 20 to 30%, and lignin 20 to 30%. The cellulose microfibrils, which include both crystalline and amorphous cellulose, have nanometric lateral dimensions (a width of 3 nm and a length of 1 μm) [2]. They are attached by lignin to provide mechanical strength and rigidity to the structure of the wood. Hemicellulose is comprised of different sugar units, such as glucose, mannose, galactose, xylose, and arabinose. Lignin is a natural glue between the wood cells and the middle lamella. The opaque, brown appearance of wood is the result of the presence of chromophore groups and the phenolic character of lignin with strong light-capturing capability [3]. Cellulose and hemicellulose are hydrophilic, whereas lignin is less hydrophilic. The removal of non-cellulosic matrices from wood is a subtractive chemical treatment that is also known as delignification. This modification treatment not only changes the chemistry composition of the natural wood but also creates nanochannels in the empty inter-fiber spacing, along with the creation of nanopores in the cell wall and the micropores at the cell wall corner, thus increasing the specific surface area from around 1 to over 40 m^2/g depending on the drying methods [4].

The details of the delignification methods reported in the literature are summarized in Table 1. One of the common techniques in pulping, the high temperature and high pH soda process fragmentates the lignin structure by cleavage of lignin-carbohydrate bonds and β-O-4 linkage using sodium hydroxide (NaOH) [6]. Alkaline sulfite pulping uses a solution system of sodium hydroxide (NaOH) and sodium sulfite (Na_2SO_3) that contains active delignifying nucleophiles, such as sulfite (SO_3^{2-}), bisulfite (HSO_3^-), and hydroxide (OH^-) anions. It is often followed by hydrogen peroxide bleaching to remove

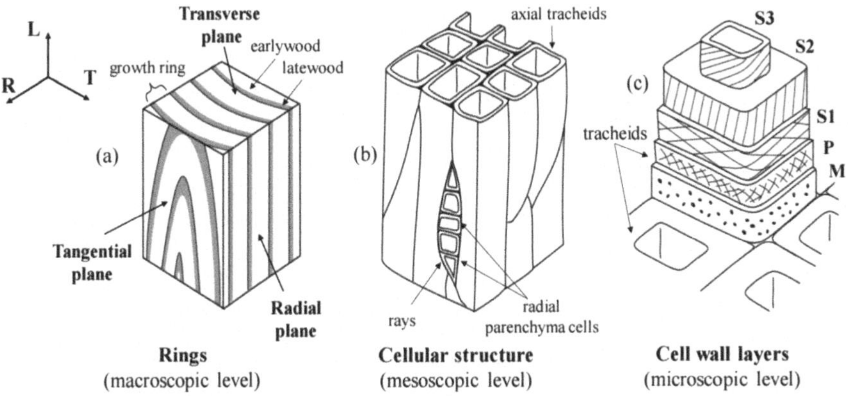

Fig. 1. Sketch of the hierarchical structure of wood [5]: (a) growth rings, layout, and main axes; (b) honeycomb-shaped cell tissue; (c) multi-layered cell wall defined by the variable orientation of cellulose microfibrils.

the unreacted chemicals and to decolorize the wood [1]. Another method identified was organosolv pulping (alcoholysis with ethylene glycol) which hydrolyses hemicellulose and removes lignin by ether linkage cleavage. It is followed by the sodium chlorite/acetic acid ($NaClO_2$/AA) bleaching that produces chlorine dioxide to oxidize the aromatic rings from lignin via electrophilic attack [7].

Table 1. Experimental details of wood modification techniques involving soda pulping, alkaline sulfite pulping and organosolv pulping reported in the literature.

Delignifying methods	Wood species & dimensions [mm^3]	Delignification protocols	Additional bleaching step	Applications & Refs.
Soda pulping	Paulownia (*Paulownia tomentosa*) 600 x 45 x 12.5	5 wt% NaOH at high temperature for 1 h	No	Thermal insulator [8]
	Spruce (*Picea abies*) 5 x 5 x 5	10 wt% NaOH at 80 °C for 8 h under constant stirring	No	Cellulose scaffold [6]
Alkaline sulfite pulping	Basswood (*Tilia americana*) 120 x 30 x 12	NaOH + Na_2SO_3 at boiling temperature for 3 h	Treatment with H_2O_2 for 5 h	Thermal insulator [1]
	Beech (*Fagus sylvatica*) 40 x 50 x 5	2.5 mol.L^{-1} NaOH + 0.4 mol.L^{-1} Na_2SO_3 at boiling temperature for 48 h	Boiling 30 wt% H_2O_2 until white without stirring	Transparent wood [9]

(*continued*)

Table 1. *(continued)*

Delignifying methods	Wood species & dimensions [mm^3]	Delignification protocols	Additional bleaching step	Applications & Refs.
Organosolv pulping	Japanese Cedar (*Cryptomeria japonica*) 10 x 10 x 10	Alcoholysis with Ethylene glycol + H_2O + 97% H_2SO_4 at 150 °C for 1 h	Wise method using 150 ml H_2O + 1 g $NaClO_2$ + 0.2 ml acetic acid at 70 °C for 8 h, add the same amount of reactant every hour	Colorless wood [7]

3 Materials and Methods

French poplar wood planks were purchased from Chagnon Carpentry, France. They were cut into a 100 mm by 10 mm section in the transversal plane and then longitudinally 100 mm along the grain (Fig. 2). Three samples from the batch were oven-dried at 103 °C to determine the average initial moisture content.

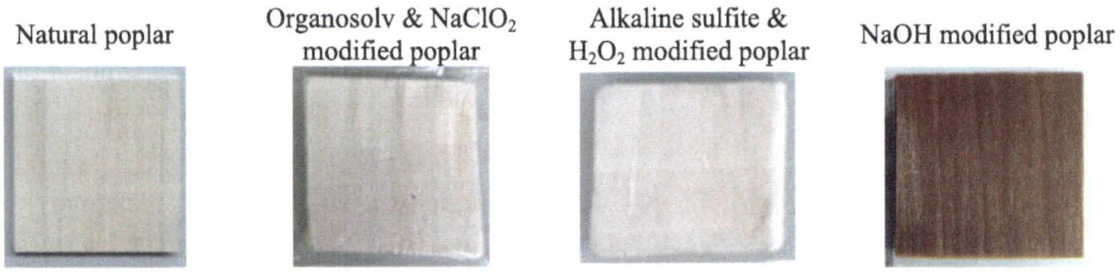

Fig. 2. Visual appearance comparison of poplar wood after different treatments.

A summary of experimental steps is shown in Fig. 3. Three kinds of wood modification techniques were used on the poplar wood samples. NaOH modified wood was prepared by soda pulping. Wood samples were treated in a stock solution of sodium hydroxide (10 wt%) at 100 °C for 8 h. Alkaline sulfite & H_2O_2 modified wood were prepared first by being introduced in a mixture of 2.5 mol.L^{-1} NaOH and 0.4 mol.L^{-1} M Na_2SO_3 at 100 °C for 6 h, followed by hydrogen peroxide (35%) bleaching at 70 °C for 4 h. Organosolv & $NaClO_2$ modified wood were subjected to two-step delignification involving alcoholysis and sodium chlorite bleaching. The solvent employed for the alcoholysis reaction was composed of ethylene glycol, 97 wt% sulfuric acid, and water in a ratio of 99:0.5:0.5. The wood was heated in the solutions at 140 °C for 4 h. After washing, they were bleached at 70 °C for 8 h, 8 g of $NaClO_2$ and 1.6 ml of acetic acid were added into the 1200 ml bleaching solution every hour [7]. Upon completion of the wood modification processes, the samples were rinsed thoroughly to remove the remaining chemical reagents and lignin fragments. Then, they were stored in ethanol for over a week to allow the exchange of solvent.

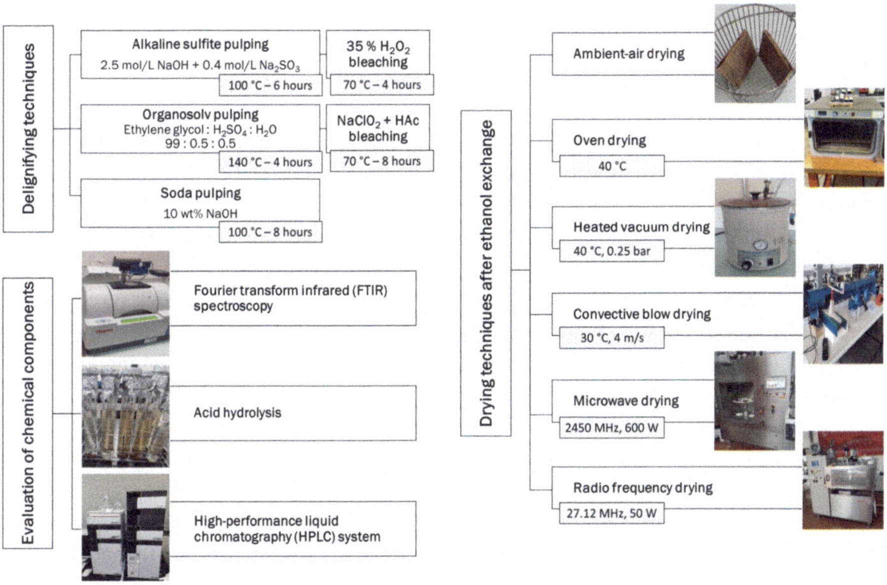

Fig. 3. Diagram summarizing each experimental step (delignifying techniques, evaluation of chemical components and drying techniques).

NaOH modified wood samples were used to study the drying kinetics of a few drying techniques. The heating was stopped when excessive shrinkage and deformation were observed. For ambient-air drying, the samples were simply left in an ambient environment. A laboratory oven (Firlabo, France) was used to oven-dry the samples at 40 °C. A heated vacuum desiccator (GCA/Precision Scientific, Chicago, USA) was connected to a vacuum pump to dry the modified wood samples at 40 °C and 0.25 bar. For convective blow drying, a tube blow dryer (Polytech Clermont, France) equipped with an air heater and a fan was used to blow air of 30 °C at a velocity of 4 m/s on a suspended sample. Dielectric drying through electromagnetic radiation was carried out using a multi-energy microwave oven Labotron MEO12 (Sairem, France) for microwave heating (2450 MHz at 600 W) and Labotron RF 3000 (Sairem, France) for radio frequency heating (27.12 MHz at 50 W).

A weighing balance (Sartorius, Germany) with a minimum sample weight of 0.1 g was used to weigh the wood samples and the size was determined with a digital caliper at a precision of +/− 0.01 mm. Fourier transform infrared (FTIR) spectroscopy (Spectrum Frontier, PerkinElmer, USA) equipped with an attenuated total reflection (ATR) accessory was used to acquire the spectra, which were then normalized with the absorbance of the band 1030 cm^{-1}. To determine the acid-insoluble lignin content, two specimens for each treatment were treated by acid hydrolysis by first heating in 72 wt% sulfuric acid at 30 °C for 1 h, then in 4 wt% sulfuric acid solution at 120 °C for 1 h. The solid residue was dried and weighed as acid-insoluble content. The monosaccharide contents were estimated by a high-performance liquid chromatography (HPLC) system (Shimadzu, Japan).

4 Results and Discussions

4.1 Degree of Lignin Removal

Three kinds of chemical modification were performed on the poplar bulk wood sample, which resulted in NaOH modified wood, alkaline sulfite & H_2O_2 modified wood and organosolv & $NaClO_2$ modified wood.

FTIR spectroscopy method was used to detect the change in the content of chemical components through the intensity change in each assigned band. When comparing the modified wood to the natural untreated wood, the overall change in peak patterns and intensities for modified wood indicates an alteration of the wood's chemical composition (Fig. 4). The decreasing intensity of the peaks at 1730 and 1240 cm^{-1} assigned to acetyl functional groups was a result of effective hemicellulose removal [10]. Organosolv & $NaClO_2$ treatment led to the disappearance of the peaks at 1508, 1465, and 1264 cm^{-1}, indicating the complete removal of lignin. In contrast, these peaks remain for the NaOH and alkaline sulfite & H_2O_2 modified wood. This is further confirmed by the remaining acid-insoluble lignin content in Table 2. A higher treatment temperature, treatment duration or delignifying solution concentration may be required for NaOH and alkaline sulfite & H_2O_2 modification method to increase the degree of lignin removal.

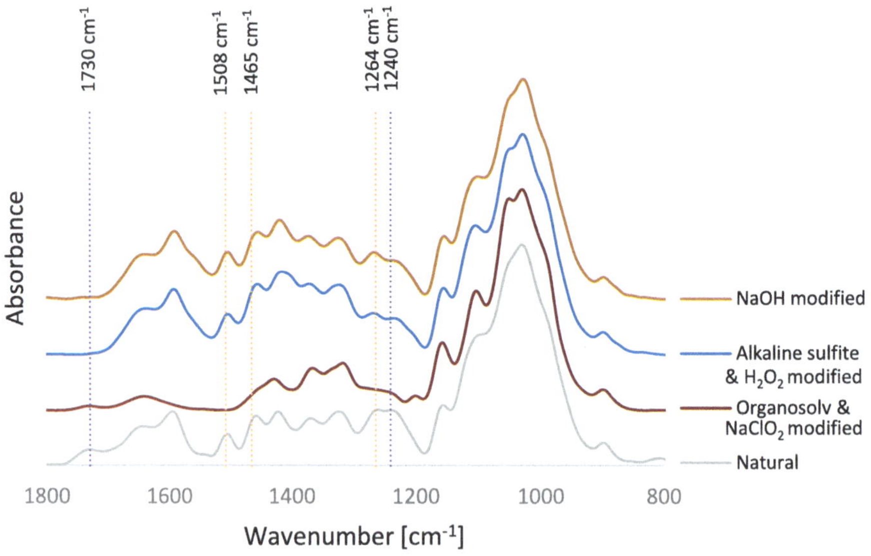

Fig. 4. FTIR spectra of poplar wood under different modification treatments.

Table 2. Percentage of chemical components in poplar wood before and after modification treatments.

Chemical components	Percentage of chemical components [%]			
	Natural poplar	Organosolv & $NaClO_2$ modified poplar	Alkaline sulfite & H_2O_2 modified poplar	NaOH modified poplar
Acid-insoluble content	19.3	0.1	12.5	16.1
Fucose	2	0.8	1.2	1.4
Xylose	15	3	9.6	9.3
Arabinose	1.1	0.5	0.8	0.7
Mannose	5.5	2.3	2.9	3.4
Glucose	52.4	59.8	58.6	59.2
Galactose	4.6	2.9	4.1	4.1
Removed substances	0	30.6	10.2	5.9

In the natural untreated poplar wood samples, the acid-insoluble lignin represents nearly 20% of its mass. Significant removal of substances, especially acid-insoluble lignin, was observed in organosolv & $NaClO_2$ modified wood. The high glucose content was maintained in all modified wood, suggesting that cellulose was not much affected by the treatments, so that only lignin and hemicellulose were preferentially altered. Organosolv & $NaClO_2$ modification, alkaline sulfite & H_2O_2 modification and NaOH modification appear to be the most to the least lignin removal efficiency respectively, based on the remaining acid-insoluble content.

4.2 Drying Kinetics

Due to the creation of capillary pressure during liquid evaporation, the modified wood would shrink when directly oven-dried. In most of the experimental studies, delignified wood samples were freeze-dried to remove water without altering the microstructure. Herein, we examined alternative drying techniques using NaOH modified wood which were solvent exchanged with ethanol, an organic solvent with a lower surface tension than water.

Ambient-air drying is the least efficient method due to its dependence on the temperature and relative humidity of the surroundings (Fig. 5). Blow drying has an initial rapid decline in moisture content because surface moisture evaporates quickly through convection, but less effective later when the liquid has to diffuse from the core to surface. Compared to oven drying, vacuum oven drying is more efficient as the lower pressure facilitates the evaporation of ethanol. Dielectric heating (microwave and radio frequency) is the fastest drying technique as it causes the moisture molecule of the whole piece of modified wood to vibrate and vaporize. For all drying methods, the drying rate decreases as the moisture content of samples reduces, as shown in Fig. 6. Only the drying rate of vacuum oven drying and dielectric heating exceeds 1%/min, other methods need longer drying duration even at high moisture content.

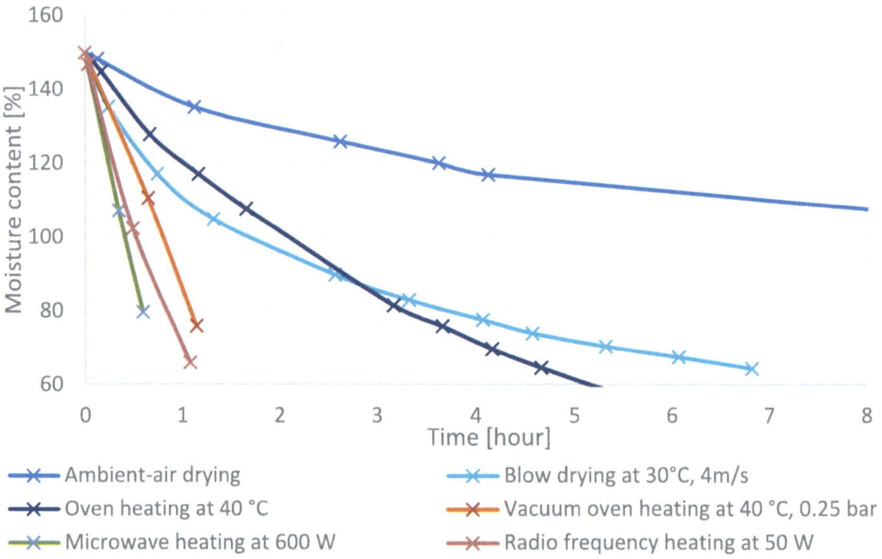

Fig. 5. Moisture content evolution of NaOH modified wood by different drying techniques.

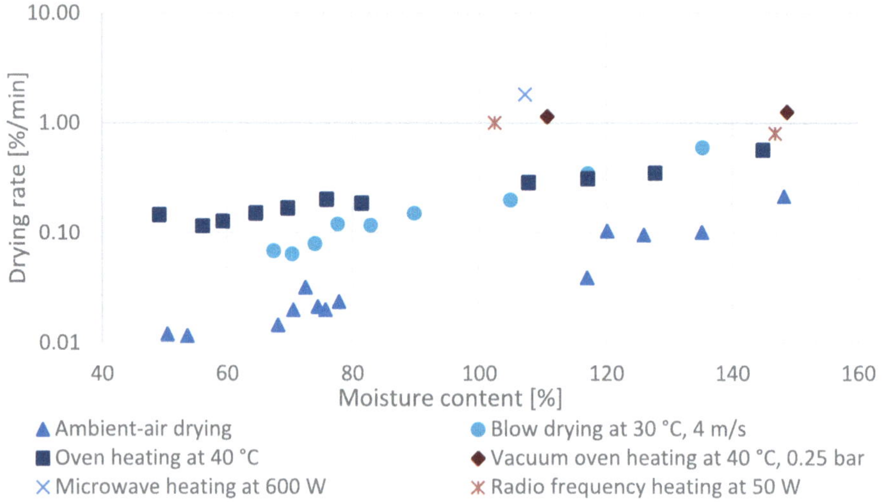

Fig. 6. Drying rate at logarithmic scale of different moisture contents for each drying technique.

All samples were finally subjected to oven drying at 103 °C to determine the dry mass and the dimensional variation. The dimension along the fiber direction was slightly affected by the wood modification and drying treatments, whereas the significant shrinkage of the sectional surface reflects the collapse of cell lumens. This study suggests that for applications requiring rapid drying, microwave or radio frequency heating would be preferable, yet these methods result in great dimensional shrinkage and densifies the samples despite organic solvent exchange. Blow drying stands out as the most adequate method in preserving the wood's dimensions, which is crucial for manufacturing functional wood scaffolds.

5 Conclusion and Perspectives

In this study, solid wood modifications and drying techniques were experimented on French poplar wood for the first time. The combination of organosolv pulping and sodium chlorite bleaching was proven to be the most effective in achieving complete removal of non-cellulosic substances towards the wood modification in producing cellulose scaffolds. Soda pulping and alkaline sulfite pulping followed by hydrogen peroxide bleaching showed their suitability in the poplar wood modification with intended lignin retention. Further optimization by employing a higher concentration of delignifying solutions, higher temperature or longer treatment duration may contribute to a more significant reduction of lignin content. Additionally, this work presented drying kinetics of possible drying techniques for upscaling applications as functionalized wood scaffolds or biobased thermal insulators. These findings empower industry professionals in decision-making about solutions best suited to their specific requirements. Future investigations should focus on the effects of delignification on the thermal, mechanical, and hygric properties of the French poplar wood.

Acknowledgments. This work is supported by the French National Research Agency (ANR) and the company Dagard, under the « France Relance » plan. The research visit to Japan was co-funded by the I-Site project CAP 20-25 and the Tokyo University of Agriculture and Technology.

References

1. Li, T., Song, J., Zhao, X., et al.: Anisotropic, lightweight, strong, and super thermally insulating nanowood with naturally aligned nanocellulose. Sci. Adv. **4**, eaar3724 (2018). https://doi.org/10.1126/sciadv.aar3724
2. Nishiyama, Y.: Structure and properties of the cellulose microfibril. J. Wood Sci. **55**, 241–249 (2009). https://doi.org/10.1007/s10086-009-1029-1
3. Li, Y., Fu, Q., Yu, S., et al.: Optically transparent wood from a nanoporous cellulosic template: combining functional and structural performance. Biomacromolecules. **17**, 1358–1364 (2016). https://doi.org/10.1021/acs.biomac.6b00145
4. Fu, Q., Medina, L., Li, Y., et al.: Nanostructured wood hybrids for fire-retardancy prepared by clay impregnation into the cell wall. ACS Appl. Mater. Interfaces. **9**, 36154–36163 (2017). https://doi.org/10.1021/acsami.7b10008

5. Phan, N.-T.: Effets de la réticulation des fibrilles de cellulose sur le comportement hygromécanique effectif du bois à différentes échelles. Ingénierie Des Matériaux, Université Clermont Auvergne (2022). http://www.theses.fr/2022UCFAC098/document
6. Segmehl, J., Studer, V., Keplinger, T., Burgert, I.: Characterization of wood derived hierarchical cellulose scaffolds for multifunctional applications. Materials. **11**, 517 (2018). https://doi.org/10.3390/ma11040517
7. Horikawa, Y., Tsushima, R., Noguchi, K., et al.: Development of colorless wood via two-step delignification involving alcoholysis and bleaching with maintaining natural hierarchical structure. J. Wood Sci. **66**, 37 (2020). https://doi.org/10.1186/s10086-020-01884-1
8. Zhao, X., Liu, Y., Zhao, L., et al.: A scalable high-porosity wood for sound absorption and thermal insulation. Nat. Sustain. (2023). https://doi.org/10.1038/s41893-022-01035-y
9. Yu, Z., Yao, Y., Yao, J., et al.: Transparent wood containing $Cs_x WO_3$ nanoparticles for heat-shielding window applications. J Mater Chem A. **5**, 6019–6024 (2017). https://doi.org/10.1039/C7TA00261K
10. Horikawa, Y., Hirano, S., Mihashi, A., et al.: Prediction of lignin contents from infrared spectroscopy: chemical digestion and lignin/biomass ratios of Cryptomeria japonica. Appl. Biochem. Biotechnol. **188**, 1066–1076 (2019). https://doi.org/10.1007/s12010-019-02965-8

Laser Incising: A Philosophical Shift from Timber Treatment to Modification

Morwenna J. Spear[1(✉)], Paul Mason[2], Geraint Williams[2], and Graham A. Ormondroyd[1]

[1] Bangor University, Bangor, Gwynedd, UK
m.j.spear@bangor.ac.uk
[2] Millennium Lasers, Neath, UK

Abstract. Timber incising has been known and investigated for many years, yet remains a very diverse field. Often simplified into three categories: mechanical incision, laser incision and bioincising, this group of technologies is often seen as the domain of timber preservation, with only mechanical incision being commercialised.

Yet laser incision offers essential differences which mark it out as suitable for wood modification and high value products, with great aesthetic appeal and high market value. Wood modification requires near-perfect distribution of treatment agent, which is achieved in laser incision through careful matching of incision patterns and species. Building on successful collaborative research projects (Laserwood and Lasercure) we have demonstrated the technology's application in the full scale resin modification process.

In this poster we present the concept of laser incision, as it applies to wood modification, and its scope for revolutionizing all fluid-based wood modification systems (esterification, furfurylation, resins, bio-polymers, polymerisation modifications) as well as other protective treatments e.g. with natural extracts .

Keywords: Laser incision · Permeability · Wood modification

1 Introduction

Wood modification is an increasingly adopted group of technologies for prolonging the service life of wood, altering the properties of wood and enhancing visual aesthetics of wood. The modification technologies can be grouped into thermal modification systems, chemical modification systems, impregnation or polymerization modification systems and physical modifications. Of these technologies, the chemical modifications and the impregnation or polymerization systems rely on excellent permeability of the timber to the treatment agents [1, 2]. This has traditionally been achieved by the use of pine sapwood, most commonly Radiata pine, but also Scots pine and Southern yellow pine for certain commercial treatments. The resulting demand on a finite resource has led to the quest to identify permeable species [3–5] or the use of veneers to increase uptake by reducing distance travelled by the modifying agent. This was most clearly demonstrated

G. Goli et al. (Eds.): ECWM 2024, SPM 86, pp. 183–189, 2025.
https://doi.org/10.1007/978-3-031-99418-0_23

in the Compreg system developed in the 1940s and 50s, in comparison to the larger pieces used in the Impreg system [6, 7]. Other technologies to enhance permeability have included microwave treatments [8, 9] and incising technologies [10–12].

Timber incising has been known and investigated for many years, generally for the preservative treatment systems used in fencing products, telegraph poles, railway sleepers and other ground contact timbers [13–15]. It remains a very diverse field, probably due to the need to match the incision method to the context and application of the treated timber. Large cross section timbers for heavy duty applications were sometimes incised to full depth using nails, spikes or spiked rollers. These mechanical incision techniques have been considerably refined in recent years, with high quality achieved through greater understanding of the beneficial effect of a controlled degree of splitting or crushing the timber to increase access points for the treatment fluid, while also minimizing large splitting and checking effects [15]. However, such systems are not suited to the high aesthetic demands prevalent in many wood modification products, where cost and intended market dictate a different level of uniformity and appearance.

Timber incising is often considered to have three categories: mechanical incision, laser incision and bioincising. The needs of wood modification are likely to be better met by either laser incision or bioincising in future technologies. Bioincising utilizes either fungi or bacteria to open up the flow pathways in the wood [16, 17]. Ponding is the oldest method, and a benefit of storing logs under wet conditions. Improvements in permeability of logs are believed to result from the action of bacteria, reducing pectin content and destroying the pit membranes [18, 19]. More controlled systems for incubating with bacteria have shown improvements [20, 21]. Other studies have used blue-stain fungi [22]. A recent study using the white rot fungus *Physisporinus viteus* showed degradation of the bordered pits in *Picea orientalis* [23]. A related approach is the use of enzyme treatments [24]. The technology is better reviewed in [17], however uniformity of the incising effect remains a challenge for practical applications, especially wood modification.

Laser incision offers a faster process than bioincising, with greater control of the distribution of the treatment fluid. Bakir [25] demonstrated laser incision for preservative treatment of Caucasian spruce for example, but had more variable results with European larch. Islam et al. [10, 26] also demonstrated the benefits of laser incising. Laser incising also offers smaller hole sizes and less damage to the wood tissue than mechanical incising, this leads to lower levels of change in mechanical properties for laser incised wood [27]. Further, it allows the pattern of holes across the surface of the plank to be adjusted, either to incorporate brand designs or patterns, or to avoid pattern altogether – generating a random appearance to be more discrete and blend with the wood grain [12]. The laser incision system is ideal for scale up to meet the need of the wood modification industry, and to support delivery of high quality or high value products, with suitable aesthetic appeal.

2 Development of the System

2.1 Context

Chemical modifications and resin modification systems for wood require near-perfect distribution of treatment agent to ensure that uneven bulking does not induce stress-related defects. This study reports work directed towards resin modification, which was achieved through laser incision. A replacement for Radiata pine was investigated, using various through careful matching of incision patterns and species. This paper demonstrates that the concept of laser incision is extremely well suited to wood modification, and has scope for revolutionizing all fluid-based wood modification systems (esterification, furfurylation, resins, bio-polymers, polymerisation modifications) as well as other protective treatments e.g. with natural extracts.

Much has been written about the better-commercialized process of laser cutting wood, and many of these principles can be transferred to the laser incision context [27, 28], although the narrow bore of the incisions may inhibit mobility of the plasma products from deep incisions in a way that is not an issue for cutting [29].

The characteristics of the laser beam itself are one aspect, e.g. wavelength, pulse rate [29]; while the properties of the timber (especially density and moisture content) are also relevant [27, 30]. The density dependence gives rise to differences observed with wood orientation (tangential versus radial or other growth ring angles) as well as species differences [31]. Thirdly the equipment itself, and processing variables must also be considered, e.g. the plank feed rate which is related to the pattern and number and depth of holes required [12].

Nath et al. [27] reviewed many studies of laser incision recording laser type and wavelength and other laser parameters, finding that most studies had used CO_2 lasers, with a wavelength of 10.6 μm. However, all other aspects varied, so a wide range of species had been investigated, and hole spacings ranging from 1600 to 6,670,000 holes per square metre (see Table 1 for summary). Where pattern information was available a wide range of approaches had been attempted, and hole depths depended on the application of the incised wood, for example in glulam treatments a shallow hole depth was preferred, while others sought deeper treatment depth.

Table 1. Laser parameters investigated in a range of literature studies.

Study	Wood species	Laser type	Laser energy (E), power (P) and wavelength (λ)	Hole density (holes per m^2)
Kamke and Peralta 1990 [32]	White oak, Red oak, Yellow poplar Loblolly pine	CO_2 laser	$\lambda = 10.6$ μm, $P = 75\text{-}115$ W	6200 13,950 37,975
Goodell et al. 1991 [33]	Red spruce	CO_2 laser	$\lambda = 10.6$ μm $P = 25\text{-}100$ W	25,000 33,000

(continued)

Table 1. (*continued*)

Study	Wood species	Laser type	Laser energy (E), power (P) and wavelength (λ)	Hole density (holes per m^2)
Hattori et al. 1997 [34]	Sugi (black heart)	CO_2 laser	$\lambda = 10.6\ \mu m$, $P = 1500$ W	10,000 20,000 30,000
Islam et al. 2009 [35]	Douglas fir	CO_2 laser	$P = 1.5$ kW	5000 7500 10,000
Kortsalioudakis et al. 2014 [36]	Fir, Spruce	Nd:YAG	$E = 301$ mJ, $\lambda = 532$ nm	5000 10,000
Ando et al. 2014 [37]	Japanese cedar	CO_2 laser	$P = 1$ kW	1600
Fukuta et al. 2016 [30]	Japanese cedar, Japanese larch, Beech	Nd:YLF and Nd:YVO_4	$\lambda = 349$ and 355 nm $P = 0.08$ and 5 W $E = 80$ and 100 μJ	6,670,000
Spear et al. (2021) [11]	Southern yellow pine, Scots pine, Tulipwood	CO_2 laser	$\lambda = 10.6\ \mu m$	8333 9091 12,500 15,625

2.2 Laserwood Project

Within the Laserwood project the initial aim was to identify candidate species with moderate permeability which could be improved sufficiently by laser incision to permit chemical modification or resin modification processes. Thus European beech, tulipwood (yellow poplar), poplar, Scots pine, Southern yellow pine were selected as candidate species. The low density of tulipwood and poplar was of particular interest, while higher uptake was expected in beech.

Many of the fundamental principles were established during this project, and Nd:YAG laser (frequency doubled and tripled to wavelengths of 533 and 355 nm) was compared with CO_2 laser (wavelength 10.6 μm) [29, 31]. The flow paths that characterize the different wood species were investigated using single hole and multi-hole experiments. Differences in conductivity between tissue types were observed, for example the poor flow in rays in beech and the role of resin canals in flow in pines [4]. A fully functional prototype laser incision rig was also built.

2.3 Lasercure Project

Within the Lasercure project the research was extended, with the development of a full commercial scale laser incision rig for preparation of planks for resin modification [12].

By way of scientific advances, the team addressed process variables and laser incision patterns suitable for higher feed rates within this commercial laser rig. These maintained the necessary level of fluid uptake and excellent dispersion of the treatment agent within the wood, as is required for wood modification.

The hole densities within the bespoke incision patterns ranged from 8333 to 12,500 holes/m^2 in softwoods and from 9091 to 15,625 holes/m^2 in hardwoods, although patterns differed between species, to reflect the differences in wood anatomy [11]. This is within the median range of laser hole densities seen in the review of such studies (Table 1), and these hole densities were easily achieved with an acceptable feed rate in the large batches of planks tested. These planks were used within full scale trials in the resin impregnation system at LIGNIA Wood Company. Trials demonstrated the incision patterns in tulipwood, southern yellow pine, and the resulting distribution of resin treatment studied. For lower permeability species requiring higher hole densities the patterns can easily be adjusted.

In a later development of the project the work was extended to consider refractory species such as spruce, in which greater numbers of holes per square metre were required to achieve full permeability. However, due to the different requirements of preservative treatment systems different hole patterns and shallower hole depth can be used to increase feed rate. Thus the laser incision system was also shown to be suitable for traditional preservative systems.

3 Conclusion

Laser incision has unique differences to the mechanical incision system, which mark it out as being highly suitable for wood modification. It also generates timber with minimal visual or tactile defects, so is ideal for wood treatments that are destined for high value products, with great aesthetic appeal and high market value.

Most importantly, wood modification requires near-perfect distribution of treatment agent, which is achieved in laser incision through careful matching of incision patterns and species. This paper sumarises findings from two successful collaborative research projects (Laserwood and Lasercure) to demonstrate the technology's application in commercial scale resin modification process.

Laser incision has the capability to revolutionize all fluid-based wood modification systems (e.g. esterification, furfurylation, resins, bio-polymers, polymerisation modifications) as well as other protective treatments.

References

1. Kupfernagel, C., Spear, M., Pitman, A., Yelle, D., Rahman, M., Brown, S.P., Ormondroyd, G.: Mapping process-related chemical variation in resin-treated wood on different spatial scales. In: Brischke, C., Buschalsky, A. (eds.) Proceedings of the 18th Annual Meeting of the Northern European Network for Wood Science and Engineering, September 21-22, Göttingen, Germany, pp. 53–55 (2022)
2. Stefanowski, B.K., Spear, M.J., Pitman, A.J.: Review of the use of PF and related resins for modification of solid wood. In: Proceedings of Timber 2018, 26-27 June 2018, London, UK, pp. 165–179 (2018)

3. Kupfernagel, C., Spear, M., Pitman, A., Ormondroyd, G.: Wood modification with phenol formaldehyde resin and its influence on the dimensional stability of homegrown and imported hardwoods. In: Proceedings of WSE 2021, Northern European Network of Wood Science and Engineering, 14-15th October 2021, Kaunas, Lithuania, pp. 12–15 (2021)
4. Spear, M.J., Holmberg, J., Nath, S., Pitman, A., Waugh, D., Mason, P., Curling, S.F., Ormondroyd, G.A.: Fluid flow in wood: investigation of the influence of laser incision parameters on uptake and flow paths in four wood species. In: Proceedings of Timber 2018, 26-27 June 2018, London, UK, pp. 137–144 (2018)
5. Kupfernagel, C., Spear, M.J., Pitman, A.J., Ormondroyd, G.A.: Wood modification with phenol urea formaldehyde (PUF) resin: the influence of wood species selection on the dimensional stability. Eur. J. Wood Wood Prod. **81**, 5–19 (2023)
6. Stamm, A.J., Seborg, R.M.: Resin-Treated, Laminated Compressed Wood. (compreg). Forest Products Laboratory 1268 (1941)
7. Stamm, A.J., Seborg, R.M.: Resin-Treated Wood. (Impreg). Forest Products Laboratory 1380 (1942)
8. Torgovnikov, G., Vinden, P.: High-intensity microwave wood modification for increasing permeability. For. Prod. J. **59**(4), 84–92 (2009)
9. Dömény, J., Koiš, V., Dejmal, A.: Microwave radiation effect on axial fluid permeability in false heartwood of beech (*Fagus sylvatica* L.). BioResources. **9**(1), 372–380 (2014)
10. Islam, M.N., Ando, K., Yamuchi, H., Kobayashi, Y., Hattori, N.: Comparative study between full cell and passive impregnation method of wood preservation for laser incised Doulas fir lumber. Wood Sci. Technol. **42**, 343–350 (2008)
11. Spear, M., Mason, P., Williams, G., Miles, C., Williams, E., Pitman, A., Bailey, R., Theobald, T., Nath, S., Waugh, D., Ormondroyd, G.: Demonstrating the suitability of laser incision for timber treatment technologies. In: Proceedings of WSE 2021, Northern European Network of Wood Science and Engineering, 14-15th October 2021, Kaunas, Lithuania, pp. 101–103 (2021)
12. Spear, M., Miles, C., Curling, S., Nath, S., Pitman, A., Mason, P., Waugh, D., Ormondroyd, G.: Mapping fluid flow in wood: the interaction of laser incisions with cell structure. In: Brischke, C., Buschalsky, A. (eds.) Proceedings of the 18th Annual Meeting of the Northern European Network for Wood Science and Engineering, 21-22 September, Göttingen, Germany, pp. 168–170 (2022)
13. Franciosi, G.F.: Effect of incising beech cross ties. For. Prod. J. **6**(7), 264–270 (1956)
14. Morris, P.I., Morrell, J.J., Ruddick, J.N.R.: A review of incising as a means of improving treatment of sawnwood. In: International Research Group on Wood Preservation Paper IRG/WP/94-40019. IRG Secretariat, Stockholm, Sweden (1994)
15. Evans, P.D.: The effects of incising on the checking of wood: a review. Int. Wood Prod. J. **7**(1), 12–25 (2016)
16. Hansmann, C., Gindl, W., Wimmer, R., Teischinger, A.: Permeability of wood—a review. Wood Res. Drevársky Výskum. **47**(4), 1–16 (2002)
17. Lehringer, C., Richter, K., Schwarze, F.W.M.R., Militz, H.: A review on promising approaches for permeability improvements in softwoods. Wood Fiber Sci. **41**(4), 373–385 (2009)
18. Unligil, H.H.: Penetrability and strength of white spruce after ponding. For. Prod. J. **22**(9), 92–100 (1972)
19. Kobayashi, Y.: Drying and anatomical characteristics of sugi wood attacked by bacteria during pond storage. J. Wood Sci. **44**(6), 432–437 (1998)
20. Kobayashi, Y., Iida, I., Imamura, Y., Watanabe, U.: Improvement of penetrability of sugi wood by impregnation of bacteria using sap – flow method. J. Wood Sci. **44**(6), 482–485 (1998)
21. Yıldız, S., Çanakçı, S., Yıldız, Ü.C., Özgenç, Ö., Tomak, E.: Improving of the impregnability of refractory spruce wood by *Bacillus licheniformis* pretreatment. Bioresources. **7**(1), 565–577 (2012)

22. Lehringer, C., Hillebrand, K., Richter, K., Arnold, M., Schwarze, F.W.M.R., Militz, H.: Anatomy of bioincised Norway spruce wood. Int. Biodeter. Biodegr. **64**(5), 346–355 (2010)
23. Bakir, D., Dogu, D., Kartal, S.N., Terzi, E.: Evaluation of pit dimensions and uptake of preservative solutions in wood after permeability improvement by bioincising. Wood Mater. Sci. Eng. **18**(1), 233–243 (2023)
24. Durmaz, S., Yıldız, U.C.: Increasing the permeability of spruce sapwood (*Picea orientalis* L.) with enzymatic treatment. Artvin Coruh Univ. J. For. Fac. **17**(1), 32–37 (2016)
25. Bakir, D.: Effects of different incising pretreatments in improving permeability of two refractory wood species. Bioresources. **17**(3), 5021–5037 (2022)
26. Islam, M.N., Ando, K., Yamuchi, H., Kobayashi, Y., Hattori, N.: Passive impregnation of liquid in impermeable lumber incised by laser. J. Wood Sci. **53**, 436–441 (2007)
27. Nath, S., Waugh, D.G., Ormondroyd, G., Spear, M., Curling, S., Pitman, A., Mason, P.: Laser incising of wood: a review. Lasers Eng. **45**, 381–403 (2020)
28. Wang, Y., Ando, K., Hattori, N.: Changes in the anatomy of surface and liquid uptake of wood after laser incising. Wood Sci. Technol. **47**, 447–455 (2013)
29. Nath, S., Waugh, D.G., Ormondroyd, G.A., Spear, M.J., Curling, S.F., Pitman, A.J., Mason, P.: Percussion Nd:YAG laser-incision of radiata pine: effects of laser wavelength, processing parameters and wood anatomy. Lasers Manuf. Mater. Process. **9**, 153–192 (2022)
30. Fukuta, S., Nomura, M., Ikeda, T., Yoshizawa, M., Yamasaki, M., Sasaki, Y.: UV laser machining of wood. Eur. J. Wood Prod. **74**, 261–267 (2016)
31. Nath, S., Waugh, D.G., Ormondroyd, G.A., Spear, M.J., Pitman, A.J., Sahoo, S., Curling, S.F., Mason, P.: CO_2 laser interactions with wood tissues during single pulse laser incision. Opt. Laser Technol. **126**, 106069 (2020)
32. Kamke, F.A., Peralta, P.N.: Laser incising for drying lumber. Wood Fiber Sci. **40**(4), 48–54 (1990)
33. Goodell, B., Kamke, F.A., Liu, J.: Laser incising of spruce lumber for improved preservative penetration. For. Prod. J. **41**(9), 48–52 (1991)
34. Hattori, N., Ando, N., Kitayama, S., Kubo, T., Kobayashi, Y.: Application of laser incising to microwave drying of sugi square lumber with black heart. For. Resour. Environ. **35**, 53–60 (1997)
35. Islam, M.N., Ando, K., Yamauchi, H., Hattori, N.: Effects of species and moisture content on penetration of liquid in laser incised lumber by passive impregnation method. Eur. J. Wood Prod. **67**(2), 129–133 (2009)
36. Kortsalioudakis, N., Petrakis, P., Moustaizis, S., Voulgaridis, E., Adamopoulos, S., Karastergiou, S., Passialis, C.: An application of laser drilling technique to fir and spruce wood specimens to improve their permeability. In: Conference: The 7th International Scientific and Technical Conference, "Innovations in Forest Industry and Engineering Design", At Nov. 13-15, 2014, Sofia-Yundola, Bulgaria (2014)
37. Ando, K., Hattori, N., Harada, T., Kamikawa, D., Miyabayashi, M., Nishimura, K., Kakae, N., Miyamoto, K.: Development of one-hour fire-resistive building materials made of combustible Japanese cedar using laser incising. Int. J. Mater. Sci. **4**(1), 31–34 (2014)

Hydrophobization of Beech Scantlings with Radiation-Curing Epoxidized Vegetable Oils for Use as Dimensionally Stable Components in Exterior Applications

Christiane Swaboda(✉) and Roger Scheffler

Institut für Holztechnologie Dresden gGmbH, Zellescher Weg 24, 01217 Dresden, Germany
christiane.swaboda@ihd-dresden.de
https://www.ihd-dresden.de

Abstract. The work investigates the influence of a treatment of European beech with curable epoxidized vegetable oils on the wood properties in order to improve them for use as construction elements (such as window frames, exterior doors, façade claddings) or for outdoor applications.

Following the political demands for enhancing the use of domestic resources and the forest conversion in Germany for growing more hardwood, it seems useful to upgrade the properties of European beech wood, which is not durable according to EN 350 (2016). Load-bearing components or components exposed to weathering made of untreated beech wood are not permitted.

The IHD has therefore developed a tempering process for beech wood, based on the use of radiation-curing epoxidized vegetable oils. By optimizing the formulation, it was possible to generate variants that guarantee optimum curing with a low γ-radiation dose to ensure a significant improvement in the material properties compared to natural beech wood. At the same time the tendency to embrittlement was reduced.

Although this type of tempering is not a wood modification in the narrower sense, i.e. no direct modification of the wood cell wall is achieved, material properties are achieved that allow the use in window and door constructions and can compete with already approved modified woods such as TMT or Kebony.

Keywords: European Beech wood · construction elements · dimensional stability · durability · epoxidized plant oils · hydrophobization

1 Introduction

1.1 Initial Situation

Wooden construction elements (such as windows, external doors, shutters, sliding shutters, conservatories, cladding) are subject to particularly fierce competition from products made of other materials or combinations of materials.

G. Goli et al. (Eds.): ECWM 2024, SPM 86, pp. 190–199, 2025.
https://doi.org/10.1007/978-3-031-99418-0_24

In the competition between the available materials, the criterion of the lowest possible care and maintenance requirements is of paramount importance (user perspective). Wood as a frame material is indiscriminately subjected to the same evaluation criteria as plastic and aluminum.

The moisture-dependent properties (swelling/shrinkage, deformation, cracking, biological durability) with their effects during use and service life are considered problematic, although there are high quality timber frame structures that can compete with non-wood materials.

Environmental protection requirements and regulations make the use of domestic resources attractive: fewer wood species are allowed to be imported into the European Union, including tropical woods, which are well suited for window construction [1]. On the other hand, the supply of European beech wood will continue to increase in the coming years because of forest restructuring in Germany.

However, beech does not fulfil the requirements for use in construction elements. Load-bearing components or components exposed to weathering made of untreated beech wood (not durable according to DIN EN 350 (2016)) are not permitted.

1.2 Objectives

The objective of this work was to develop an impregnation and tempering process for European beech based on the use of radiation-curing epoxidized vegetable oils combined with an optimized curing with a low γ-radiation dose. By significantly improving the material properties, it should be possible to use beech as frame material for windows and doors. The aim was to achieve positive effects in terms of moisture behavior and biological durability without the typical disadvantages of other effects such as loss of strength and reduced modulus of elasticity [2]. For this, laminated scantlings or profiles made of modified beech wood were produced with defined (usage) properties such as increased weathering resistance or advantageous coating properties.

2 Materials and Methods

2.1 European Beech Wood as Raw Material

The raw material used was European beech (*Fagus sylvatica*) supplied by HOLZ SCHILLER GMBH. A total of approximately 60 scantlings measuring 1850 mm x 100 mm x 50 mm were available. The quality of the scantlings met the requirements of appearance class J2 according to DIN EN 942. According to the manufacturer, the timber was kiln-dried and steamed. The timber was conditioned for at least four weeks at 20 °C and 65% relative humidity (RH). Initially, enough quantities of native beech were climatized to a standard climate of 20 °C / 65% RH and, once mass constancy had been achieved, cut to the appropriate specimen dimensions and lamellae (25 x 85 x 300 mm) for further processing.

2.2 Selection, Synthesis and Characterization of Suitable Impregnating Agents Based on Cationic Curing Epoxidized Plant Oils

Vegetable oils are suitable both for surface finishing and as impregnating agents for protecting wood from natural degradation processes. Their curing with atmospheric oxygen is a lengthy process, which is mainly used for the protection of surfaces. However, oxidative drying vegetable oils are unsuitable for consistent treating of larger wood cross-sections due to the lack of oxygen in the wood.

Epoxidizing the unsaturated carbon bonds in vegetable oils is one way of making them thermally or radiation-curable in conjunction with hardeners. The radiation curing method used in the present work involves the use of a cationic hardener, which releases a strong acid when exposed to radiation. The acid in turn initiates the cationic polymerization. Thereby the three-rings of the oxirane groups are opened and carbocations are formed, which then attack and open further oxirane groups. In this way, a polymer network is built up step by step [2].

Solvent-free or "high solid" systems are generally used here. Typically, onium salts of very strong acids such as aryl diazonium salts, diaryl iodonium salts and triaryl sulphonium salts are used as photo initiator for cationic polymerization.

The material properties of the epoxy resin are influenced by the choice of epoxy oil, hardener and curing parameters. For example, a high temperature or radiation dose will result in tighter cross-linking and therefore higher temperature or biological resistance, but also greater brittleness of the matrix, which can have a negative effect on the processing properties of the tempered wood. Additives can influence the properties of the cured resin and improve the processing properties. With the aim of developing a formulation with a sufficiently high degree of cross-linking and sufficient flexibility, a screening was carried out on a laboratory scale with various epoxidized oils, photo initiators and reactive diluents.

Various epoxidized vegetable oils, such as epoxidized linseed oil (ELO) or epoxidized soybean oil (ESBO) were considered as well as reactive diluents based on epoxidized glycidyl ether These were provided by industrial partners (Hobum Oleochemicals GmbH, Hamburg; ipox-chemicals GmbH) and characterized regarding their viscosity and reactivity by initiating the hardening process with an Haake Mars III rheometer, coupled with an UV-curing device, and measuring of the storage G' and loss modules G" by oscillation. The aim was to produce the lowest viscosity mix at the lowest elevated temperature with short gel-times. The following tables show the compositions of the test variants in the preliminary screening (Table 1) and the optimized compositions (Table 2). The Omnicat photo initiators were delivered by IGM Resins, the Deuteron variant by Deuteron GmbH.

Table 1. Variants of preliminary screening.

Variant	Oil	Hardener	Thinner
V5	50% ELO	1,5% OC 550 in 300 ml Acetonitril	50% RD 14
V6	80% ELO	3% OC 550 in 300 ml Acetonitril)	20% RD 14
V9	80% ELO	1,5% Deuteron 1242 not diluted	20% RD 14
V17	100% Merginat UVT	1,5% OC 550	–
V20	100% ELO	1,5% OC 432	–
V22	50% ELO	1,5% Deuteron 1242	50% RD 24

Table 2. Variants of optimized receipts.

Variant	Oil	Hardener	Thinner
V1	100% ELO	–	–
V2	100% ELO	1,5% Deuteron 1242	–
V3	80% ELO	1,3% Deuteron 1242	20% RD 24 (Ipox)
V4	70% ESBO	1,5% Deuteron 1242	30% RD 14 (Ipox)

2.3 Impregnation Process

The screening of different formulation variants was initially carried out using vacuum impregnation on a laboratory scale, largely based on the procedure of DIN EN 113-1:2021-02. For the subsequent production of beech wood scantlings on a larger scale, wood lamellas with the dimensions listed above were impregnated in the vacuum pressure impregnation plant at the IHD with selected formulations optimized based on previous test results.

The cut test specimens were stacked in the impregnation tank and subjected to a pre-vacuum at 50 °C for 2 h. Once this step was complete, the tank was opened, and the preheated oil was added. This was followed by another vacuum phase lasting 15 min, followed by a 16 h pressurization phase at 12 bar. Some variants were impregnated at 30 °C. After removal, the test specimens were freed from excess oil and then stored for approx. 16 h in a drying cabinet at 70 °C before being sent to irradiation.

The consistency and homogeneity of the impregnation was verified on the one hand by means of the degree of loading of the test specimens and its scattering and on the other hand by microscopic examination.

2.4 Development of a Curing Process Using Gamma Rays

The method of curing using Gamma rays was chosen because, unlike UV rays, these can penetrate larger cross-sections and thus initiate the curing reaction directly inside the wood. The test specimens were cured by BBF Sterilisationsservice GmbH. To determine the influence of the radiation dose on the curing and the final properties of the test specimens, tests were carried out with 25 kGy, 50 kGy and 100 kGy.

At higher doses, the temperature could rise causing the oil to liquefy and escape from the specimens. Some specimens, particularly those with radial cut surfaces in contact with each other, could only be separated by force after the packages had been opened. In contrast, curing of specimens with longer edge lengths, such as the lamellae, which received only 25 kGy, was unproblematic. The effect of the radiation dose on the tempering properties is discussed in detail in the following sections.

2.5 Material Tests

Table 3 contains all physical and mechanical parameters used to assess the treated wood and its suitability as window material.

Table 3. Test parameter and standards.

Parameter	Standard used
Bulk density	DIN 52182 (1976)
Flexural strength/flexural modulus	DIN 52186 (1978)
Brinell surface hardness	DIN EN 1534 (2011)
Impact energy	DIN 52189–1 (1981)
Compressive shear strength	DIN 13354 (2009)
Screw pull-out strength	DIN EN 320 (2016)
Swelling and shrinkage behavior	DIN 52184
Capillary water absorption	DIN EN ISO 15148 (2016)
Resistance to wood-destroying fungi	DIN EN 113–2 (2021)
Blue stain resistance	DIN EN 152 (2012)
Fire behaviour by means of single flame test	EN ISO 11925-2:2020
Bondability (piercing test, finger joint)	Ift-HO-10/1 (2002)

3 Results

3.1 Impregnation

During the production of scantlings using the pilot impregnation process at 70 °C, maximum loading levels of 300 kg/m^3 were determined. At 30 °C, the loading of the manufactured lamellas achieved approximately 200 kg/m^3. This is below the maximum possible load of 450 to 500 kg/m^3 for beech. Not all vessel lumens are filled with impregnant, and the parenchymal cavities are largely free of impregnant. Given the average wood ray content (15%) in the hardwood tissue, it can be assumed that most of vessel lumens are impregnated even in the inner layers of the window scantlings (Fig. 1) and only a few vessel lumens - negligible - are still free of impregnate (Fig. 1a–d).

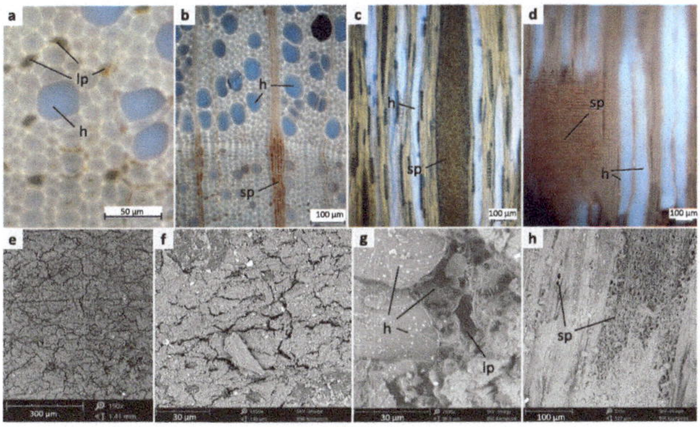

Fig. 1. Optical and electronic microscopy images of treated beech. Reflected light microscope: a, b, c, d; SEM images: e, f, g, h. Lp = Longitudinal parenchyma, h = vessel, sp. = Wood ray parenchyma.

The microscopic examination of the wood structure of oil-impregnated beech wood samples showed no influence on the cell wall thickness. Since the microscopic examination revealed only isolated impregnated areas in the parenchyma cells (Fig. 1a), it is unlikely that the impregnating agent has penetrated the cell walls. The penetration of the vessel walls is unlikely. It is more likely that the resin has mechanically reduced the expansion of the wood substance.

A lower degree of loading is seen as positive, as it contributes to a significant reduction in the amount of impregnating agent required as well as significantly improved physical and mechanical properties and a reduction in brittleness.

Achieving this was one of the main objectives of the project. Reduced brittleness is the prerequisite for better machinability and workability. Figure 1e, f shows SEM images of a brittle variant with visible cracks in the cell structure, which spread over the entire wood structure, along the center lamella and across the cells without a recognizable pattern.

3.2 Influence of Radiation Dose and Loading on Material Properties

Irradiated untreated beech wood shows a significant decrease in bulk density (−15%), flexural strength (−10%) and Brinell hardness (−17%) as the radiation dose increases from 25 kGy to 100 kGy. However, the properties of irradiated treated wood differ substantially from the latter.

Here, higher loading and radiation intensity led to higher Brinell hardness (Fig. 2a, b), compressive strength and screw pull-out strength, while the flexural strength did not change because of impregnation and irradiation. It was found that the reactivity and solubility of the photo initiator in the oil had the main influence regarding the effects of the applied radiation dose and the resulting physical-mechanical surface properties.

A formulation was found which, when using a radiation dose of 25 kGy, could guarantee a degree of hardening that ensures the necessary mechanical properties for manual processing and use as a window scantling.

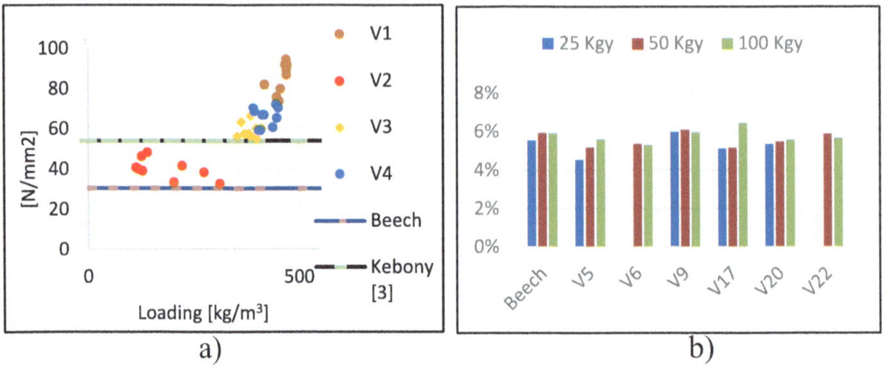

Fig. 2. (a) Brinell hardness (radial) [N/mm^2]. (b) Brinell hardness (radial) [N/mm^2] with several variants of impregnation receipt with several variants of radiation dose.

3.3 Influence of Curing on Swelling and Shrinkage Properties

Due to the chemical reaction mechanisms of the cationic curing of epoxy compounds, side reactions can lead to unreacted hydroxyl groups in the polymer, which cause increased water absorption of the corresponding material. The analysis of the resulting wood equalization moisture content (Fig. 3a) also showed major differences between the individual variants. Thus, good curing results in equilibrium moisture contents in the range of natural beech or even below, regardless of the radiation dose (V6, V9, V22), whereas poor curing results in a sharp increase in wood equilibrium moisture contents (V5, V17, V20).

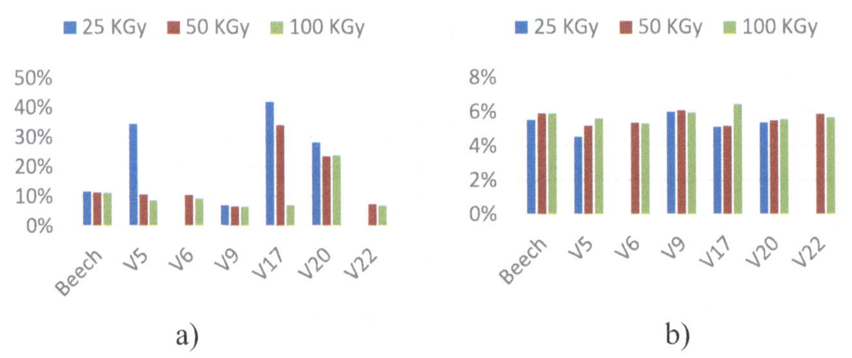

Fig. 3. (a) Wood equilibrium moisture content. (b) Maximum radial swelling in standard climate 20 °C/65% rel. hum. [%].

Figure 3b shows that despite the increased moisture absorption, relatively uniform linear swelling values between (5 and 6%) are achieved, which for the well-cured samples are mostly below the values for native red beech. The fact that the poorly cured variants V5 and V17 (25 kGy) exhibited a low maximum swelling rate despite the increased amount of water absorbed is very surprising. This could be related to the fact that OH-groups of the polymer bind the water so that it is not incorporated into the cell walls and thus does not contribute to a change in shape. Additionally, gamma irradiation may lead to additional crosslinking within the cell wall cellulose due to newly formed aldehyde groups (Figs. 4 and 5).

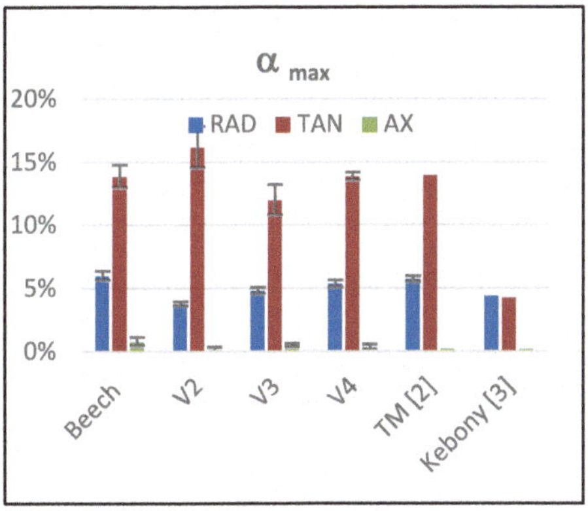

Fig. 4. Maximum swelling dimension α_{max} during the transition from the kiln-dry to the wet state.

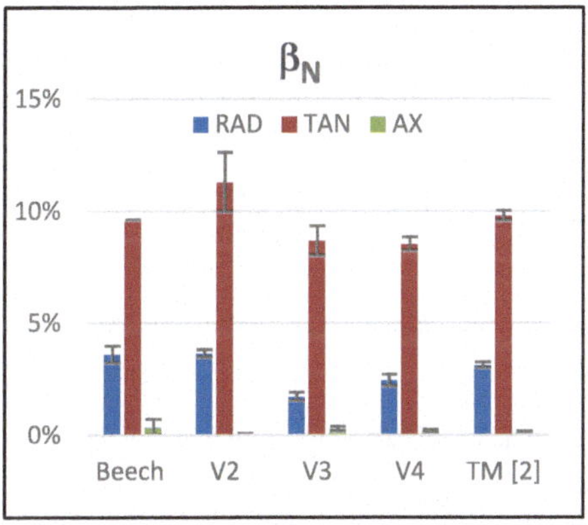

Fig. 5. Drying shrinkage β_N during the transition from the wet state to standard climate 20 °C/65% rel. hum.

In the pilot test, a slight reduction in maximum swelling was achieved with the optimized formulation V3 in each wood anatomical direction. Overall, the improvement in maximum swelling achieved with Kebony [3], particularly in the tangential direction, was not achieved with any of the epoxidized oil variants tested.

3.4 Characterization of the Material Properties and Suitability as a Window Frame Material in Optimized Variants

As can be seen from Table 4, similar material improvements can be achieved with beech treated with epoxidized vegetable oils as with thermally treated beech [2] or Kebony [3] after optimization of the impregnation formulations. In the area of physical-mechanical parameters, there are clear improvements in elastic properties. In terms of dimensional stability, improvements were also achieved compared to untreated beech. At 0.1 and 0.08 $kg/m^2/h0.5$, the capillary water absorption coefficients of V3 and V4 were even significantly lower than those of Kebony SYP (0.5 $kg/m^2/h0.5$) [3] and thermally treated beech (0.94 $kg/m^2/h0.5$) [2]. However, there was somewhat stronger swelling in the radial and tangential directions than with Kebony.

Table 4. Changes in the tested material properties of optimized formulation variants compared to untreated beech.

Improvement						Worsening								no change	
> 5%	+	> 25%	++	> 100%	+++	> 5%	-	> 10%	--	> 25%	---	> 60%	----	(±5%)	o

s.p. = not spezified

Parameter	TMT [2]	KBY [3]	V1	V2	V3	V4
Flexural strength [N/mm²]	-	---	o	++	+	+
Modulus of elasticity [N/mm²]	o	+	o	++	+	+
Brinell hardness [N/mm²]	+	++	+++	++	+	o
Impact work [kJ/m²]	--	----	---	+	-	-
Compressive shear strength [N/mm²]	++	o	++	+	++	++
Screw pull-out strength [N/mm]	-	o	++	o	+	+
Thermal conductivity l1020/65 [W/(m*K)]	o	s.p.	-	-	--	--
Thermal resistance R1020/65 [W/(m²K)]	s.p.	s.p.	-	-	--	--
Wood moisture content at 20°C/65%	+	++	++	++	+	++
Water absorption coefficient [kg/(m2*h0,5)]	o	++	++	++	++	++
Maximum radial swelling dimension	o	++	-	++	++	+
Maximum tangential swelling dimension	+	++	-	-	+	+
Maximum axial swelling dimension	o	+	++	++	++	++
Volume swelling measure αv	+	++	-	++	++	++
Volume shrinkage βv	+	++	++	++	++	++

The optimized variants V3 and V4 achieve durability class 1 or 2 and are therefore suitable for outdoor use.

The beech wood lamellas impregnated with the optimized variants exhibited higher thermal conductivities compared to the native beech due to the increased raw density, However, the thermal transmittance values (U_W)[1] of all modified variants are still below 1,3 W/m^2K, the requirement value for minimum thermal insulation from the Building Energy Act and are comparable, for example, with oak, a frequently used wood species for high-quality windows and doors.

In terms of fire behavior, all modified variants meet the requirements of class E for normal flammability, i.e. the modified beech scantling differs only slightly from natural wood which is classified to class D [4].

[1] U_W calculated with Flixo pro 8.1 analysis software. A standard window with a construction depth of 68 mm, triple insulating glass (Ug = 0.7) and EPDM seal was used as the basis.

Bonding is particularly relevant for use as window timber. As part of the project, it was confirmed that the lamellas can be bonded with adhesives commonly used for windows. The hydrophobized beech scantlings could also be coated with a typical window alkyd resin varnish and a PUR varnish on a pigmented, solvent-based primer. The adhesive strengths of the layer structures met the requirements for wood used outdoors.

The tempered beech wood can be processed with standard joinery tools. Drilling holes is more time-consuming due to the increased hardness and the associated heat development. The feed rate should be adjusted in view of the increased risk of splitting. Pre-drilling is also required for screwing in screws. The guide holes must be larger than for European beech wood. However, there were no coarse eruptions as described in [2].

4 Outlook

The project is currently being rounded off with outdoor weathering tests. In addition to window production, other applications in the field of exterior use of wood are also conceivable without a load-bearing function. Practical partners are being sought for this. Alternative curing processes that do not require gamma rays are also imaginable and should facilitate practical implementation.

References

1. Koch, G., Ohlmeyer, M., Günther, O.: Untersuchung der Verklebungseigenschaften und Dimensionsstabilität von kombinierten Kanteln mit dekorativen Ausstattungshölzern für den Fensterbau. Institut für Holztechnologie und Holzbiologie, AiF-Abschlussbericht 2012 (Projekt-Nr. 16047 N) https://openagrar.bmel-forschung.de/servlets/MCRFileNodeServlet/timport_derivate_00006492/dn050023.pdf. Zugriff (2016)
2. Klepel, C.: Charakterisierung ausgewählter Vergütungseigenschaften von Rotbuchenholz nach Behandlung mit strahlenpolymerisierbaren Lipiden hinsichtlich der Verwendung für Bauelemente. Masterarbeit, TU Dresden Fakultät Umweltwissenschaften (2017)
3. VFF-Merkblatt HO. 06-4/B2: Holzarten für den Fensterbau - Teil 4: Modifizierte Hölzer; Beiblatt 2: Kebony SYP. Verband Fenster- und Fassadenhersteller, Frankfurt (2016)
4. Wackerbauer, G.: Streitfrage: Wie brennbar dürfen Fenster und Fassaden sein? Publikation ift Rosenheim (2014)

Dimensional Stability of Scots Pine Sapwood Modified by Tannin-Based Formulas

Sheikh Ali Ahmed[1(✉)], Gianluca Tondi[2], Filippo Rizzo[3], Reeta-Maria Stöd[4], and Reza Hosseinpourpia[1,5(✉)]

[1] Department of Forestry and Wood Technology, Faculty of Technology, Linnaeus University, Växjö, Sweden
{sheikh.ahmed,reza.hosseinpourpia}@lnu.se
[2] TESAF Department, Università degli studi di Padova—University of Padua, Legnaro (PD), Italy
[3] Stora Enso Wood Products GmbH, Ybbs, Austria
[4] Stora Enso Oyj, Wood Products, Helsinki, Finland
[5] College of Forest Resources and Environmental Science, Michigan Technological University, Houghton, MI, USA

Abstract. This study was conducted to evaluate the physical and chemical properties of Scots pine (*Pinus sylvestris* L.) sapwood modified with tannin-based formulas. Wood samples were pressure impregnated with tannin solution at five different concentrations of citric acid (0, 2, 10, 20, and 40 wt/wt%). The tannin-modified samples were analyzed by mean of their weight percentage gain (WPG), leaching resistance (LR), bulking coefficient (BC), anti-swelling efficiency (ASE), Fourier-Transform Infrared (FTIR) spectroscopy, and optical microscopy, and the results were compared with unmodified sapwood and heartwood samples. It is observed that the WPG and dimensional stability of the samples increased with increasing the citric acid (CA) concentrations and the highest WPG was attained (44%) with a formulation containing 20% CA while, the BC, LR, and ASE values were considerably higher in the samples modified with 40% CA concentration as compared to other modification formulas and also the unmodified sapwood and heartwood samples. FTIR spectroscopy analyses showed higher esterification degree with higher concentrations of CA. Overall, this study suggests the potential of a fully bio-based tannin formula to improve the water-related properties of Scots pine sapwood.

Keywords: Bulking · Impregnation · Leaching · Anti-swelling efficiency

1 Introduction

Buildings and housing are the largest consumers of the wood raw material. In Europe, construction operations account for a substantial amount of wood which is approximately 37.5% [1]. With increasing environmental concerns, one important aim of the wood and building industries is to use high-quality domestic wood products that require low maintenance. Pine with a high ratio of sapwood to heartwood, the sapwood is easy

G. Goli et al. (Eds.): ECWM 2024, SPM 86, pp. 200–208, 2025.
https://doi.org/10.1007/978-3-031-99418-0_25

to impregnate, but classified into class 5, not durable [2]. Although the classical preservative impregnations, i.e. impregnation with copper chromium arsenate, amine copper quaternary, tebuconazole, boric acid, etc. enhance the durability of wood, the dimensional instability of wood could not be addressed with such approaches. In addition, their leaching problem during application would pose a threat to human health and environment safety [3].

Several emerging environmental-friendly processes of wood modification (chemical, thermal and impregnation/ polymerization) have been developed to improve wood properties and a few of them, e.g., acetylation, furfurylation, and thermal modification, have been commercialized. The most common external uses of modified wood are in claddings, windows, playground devices, gardening, and landscaping. The wood modification with renewable chemicals like tannin, poses a number of advantages, while it is not fully discovered. This is not only because of its availability in larger amounts in the future as a result of the growth in biomass refinery– biorefinery industries, but also due to its renewability and nonbiocidal characteristics [4]. Impregnation modification with tannin and citric acid (CA) provided a very good indication for its potential in the sustainable wood modification towards the circular bioeconomy.

Tannins are generally available as condensed or hydrolysable macromolecules. Condensed tannins consist of flavonoids (flavan 3-ol or flavan 3, 4-diol) without a sugar core, whereas hydrolysable tannins are comprised of ellagic and gallic acids with a sugar core, primarily glucose [5]. The effectiveness of tannin for wood protection has been widely studied previously, but limited studies have been directed at addressing its leachability challenge from wood tissue [6–8]. Efforts to improve the leaching resistance (LR) of tannin involve crosslinking with chemicals like hexamine and formaldehyde; however, the release of toxic formaldehyde gases raises environmental concerns [9]. Tondi et al. [10] reported that tannin-hexamine formulation reticulated into the wood structure, resulting in improved strength and LR properties.

This study seeks to introduce innovation to contemporary wood impregnation modification practices, promoting more sustainable production by eliminating the use of biocidal and metal-based chemicals and enhancing dimensional stability. Therefore, the physical and chemical properties of Scots pine sapwood modified with tannin-CA formulas at various CA concentration levels were evaluated by means of weight/dimensional changes, FTIR and microscopy analyses.

2 Material and Methods

Kiln-dried Scots pine (*Pinus sylvestris* L.) lumbers (380 × 40 mm) were collected from Hanåsa Såverket AB (Högsby, Sweden). Working samples both from sapwood and heartwood of 25 (R) × 25 (T) × 10 (L) mm^3 were converted from the lumbers. Total 20 samples were used in each treatment. Untreated heartwood was used as reference. Samples were then oven-dried at 103 ± 2 °C for 18 h and were used for impregnation with different tannin solutions.

Quebracho Tannin (moisture content 6.8%, flash point > 190 °C) in powder from was used (Silvateam SpA, San Michele di Mondovi, Italy). Tannin solution was prepared at 20 wt% in water and then mixed with CA (Sigma-Aldrich Chemie GmbH, Steinheim,

Germany) at different concentrations of 2, 10, 20 and 40 wt% (Table 1). Scots pine sapwood samples were impregnated with the tannin-CA formulas by initially exposing to vacuum at −91 kPa for 60 min and then pressure at 700 kPa for an additional 60 min. The samples were then cured at 140 °C for 10 h.

Table 1. Tannin-citric acid formulation.

Sample description	Code	Tannin (%)	Cross-linker[a] (%)	Curing temp. (°C)	Curing time (h)
Heartwood	Heart	–	–	–	–
Sapwood	Sap	–	–	–	–
Sapwood	CA0%	20	–	140	10
Sapwood	CA2%	20	2	140	10
Sapwood	CA10%	20	10	140	10
Sapwood	CA20%	20	20	140	10
Sapwood	CA40%	20	40	140	10

[a]Cross-linker (citric acid, CA) concentrations are based on the total tannin solid

Tannin uptake was measured as weight percentage gain (WPG) and calculated using the oven-dried weight before and after impregnation. On the contrary, bulking coefficient (BC) was measure based on the oven-dried volume. Total 18 samples were used in each treatment group. Leaching resistance (LR) was measured following the standard SS-EN 84 [11], with slight modification to the immersion time. The samples were immersed in distilled water for 5 days, instead of the standard 14 days, with the water being replaced at two-day intervals. After leaching, samples were gradually dried and the remaining tannin in the sample was calculated using the equation below:

$$LR\ (\%) = \frac{\mathrm{OD}_{wt}\ \text{after leaching} - \mathrm{OD}_{wt}\ \text{before impregnation}}{\mathrm{OD}_{wt}\ \text{after impregnation} - \mathrm{OD}_{wt}\ \text{before impregnation}} \times 100 \quad (1)$$

Anti-swelling efficiency (ASE) was measured on the samples after leaching. Samples were vacuum impregnated with distilled for 30 min at 13 kPa followed by storage in water for 24 h. Volumetric swelling (S) and ASE were measured according to the steps followed by Ahmed and Morén [12]. After every cycle, samples were gradually dried at 35, 45, 65, 85 and finally at 103 °C for 18 h in each step to avoid cracking.

The chemical structure of tannin impregnated cured samples, unimpregnated sapwood and heartwood were analyzed by a Frontier FTIR spectrophotometer (PerkinElmer, Waltham, MA, USA). The analysis was performed averaging 30 scans with a resolution of 4 cm^{-1} in a wavelength range between 4000 and 650 cm^{-1}.

Thin sections of 15 μm thick were prepared from small blocks of 5 (R) × 5 (T) × 10 (L) mm^3 using a sledge microtome and were observed under a motorized Olympus BX63F light microscope (Olympus, Tokyo, Japan) connected with the software program OLYMPUS cellSens Dimension, version 1.18 (Olympus, Tokyo, Japan).

The IBM SPSS Statistics software package, Version 29.0.1.0 (IBM Corporation, New York, USA) was used for one-way analysis of variance (ANOVA) at 0.05 significance level. When significant differences were found, Duncan's multiple-range test was performed to separate the specific effects of the cross-linker utilized.

3 Results and Discussion

Figure 1a shows that the WPG of the tannin modified samples increased with the increasing of CA concentration. Significantly the highest WPG of 44% was achieved at a CA concentration of 20 wt%, and then decreased to 39% at a concentration of 40 wt%. Sommerauer et al. [13] used different types of cross-linkers in tannin solution (10 wt%) viz. formaldehyde, hexamine, glyoxal, maleic anhydride, furfural and furfural alcohol. They quoted the WPG values in a range of 70% to 130% depending on the size of samples used. In this study, irrespective of CA concentration, the formulations demonstrated complete impregnation, suggesting a thorough and uniform distribution of the tannin-CA solution within the smaller sample matrix.

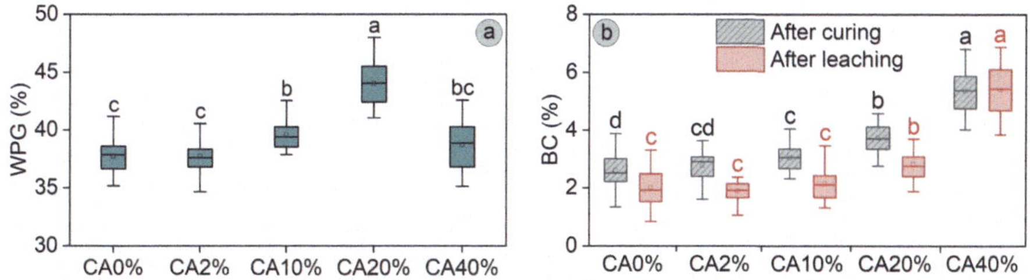

Fig. 1. Weight percentage gain (WPG) and bulking coefficient (BC) of sapwood samples modified with tannin and citric acid (CA) solution. Each treatment represents CA concentrations of 0, 2, 10, 20 and 40 wt% of total tannin solid. Mean values followed by different letter within each group (marked by same color) indicate that there is a significant difference ($P \leq 0.05$) as determined by ANOVA and Duncan's multiple range test. The error bars in columns indicate the standard deviations.

After five days of leaching, the BC values slightly decreased in CA concentrations between 0%–20%, but no change at 40% of CA. This suggests that reduced leaching loss might be due to fixation of tannin-based formulas through formation of crosslinked structure in the wood cells. An increased volume of wood samples after chemical treatment indicates a cell wall bulking due to the incorporation of chemicals into the cell wall. In Fig. 1b, the behavior of BC after both curing and leaching is illustrated. Notably, BC exhibits an upward trend with increasing concentrations of CA. Specifically, it is evident that an increase of BC is approximately 1.1, 1.2, 1.4 and 2.1 times when CA is introduced at concentrations of 2, 10, 20, and 40 wt% of total tannin solid, respectively. Despite a subsequent decline in BC levels following a 5-day water leaching process, an intriguing resilience is observed, particularly at the elevated CA concentration of 40 wt% of the total tannin solid. In this case, the BC remains nearly constant even after the leaching process. This might be due to the fixation of tannin-based formulas by forming crosslinked structures in the wood cells [13].

The amount of residual tannin after leaching for 5 days in water was measured and presented in Fig. 2a. Higher percentage of tannin remained in the wood sample at higher concentrations of CA added in the formula. This suggests that the incorporation of CA contributes to the retention of tannin within the wood matrix during the leaching process. The observed increase in residual tannin with higher CA concentrations implies a potential cross-linking effect, enhancing the tannin's adherence to the wood substrate and thus improving dimensional stability. Dimensional stability in terms of ASE of different treated wood samples was measured and presented in the Fig. 2b. The highest dimensional stability was achieved with CA at 40 wt% of total tannin solid and the differences were found statically significant. Tannin without CA did not improve the dimensional stability compared to the heartwood. When CA was used as a crosslinker, it ensured tannin stayed inside the wood and improved the dimensional stability. With increasing amount of CA increased the cross-linking and thus significantly improved the ASE properties of treated wood, especially for CA at 40 wt% of total tannin solid. Cyclic tests also showed that even at the fourth cycle of soaking, ASE remained almost the same for all tannin-CA-modified wood.

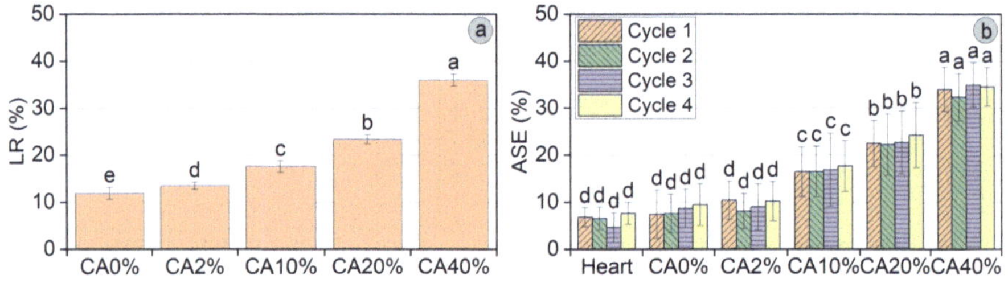

Fig. 2. Leaching resistance, LR (a) after 5 days of leaching in water and anti-swelling efficiency, ASE of different treated samples at 4 cycles of soaking and drying (b). Legends are the same as in Fig. 1. Heart: untreated heartwood.

Lower LR of only tannin might result from tannins' ability to form reversible bonds with saccharides [14], particularly hemicelluloses that are spatially accessible. Upon exposure to water, the entire structure partially dissolves, aligning with findings from Jakobek [15]. Cross-linkers are used in tannin formulations for improved leaching resistance [9, 13]. In this research, CA was used as a cross-linker and higher LR of wood modified was found with higher CA concentration. This could be due to the -OH groups on the polyphenols in tannin being effectively crosslinked with CA [16] and/or esterification of aliphatic OH groups. Furthermore, the experimental outcome also indicated that the addition of CA was favorable to improving the LR of tannin most probably due to the esterification of -COOH and -OH groups. Teixeira et al. [17] stated that polyesterification between CA, tannin and the -OH groups of wood biopolymers is more likely to react between tannin and CA molecules since the tannin is more accessible than wood biopolymers. It was observed that a lower concentration of CA did not contribute to improved LR, indicative of insufficient cross-linking of tannins. Notably, tannin-CA can form covalent interaction with the wood polymer by the esterification between -COOH groups of CA and -OH groups on the surface of the wood [16], which is favorable to improve water resistance and dimensional stability at higher concentration of CA (Fig. 2b). This finding aligns with the findings of Zhao et al. [18] and Li et al. [16] indicated that higher CA concentration improved the water resistance of wood composites. With the increase of CA, the furan ring content increases, and more chemical bonds derived from tannin participate in the reaction [18]. These findings highlight the intricate interplay between tannin and CA, emphasizing their role in forming covalent interactions that significantly contribute to the improvement of water resistance and dimensional stability of modified wood.

Figure 3 shows the chemical changes of wood due to the modification by tannin-CA formulas analyzed by FTIR spectroscopy. It is evident that the chemical structure of wood samples was altered by modification. The region at around 1609 cm^{-1} is typically due to aromatic C=C stretching [19] and increase the peak with addition of CA. Increasing the concentration of CA resulted in higher absorption peaks in both spectra 1374 cm^{-1} and 1736 cm^{-1}, which are assigned to ester bonds [20]. Together with the formation of the ester bond, there are recorded changes in wood structure, such as a broadening absorption at around 1236 cm^{-1}, which is attributed to asymmetric stretching of C–O–C of acetyl group assuming a modification of hemicellulose [21]. The peak at 1420 cm^{-1} associated with the bending vibration of –COOH has disappeared by modification. This might be related to the involvement of carboxylic acids in the reaction [21]. Also, a peak at ca. 1160 cm^{-1} (symmetric stretching vibration C-O-C in esters, Socrates [22]) increased in comparison to untreated wood, following the same order. Both peaks are indicators of the tannin-CA polyester formation. Since carboxyl groups of CA are not equally reactive (different pKa values, Max et al. [23]), some of them will not covalently bond with tannin. These unreacted (free) C=O groups of CA may form hydrogen bonds with hydroxyl groups of tannin or wood polymer constituents. Similar observations regarding characteristics for ester formation peaks at ca. 1730 cm^{-1} (C=O) and ca. 1160 cm^{-1} (C-O) have been made for wood modified with CA and sorbitol [21], glycerol [24] and sucrose [25]. It is then concluded that the modification with tannin at 40 wt% CA resulted in the most efficient esterification reaction of wood.

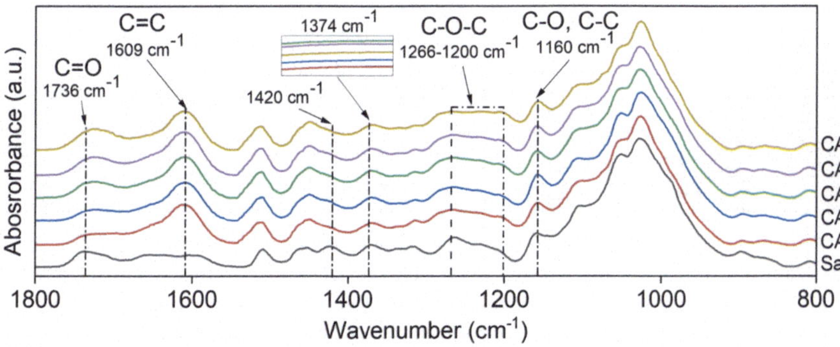

Fig. 3. Fourier-transform infrared (FTIR) spectroscopy spectra for each treatment with tannin formulations. Treatment represents with unimpregnated sapwood (sap), tannin impregnated sapwood with citric acid (CA) concentrations of 0, 2, 10, 20 and 40 wt% of total tannin solid.

Longitudinal tracheids were the main avenue for liquid impregnation. Full impregnation was observed for the most part of the latewood, while partial impregnation was in the earlywood (Fig. 4b, c). Tracheids are impregnated mostly when they are in an adjacent position to parenchyma rays. The impregnation liquid was conducted also by the ray parenchyma and ray tracheids (Fig. 4d, e, h). Almost all of them are entirely filled and remained by the impregnation solution when higher concentration of CA was used. It is evident that liquid also passed through the bordered pit membrane of longitudinal tracheids (Fig. 4f, g). The longitudinal penetration in dried wood through tracheids is easier for latewood because its bordered pits are rarely closed [26]. On the contrary, bordered pits in earlywood are often closed and only wet conditioning of the samples or strong vacuum-pressure cycles would allow passing of the liquid through these pits [27]. Studies on Scots pine showed that rays play a vital role in the radial conduction of liquid [28]. It is evident from the microscopical study that tannin solutions can enter the cell walls and accumulate in the cell lumen. Interactions with lignin at the cell wall interface are more likely.

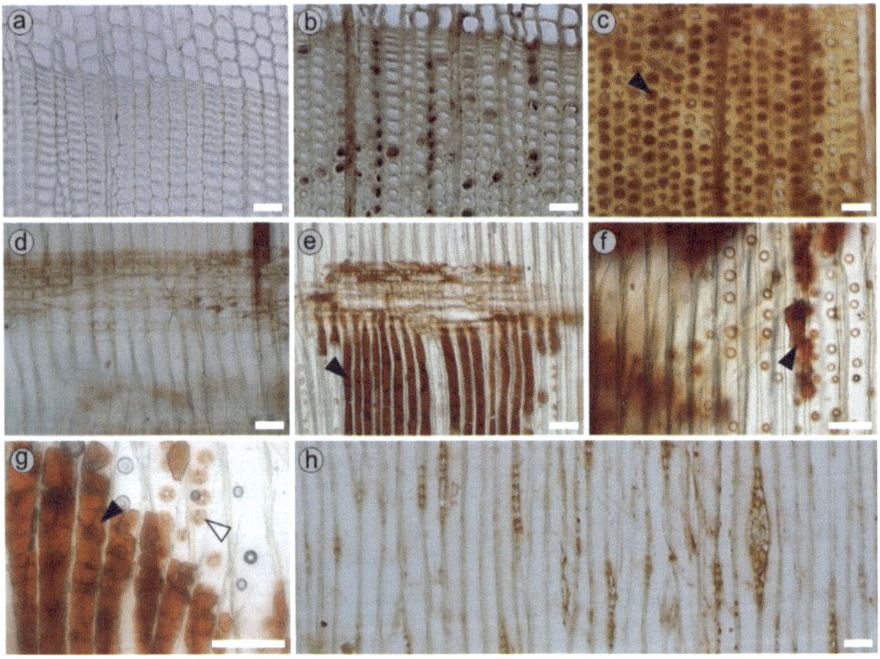

Fig. 4. Tannin penetration as observed under optical microscope. Transverse sections of unimpregnated sapwood (a), tannin impregnated wood without citric acid, CA (b) and with 40 wt% CA (c). Radial sections of tannin impregnated wood without CA (d), with 40 wt% CA (e-g). The tangential section of tannin impregnated earlywood with 40 wt% CA (h). Filled arrow heads show the cured tannin inside the cell lumen of tracheids. Empty arrowhead shows the tannin impregnation through the bordered pit. Scale bar = 50 μm.

4 Conclusion

Wood modification based on the cross-linking of condensed polyflavonoid tannins by CA as a non-formaldehyde-yielding compound, markedly slowed down the leaching of tannin especially at higher concentrations of CA. This study showed that CA concentrations as a crosslinker, significantly impacted the properties of tannin-treated wood by changing the chemical properties and considerably improving the leaching resistance, bulking coefficient and anti-swelling efficiency values. Microscopic analysis also indicated the longitudinal penetration of tannin-based formulas in Scots pine tracheids with open bordered pits and across radial direction through parenchyma rays. This underscores the pivotal role of CA, especially at 40 wt%, in optimizing wood treatment for superior performance. Further research on the durability and weathering performance of tannin-CA modified wood needs to be extended.

References

1. Chen, L., Zhang, Y., Chen, Z., et al.: Biomaterials technology and policies in the building sector: a review. Environ. Chem. Lett. **22**, 715–750 (2024)
2. EN 350-2: Durability of wood and wood-based products. Natural durability of solid wood. Part 2: Guide to natural durability and treatability of selected wood species of importance in Europe. European Committee for Standardization Central Secretariat, Brussels (1994)
3. Wang, Y., Zhang, R., Yang, et al.: Improvement on dimensional stability and mold resistance of wood modified by tannin acid and tung oil. Holzforschung. **76**(10), 929–940 (2022)
4. Broda, M.: Natural compounds for wood protection against fungi—a review. Molecules. **25**(15), 3538 (2020)
5. Sharma, K.P.: Tannin degradation by phytopathogen's tannase: a plant's defense perspective. Biocatal. Agric. Biotechnol. **21**, 101342 (2019)
6. Yamaguchi, H., Okuda, K.I.: Chemically modified tannin and tannin-copper complexes as wood preservatives. Holzforschung. **52**(6), 596–602 (1998)
7. Mubarok, M., Azadeh, E., Obounou, et al.: Effect of tannins addition on thermal stability of furfurylated wood. Polymers. **15**(9), 2044 (2023)
8. Scalbert, A., Cahill, D., Dirol, D., et al.: A tannin/copper preservation treatment for wood. Holzforschung. **52**(2), 133–138 (1998)
9. Thevenon, M.F., Tondi, G., Pizzi, A.: High performance tannin resin-boron wood preservatives for outdoor end-uses. Eur. J. Wood Prod. **67**(1), 89–93 (2009)
10. Tondi, G., Wieland, S., Wimmer, T., et al.: Tannin-boron preservatives for wood buildings: mechanical and fire properties. Eur. J. Wood Prod. **70**(5), 689–696 (2012)
11. SS-EN 84: Durability of wood and wood-based products– Accelerated ageing of treated wood prior to biological testing–Leaching procedure. European Committee for Standardization Central Secretariat, Brussels (2020)
12. Ahmed, S.A., Morén, T.: Moisture properties of heat-treated Scots pine and Norway spruce sapwood impregnated with wood preservatives. Wood Fiber Sci. **44**(1), 85–93 (2012)
13. Sommerauer, L., Thevenon, M.-F., Petutschnigg, A., et al.: Effect of hardening parameters of wood preservatives based on tannin copolymers. Holzforschung. **73**(5), 457–467 (2019)
14. Haslam, E.: Plant polyphenols: vegetable tannins revisited. Cambridge University Press, Cambridge (1989)
15. Jakobek, L.: Interactions of polyphenols with carbohydrates, lipids, and proteins. Food Chem. **175**, 556–567 (2015)

16. Li, J., Lei, H., Xi, X., et al.: A sustainable tannin-citric acid wood adhesive with favorable bonding properties and water resistance. Ind. Crop Prod. **201**, 116933 (2023)
17. Teixeira, D.E., do Carmo Pereira, D., Nakamura, A.P.D., et al.: Adhesivity of bio-based anhydrous citric acid, tannin-citric acid and ricinoleic acid in the properties of formaldehyde-free medium density particleboard (MDP). Drvna Industrija. **71**(3), 235–242 (2020)
18. Zhao, Z., Umemura, K., Kanayama, K.: Effect of the addition of citric acid on tannin-sucrose adhesive and physical properties of particleboards. Bioresources. **11**(1), 1319–1333 (2016)
19. Moosavinejad, S.M., Madhoushi, M., Vakili, M., et al.: Evaluation of degradation in chemical compounds of wood in historical buildings using FT-IR and FT-Raman vibrational spectroscopy. Maderas: Cienc. Tecnol. **21**(3), 381–392 (2019)
20. Crisostomo, M., Menezzi, C.D., Militz, H., et al.: Effect of citric acid on the properties of sapwood of *Pinus sylvestris* submitted to thermomechanical treatment. Forests. **14**, 1839 (2023)
21. Mubarok, M., Militz, H., Dumarçay, S., et al.: Beech wood modification based on in situ esterification with sorbitol and citric acid. Wood Sci. Technol. **54**(3), 479–502 (2020)
22. Socrates, G.: Infrared and Raman characteristic group frequencies: tables and charts, pp. 340–348. John Wiley and Sons, Hoboken, NJ (2004)
23. Max, B., Salgado, J.M., Rodríguez, N., et al.: Biotechnological production of citric acid. Braz. J. Microbiol. **41**, 862–875 (2010)
24. Essoua Essoua, G.G., Blanchet, P., Landry, V., et al.: Pine wood treated with citric acid and glycerol mixture: biomaterial performance improved by a bio-product. Bioresources. **11**(2), 3049–3072 (2016)
25. Sun, S., Zhao, Z.: Influence of acid on the curing process of tannin-sucrose adhesives. Bioresources. **13**(4), 7683–7697 (2018)
26. Dinwoodie, J.M.: Timber: Its nature and behviour, 2nd edn. E & FN Spon, Taylor & Francis, London (2000)
27. Tondi, G., Thevenon, M.F., Mies, B., et al.: Impregnation of Scots pine and beech with tannin solutions: effect of viscosity and wood anatomy in wood infiltration. Wood Sci. Technol. **47**(3), 615–626 (2013)
28. Ahmed, S.A., Sehlstedt-Persson, M., Karlsson, O., et al.: Uneven distribution of preservative in kiln-dried sapwood lumber of Scots pine: impact of wood structure and resin allocation. Holzforschung. **66**(2), 251–258 (2012)

Development of Novel Guitar Fretboards by Thermal Modification and Impregnation with PF-Resin of Beech (*Fagus sylvatica*) and Maple (*Acer ssp.*) Wood

Christina Zwanger(✉) and Marcus Müller

University of Applied Forest Sciences Rottenburg, Rottenburg, Germany
{zwanger,mueller}@hs-rottenburg.de

Abstract. Tropical woods like ebony (*Diospyrus ssp.*) or rosewood (*Dalbergia ssp.*) are typical tonewoods used to produce guitar fretboards. Trade of these woods is strongly restricted by CITES (Convention of International Trade in Endangered Species). Furthermore, there is an increasing demand for guitars built without tropical woods. Therefore, it is necessary to search for alternatives. Most native woods in Europe do not meet the requirements that are important for usage in guitar fretboards. These requirements are a sufficiently high hardness and density, dimensional stability and a dark wood colour. To achieve these properties a two-step modification was applied. In a first step wood samples of beech (*Fagus sylvatica*) and maple (*Acer ssp.*) were heat-treated at 220 °C for 4 h under nitrogenous atmosphere, followed by an impregnation with a 15% (solid content) solution of phenolic resin (PF). After treatment, the wood was dried stepwise closing with a curing step. The effect of the two-step treatment on dimensional stability, Brinell-hardness and L*a*b*-values were measured. Thermally treatment of beech and maple wood at 220 °C for 4 h changed the light wood colour to dark brown/black. Impregnation of the thermally treated wood with a 15% solution of PF resin resulted in an increase of hardness of 8% for beech and a slight decrease of 5% for maple. Density could be increased by 1.6% for beech and 11.6% for maple wood. The ASE (Anti-Swelling-Efficiency) values of thermally and PF-resin impregnated (TPF)-beech (51% in tenth cycle) and TPF-maple (52% in tenth cylce) were higher than for the thermally treated (T) and the resin impregnated (PF) wood. The ASE values of TPF treatment are stable over ten cycles. The properties of TPF-treated beech meet those of rosewood.

Keywords: guitar fretboard · thermal modification · PF-resin

1 Introduction

Guitar fretboards are usually made of tropical woods like ebony (*Diospyrus ssp.*), grenadill (*Dalbergia melanoxylon*) or rosewood (*Dalbergia ssp.*). These woods have a high dimensional stability, density, hardness and a dark colour. Because of their bright wood colour, high swelling and shrinking behavior and rather low hardness domestic

G. Goli et al. (Eds.): ECWM 2024, SPM 86, pp. 209–216, 2025.
https://doi.org/10.1007/978-3-031-99418-0_26

woods like beech or maple are not suitable for using them in guitar fretboards. A way to change wood properties is wood modification. In the recent decades different kinds of wood modifications have been developed. One of these treatments is a modification by thermal treatment. Thermal modification usually takes place at temperatures between 150 and 240 °C under dry (air, vacuum, inert atmosphere, superheated steam) or wet (saturated steam, water) conditions [1]. During thermal modification a change in chemical composition and structure of the wood occurs. Degradation of hemicelluloses and volatiles/extractives from solid into volatile compounds takes place [2, 3]. These changes in chemical composition and structure affect physical and optical properties. Lower density, L-value, moisture content thus higher dimensional stability, bending strength/MOR (Modulus of Rupture), increased brittleness and durability are results of this treatment. Thermal treatment has no important impact on MOE (Modulus of Elasticity) [2, 3, 4]. Higher dimensional stability is due to degradation of hemicelluloses, which are the main reason for swelling and shrinking behaviour of wood [2, 5]. While several properties are improved hardness and density of thermally treated wood is decreasing [6]. Another common wood modification is the impregnation of wood with resins. This kind of modification has similar effects on wood as thermal modification. Meanwhile a big variety of resins is available, which are used to modify wood. One of these resins is based on phenolic formaldehyde (PF). By impregnating a resin into the cell wall and polymerization afterwards, it is possible to reduce swelling and shrinking behaviour. This is due to a bulking effect. Dimensional stability is improved as well as mechanical properties like bending strength, MOE, MOR and hardness [7–10].

Separately the different kinds of modifications are well researched. But until now there are only a few studies which have examined the combination of different wood modifications. Nevertheless a combination of different modification systems has already been performed as the studies of [11–13] show. In a study conducted by [12] modified Eucalypt wood was impregnated with melamine-urea-formaldehyde (MUF) resin and then heat-treated at different temperatures. Investigated properties were dimensional stability and mechanical properties (MOE and MOR were determined by a three-point bending test). [14] already showed early that impregnating MHP (Moisture-Heat-Pressure)-treated pine wood results in a high dimensional stability. [13] first heat-treated native hardwoods followed by an impregnation with melamine resin. Properties like dimensional stability, Brinell-hardness, bending strength were investigated. [11] combined a modification by acetylation with an impregnation with melamine-formaldehyde-resin. The aim of this research was to create a new material out of domestic woods (with good availability in central Europe) that is dark in colour, has less swelling and shrinking behaviour, high density and hardness. Depending on which modification method is used some properties will be improved while others decrease. The idea was to combine thermal modification and impregnation with phenolic resin to use the advantages of both treatments and to equalize the disadvantages caused by thermal modification

2 Materials and Methods

Wood samples of beech (*Fagus sylvatica*) and maple (*Acer ssp.*) were treated by combining two wood modifications. In a first step the wood was heat-treated, followed by an impregnation with phenolic resin. For determining the single effects of thermal and chemical treatment, specimens with only one treatment were tested as well. References in any case were untreated beech and maple specimens. Specimens were labelled as followed: Reference (R), thermal modification (T), PF-resin impregnation (PF), thermal modification + PF resin impregnation (TPF).

Thermal modification was carried out at 220 °C for 4 h under nitrogenous atmosphere. The efficiency of thermal modification was measured by weight loss. In a second step the thermal modificated wood was impregnated with a 15% (solid content) solution of phenolic resin (prefere resins GmbH). In a pressure apparatus the wood was put under vacuum at 100 mbars for 30 min and then set under pressure at 9 bars for 120 min. After impregnation, the wood was dried stepwise at 40, 60, 80, 103 °C for 24 h at each temperature. At last a curing step at 120 °C for 180 min was applied.

The effect of double treatment on swelling and shrinking behaviour was investigated according to DIN 52184 (1979) and [15]. Therefore, specimens of the size 25 x 25 x $10mm^3$ (radial x tangential x axial) were cut and planed out of the heat-treated beech and maple wood, impregnated and dried as explained in the section before. For determining changes in swelling and shrinking behaviour Anti-Swelling-Efficiency (ASE) of the specimens was calculated as shown in Eq. (1).

$$\mathrm{ASE}\ [\%] = \Big((S_u - S_m/S_u)\ \mathrm{x}\ 100 \text{ with S.}$$

$$S = \Big((V_{ws} - V_{kd}/V_{kd})\ \mathrm{x}\ 100 \qquad (1)$$

S_u is the volumetric swelling coefficient of the unmodified wood sample, S_m is the volumetric swelling coefficient of the modified wood sample, V_{ws} is the volume of the water-saturated wood sample and V_{kd} is the volume of the kiln-dried wood sample. Besides ASE, Weight-Percent-Gain (WPG) and bulking were detected.

For establishing the effect on hardness of the wood, samples following DIN EN 1534 (2019) are used. Five specimens of the size 50 x 25 x $150mm^3$ (radial x tangential x axial) were cut and then conditioned at 20 °C and 65% RH until equilibrium moisture content was reached. Testing was carried out at a Zwick Roell testing machine (Z100). Three tests were performed on radial section of each specimen. Applied load here was 1 kN. As Brinell-hardness (HB) is calculated with Eq. (2) it was necessary to measure the diameter of the imprint. d_1 and d_2 were determined by means of a camera with a measuring system (Ash Inspex HD 1080p (Vesa)).

$$\mathrm{HB}\ \left[N/mm^2\right] = 2F/\Big(\pi\ \mathrm{x}\ D\ \mathrm{x}\ \Big(D - \sqrt{}D^2\text{--}d^2\Big)\Big).$$

$$d = (d_1 + d_2)/2 \qquad (2)$$

F is the applied load in N, π is the factor "pi", D is the diameter of the indenter in mm, d is the mean-value of d_1 and d_2, where d_1 and d_2 are the diameter of indentation in mm.

To document colour changes, L*a*b-values according to CIELab (Comission Internationale d'Eclairage) standard of wood samples following DIN EN 927-6 were measured. Five samples of each type of modification were cut and planed. For each specimen five measurements were made. Measuring device for this was a colorimeter (spectro-guide 45/0 BYK). Equipment calibration was carried out with standards provided by the supplier. Measurements were made using D65 illuminant and 10-degree standard observer. The color system is built up as three axes system where L* defines lightness from 0% (black) and 100% (white), a* the colors from green (−a) to red (a+), and b* from blue (−b) to yellow (+b).

3 Results and Discussion

3.1 Dimensional Stabilization

As thermal modification usually is carried out at 150 to 240 °C, the applied temperature of 220 °C is considered as high treatment intensity. The thermal modification caused a mass loss of 10.7% for beech and 10.2% for maple wood (Table 1). Beech and maple wood have nearly the same WPG for both impregnations (PF and TPF). However, what can be seen is that priorly thermally treated wood (TPF) has a slightly higher WPG compared to PF-treated wood (PF). The results for bulking differ between beech and maple. Bulking effect of TPF-beech was twice as high as for PF-beech. For maple thermal treatment prior to impregnation (TPF) resulted in slightly less bulking than wood impregnated with PF-resin alone. [16] found that porosity and average pore diameter of maple increased with higher degree of thermal treatment due to hemicellulose degradation. This could explain higher WPG and bulking of beech and maple TPF-samples compared to PF-samples, but not the contradictory results of maple regarding bulking. As Table 1 shows, swelling and shrinking behaviour of beech and maple samples could be successfully improved. Thermal treatment increased dimensional stability up to 32% (tenth cycle) for beech and 38% (tenth cycle) for maple. This is in accordance with results found by [14, 17]. There dimensional stability of different wood species could be increased from 50% up to 80% respectively due to thermal modification. [17] found for beech an improvement of dimensional stabilization of 60% after heat-treatment at 160 °C for 15 h. The results of treatment with phenolic resin showed an improvement of dimensional stability of 43% (tenth cycle) for beech and 37% (tenth cycle) for maple. [8] noticed that a 30% solution of phenolic resin can reduce swelling and shrinking of Jabon wood (*Anthocephalus cadamba*) about 70.59%. [18] attained ASE values of 60% by impregnating Japanese cedar wood with a 30% solution of phenolic resin. The ASE values of TPF-beech (51% in tenth cycle) and TPF-maple (52% tenth cylce) are higher than for the single treatments. [19] reached ASE values of 60% for beech wood impregnated with a 25% solution of PF-resin. Until now there are no comparable results for a combined thermal treatment and impregnation modification with phenolic resin. However, looking at a combined treatment of thermal modification and secondary impregnation with melamine resin similar results can be found [13]. A thermal treatment at 210 °C for 7.5 h and impregnation with a 10% melamine resin solution of beech caused an improvement of swelling and shrinking behaviour of 52%.

Table 1. Characteristics of specimens and modifications: Oven-dry density (OD), thermal modification (TM), solid content (SC) of the PF-resin impregnation solution, Weight-Percent-Gain (WPG), bulking and Anti-Swelling-Efficiency (ASE). Mean values (Standard deviation).

Wood species	Label	OD [kg/m^3]	TM [C°]/[h]	SC [%]	WPG [%]	Bulking [%]	ASE [%] tenth cycle
Beech	R	687 (37)			–	–	–
	T	612 (50)	220/4		−10.7 (0.4)	–	32.18 (6.38)
	PF	762 (18)		15	21.3 (1.4)	2.0 (0.6)	43.75 (3.52)
	TPF	698 (53)	220/4	15	24.5 (3.6)	4.2 (0.3)	50.77 (3.78)
Maple	R	587 (8)			–	–	–
	T	568 (12)	220/4		−10.2 (0.5)	–	38.01 (2.22)
	PF	656 (8)		15	21.5 (0.5)	4.5 (0.6)	37.95 (1.53)
	TPF	655 (13)	220/4	15	23.5 (0.7)	3.14 (0.4)	52.09 (1.53)

3.2 Brinell Hardness

Brinell hardness of native beech and maple wood measured in this study are rather low, but can be found in the lower third of the values represented in the literature [20, 21]. Thermal treatment decreased hardness of beech wood by 25% and for maple wood even higher by 43% (Table 2). Similar effects were observed by [14] who showed that a thermal treatment (180–200 °C, moisture-heat-pressure-treatment) of different wood species caused a decrease in hardness up to 30%. [22] examined the effects of heat treatment under various conditions (120, 150, 180 °C; 2, 6, 10 h) on Red-bud maple wood and showed that hardness on the radial section decreased by 53% at 180 °C for 6 h treatment. For beech wood we can find similar results. [6] measured a decrease of hardness up to 20% for beech samples treated at 220 °C for 2 h. Changes of mechanical properties are strongly dependent on composition of polymeric components of the cell wall. Thermal treatment causes a degradation of hemicelluloses, degradation and crystallization of amorphous cellulose and polycondensation reactions of lignin. As hemicellulose content is reduced the load sharing capacity of lignin-hemicellulose matrix of the cell wall is affected negatively regarding compressive strength and hardness. Small radial fissures can also be a reason for decreased compressive strength in radial direction [23, 24]. For PF-samples hardness could be increased by 20% (maple) and 65% (beech), respectively. [7] impregnated sugar maple with PF-resin, which resulted in a 23.1% WPG and increased hardness up to 35.3–41.7%. Compared to the results for maple in this study (12–24%) we can see lower improvement with a slightly lower WPG (21.5%). During the curing step PF-resin reacts in a polycondensation reaction and forms a polymer network within the cell wall and the lumen. This results in a higher hardness [7]. A two-step modification combines the effect of thermal modification and PF-resin impregnation what causes a slight increase of hardness for beech (8%), whereas we can see a slight decrease (−5%) for maple wood. For TPF-maple the impregnation with PF-resin cannot compensate the decrease of hardness caused by thermal modification.

Table 2. Brinell hardness. Mean values (Standard deviation).

Wood species	R	T	PF	TPF
Hardness beech [N/mm^2]	31.7 (1.9)	24.9 (2.7)	52.2 (7.6)	35.3 (6.3)
Hardness maple [N/mm^2]	29.3 (1.7)	19.0 (1.0)	34.6 (1.7)	27.9 (4.8)
Hardness rosewood	37.6 (5.1)	–	–	–

3.3 Colour Changes

Thermal treatment is known to darken the colour of wood and this is what could be observed for thermal treatment of beech and maple wood (Fig. 1). Treatment with PF-resin caused only a slight decrease of the L*-value of both wood species. Thermal treatment of beech reduced L* by 51% and the combined modification (TPF) reduced L* by 60% (Table 3). For maple wood thermal treatment caused a loss of L* of 54% and the combined modification of 53%. These results agree with those of several other studies [25–28]. The reason for a darker colour after heat treatment is the degradation of hemicelluloses, resulting in a proportionally higher lignin content [6, 25, 29]. Formation of oxidation products such as quinones and changes in soluble phenolic compounds are other reasons for a darker wood colour after heat treatment [10, 27, 31].

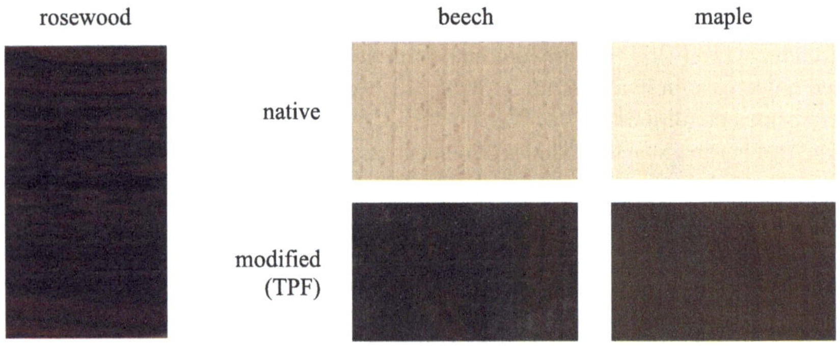

Fig. 1. Change of wood colour after a two-step modification (thermal modification and impregnation with PF-resin) and comparison to the wood colour of rosewood.

Table 3. L*-values of beech and maple. Mean values (Standard deviation).

Wood species	R	T	PF	TPF
Beech	74.02 (3.17)	33.23 (1.04)	60.98 (2.65)	29.27 (1.00)
Maple	81.67 (0.51)	38.06 (5.17)	71.75 (1.91)	37.36 (1.08)
Rosewood	34.22 (1.49)	–	–	–

4 Conclusions

1. To achieve a swelling and shrinking behavior that is similar to that of rosewood, ASE-values should be between or above 34–49% for beech and 28–44% for maple. By double treatment these values were achieved for both wood species.
2. Impregnation with 15% resin sufficiently improved the hardness of beech, but for maple the resin content was not high enough to achieve values similar to those of rosewood.
3. Treatment of beech and maple wood at 220 °C for 120 min darkened their colour sufficiently.

With a combined modification of beech as carried out in this research it is possible to create a material similar to rosewood. For TPF-maple hardness couldn't be improved sufficiently with a 15% PF-resin. Impregnation with a higher solid content could be a solution to reach a higher hardness for maple wood.

References

1. Hill, C., Altgen, M., Rautkari, L.: Thermal modification of wood—a review: chemical changes and hygroscopicity. J. Mater. Sci. **56**(11), 6581–6614 (2021)
2. Burmester, A.: Zur Dimensionsstabilisierung von Holz. Holz. Roh. Werkst. **33**, 333–335 (1975)
3. Niemz, P., Hofmann, T., Rétfalv, T.: Investigation of chemical changes in the structure of thermally modified wood. Maderas. Cienc. Tecnol. **12**(2) (2010)
4. Hakkou, M., Pétrissans, M., Zoulalian, A., Gérardin, P.: Investigation of the reasons for the increase of wood durability after heat treatment based on changes of wettability and chemical composition. In: Second European Conference on Wood Modification, pp. 37–46 (2005)
5. Kollmann, F., Schneider, A.: Über das Sorptionsverhalten wärmebehandelter Hölzer. Holz. Roh. Werkst. **21**(3), 77–85 (1963)
6. Kapidani, A., Ajdinaj, D., Beqo, S.: The effect of heating temperature on static hardness of beech (*Fagus sylvatica* L.) wood. IJCET. **9**(03) (2019)
7. Stamm, A.J., Seborg, R.M.: Minimizing wood shrinkage and swelling. Treating with synthetic resin-forming materials. Ind. Eng. Chem., 1164–1169 (1936)
8. Deka, M., Saikia, C.N.: Chemical modification of wood with thermosetting resin: effect on dimensional stability and strength property. Bioresour. Technol. **73**(2), 179–181 (2000)
9. Rowell, R.M.: The Chemistry of Solid Wood: Penetration and Reactivity of Cell Wall Components. American Chemical Society, Washington, DC (1984)
10. Wan, H., Yang, D., Zhang, C.: Impact of biological incising to improve resin retention and hardness of various wood species. For. Prod. J. **56**(4), 61–67 (2006)
11. Hansmann, C., Weichslberger, G., Gindl, W.: A two-step modification treatment of solid wood by bulk modification and surface treatment. Wood Sci. Technol. **39**(6), 502–511 (2005)
12. Sun, B., Wang, X., Liu, J.: Changes in dimensional stability and mechanical properties of Eucalyptus pellita by melamine–urea–formaldehyde resin impregnation and heat treatment. Holz. Roh. Werkst. **71**(5), 557–562 (2013)
13. Behr, G.: The Influence of Melamine Treatment in Combination with Thermal Modification on the Properties and Performance of Native Hardwoods. Georg-August-Universität Göttingen, Göttingen (2019)
14. Giebeler, E.: Dimensionsstabilisierung von Holz durch eine Feuchte/Wärme/Druck-Behandlung. Holz. Roh. Werkst. **41**, 87–94 (1983)

15. Rowell, R.M., Ellis, W.D.: Determination of dimensional stabilization of wood using the water-soak method. Wood Fiber, 104–111 (1978)
16. Pfriem, A., Zauer, M., Wagenführ, A.: Alteration of the pore structure of spruce (Picea abies (L.) Karst.) and maple (Acer pseudoplatanus L.) due to thermal treatment as determined by helium pycnometry and mercury intrusion porosimetry. Holzforschung. **63**(1) (2009)
17. Burmester, A.: Einfluss einer Wärme-Druck-Behandlung halbtrockenen Holzes auf seine Formbeständigkeit. Holz. Roh. Werkst. **31**, 237–243 (1973)
18. Furuno, T., Imamura, Y., Kajita, H.: The modification of wood by treatment with low molecular weight phenol-formaldehyde resin: a properties enhancement with neutralized phenolic-resin and resin penetration into wood cell walls. Wood Sci. Technol. **37**(5), 349–361 (2004)
19. Kielmann, B., Butter, K., Mai, C.: Modification of wood with formulations of phenolic resin and iron-tannin-complexes to improve material properties and expand colour variety. Holz. Roh. Werkst. **76**(1), 259–267 (2018)
20. Kollmann, F.: Technologie des Holzes und der Holzwerkstoffe. 2nd book, 2nd edn. Springer, Berlin, Heidelberg (1955)
21. Sell, J.: Eigenschaften und Kenngrössen von Holzarten. In: Lignum, Schweizerische Arbeitsgemeinschaft für das Holz, 4th edn. Baufachverl, Dietikon (1997)
22. Korkut, S., Kök, M.S., Korkut, D.S., Gürleyen, T.: The effects of heat treatment on technological properties in Red-bud maple (Acer trautvetteri Medw.) wood. Bioresour. Technol. **99**(6), 1538–1543 (2008)
23. Boonstra, M.J., Rijsdijk, J.F., Sander, C., Kegel, E., Tjeerdsma, B., Militz, H., et al.: Microstructural and physical aspects of heat treated wood: Part 2. Hardwoods. Maderas. Cienc. Tecnol. **2**(5), 99–110 (2006)
24. Boonstra, M.J., Acker, J., Tjeerdsma, B.F., Kegel, E.V.: Strength properties of thermally modified softwoods and its relation to polymeric structural wood constituents. Ann. For. Sci. **64**(7), 679–690 (2007)
25. Huang, X., Kocaefe, D., Kocaefe, Y., Boluk, Y., Pichette, A.: A spectrocolorimetric and chemical study on color modification of heat-treated wood during artificial weathering. Appl. Surf. Sci. **258**(14), 5360–5369 (2012)
26. Bekhta, P., Niemz, P.: Effect of high temperature on the change in color, dimensional stability and mechanical properties of spruce wood. Holzforschung. **57**(5), 539–546 (2003)
27. Sundqvist, B., Morén, T.: The influence of wood polymers and extractives on wood colour induced by hydrothermal treatment. Holz. Roh. Werkst. **60**(5), 375–376 (2002)
28. Militz, H., Altgen, M.: Processes and properties of thermally modified wood manufactured in Europe. American Chemical Society, Washington, DC (2013)
29. Bourgois, J., Guyonnet, R., Janin, G.: La Mesure de Couleur: Une Méthode d'Etude et d'Optimisation des Transformations Chimiques du Bois Thermolysé. Holzforschung. **45**(5), 377–382 (1991)
30. Tjeerdsma, B.F., Boonstra, M., Pizzi, A., Tekely, P., Militz, H.: Characterisation of thermally modified wood: molecular reasons for wood performance improvement. Holz. Roh. Werkst. **56**(3), 149–153 (1998)
31. Burtin, P., Jay-Allemand, C., Charpentier, J., Janin, G.: Modifications of hybrid walnut (Juglans nigra 23 x Juglans regia) wood colour and phenolic composition under various steaming conditions. Holzforschung. **54**(1), 33–38 (2000)

Mechanical Properties and Biological Durability of Beech Wood (*Fagus sylvatica*) Modified with PEG and Various Carboxylic Acids

Melissa Christ(✉), Nicole Flaig, and Marcus Müller

University of Applied Forest Sciences, 72108 Rottenburg am Neckar, Germany
{christ,mueller}@hs-rottenburg.de

Abstract. The goal of this study was to investigate the influence on physical and mechanical properties and on biological durability of beech wood (*Fagus sylvatica*) modified with polyethylene glycol (PEG) 400 as an alcohol and various carboxylic acids (citric acid, malic acid, 1,2,3,4-butanetetracarboxylic acid). PEG 400, chosen for its large and flexible structure, was expected to minimize negative effects on mechanical properties. However, contrary to expectations, the tensile and impact bending strength decreased – similar to other modifications. The tensile strength decreased (36.0% to 46.3%), while the modulus of rupture did not show a strong decrease (2.4% to 15.2%). The compression strength (11.7% to 40.6%) and hardness (6.0% to 17.1%) increased. The impact bending strength reduced strongly (43.0% to 69.3%). The modifying agents, fixed inside the wood, demonstrated a high anti-swelling efficiency during leaching cycles (27.1% to 45.9%). Decay resistance against brown, white and soft rot fungi was improved. Modification with PEG and citric acid falls into DC 1-2, modification with PEG and malic acid into DC 2-4 and modification with PEG and 1,2,3,4-butanetetracarboxylic acid in DC 1 for brown rot. In case of white rot, all modifications can be classified as DC 1.

Keywords: Polyethlyene Glycol · Carboxylic Acids · Physical and Mechanical Properties · Biological Durability · Wood Modification

1 Introduction

Chemical modification is an alternative method of increasing the biological durability, dimensional stability and mechanical properties of wood as well as reducing the moisture absorption without the use of biocides. Commercial chemical modification processes are increasing in market share [1, 2], but growth is restrained by higher production costs compared to conventional wood preservatives. For this reason, there is a need for a cost-effective wood preservative system with non-toxic mode of action [3]. In the current study, polyethylene glycol (PEG) 400 and various carboxylic acids (malic acid (MA), 1,2,3,4-butanetetracarboxylic acid (BTCA), and citric acid (CA)) were used for modification.

G. Goli et al. (Eds.): ECWM 2024, SPM 86, pp. 217–223, 2025.
https://doi.org/10.1007/978-3-031-99418-0_27

PEG finds diverse applications in pharmaceuticals, cosmetics, textiles, ceramics, and the conservation of wet archaeological wood [4, 5]. It can also lead to a considerable bulking-effect [6]. However, PEG is easily washed out of substances impregnated with it due to its water-soluble character [7]. There are already studies on dimensional stability, mechanical properties and biological durability [8–12]. The selected carboxylic acids should fix PEG 400 inside the wood. The mechanism that takes place is supposed to be a two-step esterification: in the first step, an anhydride is formed, while in the second step this cyclic anhydride reacts with the OH groups [13]. Of the three acids, CA is the most studied substance for wood modification. There are many studies on biological durability and on physical and mechanical properties [3, 14–26]. MA is predominantly utilized in the food and beverage industry, constituting 85% to 90% of its current production [27]. Limited studies exist on wood modification with MA – the only study found is by [28]. Recognized as the most efficient crosslinking agent for cotton fabrics among polycarboxylic acids [29], BTCA's application in wood modification is minimally explored. Despite this, some studies exist that analyzed the mechanical properties of BTCA treated wood or monocotyledons [13, 30, 31].

The aim of this study was to investigate whether the modification with PEG 400 and carboxylic acids, which is a novel combination, reduces the embrittlement of the wood (lower losses in tensile and impact bending strength). Resistance to fungal attack was also investigated.

2 Materials and Methods

Beech wood (*Fagus sylvatica*) was used for the investigations. The type of test as well as number and dimensions of the samples were based on the respective standards (Table 1). The test fungi *Coniophora puteana* (*C. puteana*) and *Trametes versicolor* (*T. versicolor*) were used for the decay resistance test against brown and white rot fungi

Three different formulations were tested: 20% (w/v) PEG 400 with 19% (w/v) CA; or with 13.4% (w/v) MA as well as 8.5% (w/v) PEG 400 with 10% (w/v) BTCA and 3% (w/v) sodium hypophosphite monohydrate (SHP) as catalyst. The molar ratio of acid to PEG was always 2:1. Untreated beech wood served as control. The specimens were immersed in the impregnation solution and vacuum was applied for 30 min in an autoclave. Then, pressure was applied for 120 min. After impregnation, the specimens were dried and cured at 160 °C for 5 h. The selected parameters are based on previous investigations [33].

Table 1. Type of test with associated standard and dimensions of the specimens.

Type of test	Longitudinal [mm]	Radial [mm]	Tangential [mm]	Standard
Weight-percent-gain, bulking, anti-swelling efficiency	10	25	25	[32]
Tensile strength	400	20	15	DIN 52188
Compression strength	30	20	20	DIN 52185
Modulus of rupture	360	20	20	DIN 52186
Impact bending strength	300	20	20	DIN 52189-1
Hardness	150	50	25	DIN EN 1534
Biological durability	50	25	15	EN 113-2
Biological durability	100	10	5	ENV 807 & CEN/TS 15083-2

3 Results and Discussion

Table 2 shows the results of the investigations. The values after 10 leaching cycles are shown for weight-percent-gain (WPG), bulking and anti-swelling efficiency (ASE). A clear bulking effect and an increase in dimensional stability can be seen for all formulations. In case of PEG/CA, the swelling and shrinkage behaviour was reduced strongly. All parameters indicate that the substances are fixed in the wood and that a reaction between PEG, acid and wood constituents has taken place. PEG/MA appears to be less efficient than PEG/CA and PEG/BTCA, as high solution concentrations were selected, which led to less promising results. These statements are consistent with [33]. [30] also found an increase of WPG, bulking and ASE in bamboo strips modified with CA and BTCA.

The tensile strength of the modified wood has been reduced strongly. This could be due to the carboxylic acids, which can cause hydrolysis of cellulose and hemicellulose. Hydrolysis of cellulose in an acidic environment is characterized by the cleavage of the β-1,4-glycosidic bonds of the cellulose molecules due to an addition of water molecules. Based on this addition, fragments with smaller chain lengths are formed, while the basic structure remains. Hydrolysis can very quickly reduce the average degree of polymerization of cellulose to a certain limit [34, 35].

The compression strength of the modified wood increased slightly. Most likely the composite resulting from carboxylic acids, PEG and wood constituents is responsible for the increased values. [19] also found that compression strength of CA-treated beech wood increases. In the current study, the incorporation of PEG and acids into the wood may have created a new polymer network. This may have formed new bonds and caused an altered transfer of forces.

Table 2. Results of the respective investigations and modifications (standard deviation in parentheses).

Type of test	Control	PEG/CA	PEG/MA	PEG/BTCA	Unit
Weight-percent-gain	–	23.0 (0.5)	13.3 (1.0)	12.0 (0.4)	%
Bulking	–	12.4 (0.9)	8.8 (0.6)	7.5 (0.4)	%
Anti-swelling efficiency	–	45.9 (1.6)	33.8 (3.6)	39.4 (2.0)	%
Tensile strength	136 (19.8)	78 (13.2)	83 (16.9)	85 (14.6)	N/mm^2
Compression strength	64 (7.8)	73 (5.8)	63 (7.7)	90 (5.1)	N/mm^2
Modulus of rupture	129 (7.2)	110 (15.6)	117 (22.8)	111 (28.4)	N/mm^2
Impact bending strength	114 (26.1)	39 (18.6)	65 (27.8)	35 (6.5)	kJ/m^2
Hardness	36 (5.3)	38 (5.3)	38 (4.7)	42 (4.4)	N/mm^2
Mass loss *C. puteana*	34.5 (5.6)	4.6 (3.2)	11.6 (5.4)	−0.3 (1.5)	%
Mass loss *T. versicolor*	29.4 (2.4)	1.3 (1.1)	2.5 (0.9)	2.9 (1.7)	%
Mass loss (soft rot)	35.1 (3.2)	5.1 (0.5)	13.9 (2.8)	2.8 (0.9)	%

Modulus of rupture (MOR) of the modified wood has slightly decreased. Since the values for tensile strength showed strong decrease, but an increase in compression strength, it is obvious that the MOR is in between.

The impact bending strength of the modified wood showed a strong decrease. PEG/MA showed a higher value than PEG/CA and PEG/BTCA. This finding possibly indicates that fewer reactions occurred with this modification. [36] found that the three-dimensional crosslinked structure results in increased stiffness of the wood. This hinders the free movement of the microfibrils, which in turn may result in a decrease in impact bending strength [36].

The hardness of the modified wood has increased slightly. [36] explain that the incorporation of chemicals into the wood structure increases the density of the wood, which leads to an increase in hardness. The surface hardness of lodgepole pine and white pine modified with Glycerol/CA also improved in studies by [22, 37].

The loss of mass due to brown rot and white rot fungi after 16 weeks of incubation was strongly reduced in the modified wood. The biological durability was increased, possibly due to the bulking effect of PEG. The bulking can reduce the cell wall porosity, creating a physical barrier [38]. This would prevent the growth of fungus, as there is no longer enough space. In addition, the reaction of the acids with the OH groups of the wood components can lead to an altered chemical structure. In this respect, the fungus would no longer recognize the wood as its food source. According to EN 113-2, PEG/CA falls into DC 1-2, with PEG/MA in DC 2-4 and with PEG/BTCA in DC 1 for brown rot. In case of white rot, all modifications can be classified as DC 1. The loss of mass due to soft rot fungi after 32 weeks of incubation was also strongly reduced in the modified wood. The biological durability has been strongly increased and use in contact with ground would be possible. Reasons for the improved fungal resistance have already been discussed for brown and white rot.

4 Conclusions

The modification of beech wood with PEG 400 and the various carboxylic acids can be regarded as successful. The substances are fixed in the wood and achieve high dimensional stabilization as well as good bulking effects. Modification with PEG/CA and PEG/BTCA is more efficient than with PEG/MA. The changed structure through reactions led to a decrease in tensile strength and impact bending strength. The hypothesis that the use of PEG does reduce these properties to a less amount could not be confirmed. It is possible that the carboxylic acids cause hydrolysis of the cellulose, which shortened the chains and lead to reduced force transmission. The compression strength and hardness increased. Maybe the incorporation of PEG and acids into the wood may have created a new polymer network. This may have formed new bonds and caused an altered transfer of forces. The MOR remained almost unchanged. This is because the compression strength and tensile strength decreased. Decay resistance against brown, white and soft rot fungi has increased strongly. In case of brown rot, PEG/CA was classified as DC 1-2, PEG/MA as DC 2-4 and PEG/BTCA as DC 1. In case of white rot, all modifications could be classified as DC 1.

With regard to industrial application, further investigations have shown that complete impregnation is also possible with planks whose dimensions correspond to the industrial scale. An advantage compared to other established modification methods is that the process management in terms of impregnation requires less effort. Except for BTCA, the modifying substances used are available in large quantities. For these reasons, and due to the fact that there have been few studies on the mechanical properties so far, the potential for future studies in this field is very high.

References

1. Sandberg, D., Kutnar, A., Mantanis, G.: Wood modification technologies—a review. IForest Biogeosci. For. **10**(6), 895–908 (2017)
2. Heräjärvi, H., Kunttu, J., Hurmekoski, E., Hujala, T.: Outlook for modified wood use and regulations in circular economy. Holzforschung. **74**(4), 334–343 (2020)
3. Beck, G.: Leachability and decay resistance of wood polyesterified with sorbitol and citric acid. Forests. **11**(6), 1–20 (2020)
4. Morén, R., Centerwall, B.R.: Use of Polyglycols in the Stabilizing and Preservation of Wood. The Historical Museum of Lunds University, Meddelande (1961)
5. Norimoto, M., Gril, J., Rowell, R.M.: Rheological properties of chemically modified wood: relationship between dimensional and creep stability. Wood Fiber Sci. **24**(1), 25–35 (1992)
6. Kollmann, F., Côté, W.A.: Principles of wood science and technology. Allen Unwin, Springer, London, Berlin (1968)
7. Schneider, A.: Beiträge zur Dimensionsstabilisierung des Holzes mit Polyäthylenglykol – Erste Mitteilung: Grundlegende Untersuchungen zur Dimensionsstabilisierung des Holzes mit Polyäthylenglykol. Holz. Roh. Werkst. **27**(6) (1969)
8. Stamm, A.J.: Effect of polyethylene glycol on the dimensional stability of wood. For. Prod. J. **9**, 375–381 (1959)
9. Schneider, A.: Beiträge zur Dimensionsstabilisierung des Holzes mit Polyäthylenglykol – Zweite Mitteilung: Untersuchungen über Eigenschaftsänderungen des Holzes durch Tränkung mit Polyäthylenglykol und über die Wirksamkeit verschiedener Tränkverfahren. Holz. Roh. Werkst. **28**(1) (1970)

10. Bjurhager, I., Ljungdahl, J., Wallström, L., Gamstedt, E.K., Berglund, L.A.: Towards improved understanding of PEG-impregnated waterlogged archaeological wood: a model study on recent oak. Holzforschung. **64**(2), 243–250 (2010)
11. Müller, L.: Einfluss einer Polyethylenglycol-Behandlung auf die mechanischen Eigenschaften von Buchenholz (Fagus sylvatica). Bachelorarbeit. Hochschule für Forstwirtschaft Rottenburg, Rottenburg am Neckar (2017)
12. Repák, M., Reinprecht, L.: Physio-mechanical properties of thermally modified beech wood affected by its pre-treatment with polyethylene glycol. Acta Facultatis Xylologiae Zvolen. **62**(1), 67–78 (2020)
13. Vukušić, S.B., Katović, D., Schramm, C., Trajković, J., Šefc, B.: Polycarboxylic acids as non-formaldehyde antiswelling agents for wood. Holzforschung. **60**(4), 439–444 (2006)
14. Larnøy, E., Karaca, A., Gobakken, L.R., Hill, C.A.S.: Polyesterification of wood using sorbitol and citric acid under aqueous conditions. Int. Wood Prod. J. **9**(2), 66–73 (2018)
15. L'Hostis, C., Thévenon, M.-F., Fredon, E., Géradin, P.: Improvement of beech wood properties by in situ formation of polyesters of citric and tartaric acid in combination with glycerol. Holzforschung. **72**(4), 291–299 (2018)
16. Mubarok, M., Militz, H., Dumarçay, S., Géradin, P.: Beech wood modification based on in situ esterification with sorbitol and citric acid. Wood Sci. Technol. **54**(3), 479–502 (2020)
17. Samani, A., Ganguly, S., Hom, S.K.: Effect of chemical modification and heat treatment on biological durability and dimensional stability of Pinus roxburghii Sarg. N. Z. J. For. Sci. **51**(5), 1–10 (2021)
18. Hasan, M., Despot, R., Jug, M.: Modification of wood with citric acid for increasing biological durability of wood. In: New Technologies and Materials in Industries Based on the Forestry Sector. International Symposium, pp. 85–89 (2007)
19. Hasan, M., Despot, R., Šefc, B., Ištok, I., Sedlar, T., Lacić, R.: Optimisation of modification of beech wood by citric acid. In: Jones, D., Militz, H., Petrič, M., Pohleven, F., Humar, M., Pavlič, M. (eds.) The Sixth European Conference on Wood Modification. Proceedings, Ljubljana, Slovenia (2012)
20. Despot, R., Hasan, M., Jug, M., Šefc, B.: Biological durability of wood modified by citric acid. Drvna Industrija. **59**(2), 55–59 (2008)
21. Šefc, B., Hasan, M., Trajkovic, J., Despot, R., Jug, M., Katovic, D., Bischof Vukušić, S.: Selected properties of beech wood modified by citric acid. In: European Conference on Wood Modification, pp. 425–428 (2009)
22. Essoua, G., Blanchet, P., Landry, V., Beauregard, R.: Pine wood treated with a citric acid and glycerol mixture. biomaterial performance improved by a bio-byproduct. Bioresources. **11**(2), 3049–3072 (2016)
23. Feng, X., Xiao, Z., Sul, S., Wang, Q., Xie, Y.: Esterification of wood with citric acid. The catalystic effects of sodium hypophosphite (SHP). Holzforschung. **68**(4), 427–433 (2014)
24. Guo, W., Xiao, Z., Wentzel, M., Emmerich, L., Xie, Y., Militz, H.: Modification of Scots pine with activated glucose and citric acid. Physical and mechanical properties. BioResources. **14**(2), 3445–3458 (2019)
25. Kurkowiak, K., Emmerich, L., Militz, H.: Biological durability and wood-water interactions of sorbitol and citric acid (SorCA) modified wood. J. Wood Sci. **69**(1), 34 (2023)
26. Crisostomo, M., Del Menezzi, C., Militz, H., Kurkowiak, K., Mayer, A., Carvalho, L., Martins, J.: Effect of citric acid on the properties of sapwood of Pinus sylvestris submitted to thermomechanical treatment. Forests. **14**(9), 1839 (2023)
27. Kövilein, A., Kubisch, C., Cai, L., Ochsenreither, K.: Malic acid production from renewables: a review. J. Chem. Technol. Biotechnol. **95**, 513–526 (2019)
28. Chabert, A.J., Fredon, E., Rémond, R.: Improving the stability of beech wood with polyester treatment based on malic acid. Holzforschung. **76**(3), 268–275 (2022)

29. Alay, S., Genç, E.: Functionalization of cotton fabrics by esterification cross-linking with 1,2,3,4-butanetetracarboxylic acid (BTCA). Cellul. Chem. Technol. **49**(5–6), 405–413 (2015)
30. Dong, Y., Liu, X., Liu, J., Yan, Y., Liu, X., Wang, K., Li, J.: Evaluation of anti-mold, termite resistance and physical-mechanical properties of bamboo cross-linking modified by polycarboxylic acids. Construct. Build Mater. **272**, 121953 (2021)
31. Dong, Y., Gao, S., Wang, K., Zhan, X., Li, Z., Zhao, Y., et al.: The effect mechanism and properties of poplar wood cross-linking modified with polyols and polycarboxylic acid. Wood Mater. Sci. Eng. **18**, 1630 (2023)
32. Hill, C.A.S., Jones, D.: The dimensional stabilisation of Corsican pine sapwood by reaction with carboxylic acid anhydrides. The effect of chain length. Holzforschung. **50**(5), 457–462 (1996)
33. Flaig, N., Christ, M., Müller, M.: Influence of wood modification with polyethylene glycol and various carboxylic acids on the dimensional stability of beech wood (*Fagus sylvatica*). Int. Wood Prod. J. **14**, 120–134 (2023)
34. Burchard, W. (ed.): Polysaccharide. Eigenschaften und Nutzung. Institut für Makromolekulare Chemie der Universität Freiburg. Springer-Verlag, Berlin, Heidelberg, New York, Tokyo (1985)
35. Sipahi Saglam, E.: Synthese und Charakterisierung spezieller Cellulose- Pfropfcopolymerer. Dissertation. Technische Universität Darmstadt, Fachbereich Chemie. Darmstadt (2003)
36. Xie, Y., Fu, Q., Wang, Q., Xiao, Z., Militz, H.: Effects of chemical modification on the mechanical properties of wood. Eur. J. Wood Wood Prod. **71**(4), 401–416 (2013)
37. Schorr, D., Blanchet, P., Essoua, G.: Glycerol and citric acid treatment of lodgepole pine. J. Wood Chem. Technol. **38**(2), 123–136 (2018)
38. Mai, C.: Chemical modification of wood and determination of the resulting wood properties. In: Schriftliche Habilitationsleistung. Georg-August-Universität Göttingen, Burckhardt-Institut, Abteilung Holzbiologie und Holzprodukte. Fakultät für Forstwissenschaften und Waldökologie, Göttingen (2010)

Acetylation of European Hornbeam Wood (*Carpinus betulus* L.): An 8-Year-Long Study

Fanni Fodor(✉) and Róbert Németh

University of Sopron, Sopron 9400, HU, Hungary
fodor.fanni@uni-sopron.hu

Abstract. In this work, European hornbeam wood (*Carpinus betulus* L.) was industrially acetylated in order to improve its properties and widen its usage.

Chemical analysis was carried out to determine the chemical components, pH, buffering capacity, and content of organic extractives. These findings were evaluated by Fourier-transform infrared spectroscopy (FTIR) and high-performance liquid chromatography (HPLC). After acetylation, changes in microscopic structure were examined by bright-field microscopy and scanning electron microscopy. Since 2016, acetylated hornbeam has been exposed to soil along with hornbeam, beech and pine sapwood control stakes. To evaluate its product-related properties, contact angle, surface energy, shear strength, bonding strength, and color stability against UV light with or without coating was tested as well.

Finally, according to the results, use classes, product groups, market share of acetylated hornbeam were determined, supplemented by SWOT analysis and related literature, in order to give a full picture of it as a product.

Keywords: Durability · Microscopy · Chemical analysis

1 Introduction

Many European wood species, including hornbeam (*Carpinus betulus* L.), have low natural durability which makes them unable to be used for exterior applications without additional protection. For decades, this has been done using biocide preservatives which are toxic for both environment and humanity. Today, non-toxic methods are preferred for wood preservation, among which are the wood modification methods, including acetylation [1].

Presently, acetylated wood is industrially produced by Accsys Technologies in Arnhem, the Netherlands. As the value of weight percentage gain (WPG) cannot be precisely calculated from its dry weight for industrially treated wood, the maximum degree of acetylation and acetyl content is determined with chemical analysis methods. Its main uses are typically in the outdoors, above or in ground contact. Because of its cost, it is mainly used in value-added products in transportation, sports equipment, military, and construction [2].

European hornbeam (*Carpinus betulus* L.) can be found all over Europe, except for the Mediterranean. Besides Hungary, hornbeam can be found in fair amounts in

G. Goli et al. (Eds.): ECWM 2024, SPM 86, pp. 224–231, 2025.
https://doi.org/10.1007/978-3-031-99418-0_28

various countries, the most in France, Romania, and Iran. Hornbeam trees are of medium height (15–25 m) and usually grow in hill and mountain forests as there are favorable climatic conditions for it. If there is optimal climate, it can also grow on planes. The stems are often crooked, buttressed, fluted, and twisted. There are also populations with smooth, cylindrical, straight trunks, but these require more planning and work. They have diameters around 35–50 cm, rarely up to 90 cm. Due to climate change, its territory will shift to northern areas, or from plains to higher altitudes, where soil conditions do not limit its spread [3]. Its potential habitat will decrease in southern territories of its area, including Hungary. On the other hand, its standing stock can still increase in the short-term depending on forest political decisions, forest management changes, and unexpected events, like new wood pests. In the future, hornbeam will be a good option as mix species in regions where its area will grow. It will appear in the areas of beech, spruce and fir forests, which will need to be exchanged to oak hornbeam mixed forests due to climate change-induced drought and dry climate [4].

It is a diffuse-porous wood species, with dull white to grayish color, high density, high tendency to warping, moderately fast moisture uptake, and high fiber saturation point. It has low durability (Durability Class 5), high compression strength, bending strength, Brinell hardness and modulus of elasticity. It is mainly used indoors, about half of the felled timber is fuelwood [5]. The thermal modification of hornbeam was the subject of several research projects in Hungary and abroad, where all the papers identified change in mechanical properties, reduction of physical properties, mass loss, color darkening and improvement of durability. In addition, acetylation is also a way to improve its properties which is the base of this research [4].

2 Materials and Methods

2.1 Material

In a previous work [6], twenty pieces of edged and air-dry hornbeam boards from the southwest part of Hungary were ordered from a Hungarian sawmill (BOPAÁR Ltd.). The boards' dimensions were 28 $\times$ 160 $\times$ 2500 mm^3 (t $\times$ w $\times$ l), the average moisture content was 15.2 $\pm$ 0.6% and the average density was 789 $\pm$ 64 kg/m^3. Half of the boards were acetylated under industrial conditions at Accsys Technologies. The WPG ranged between 13.6% and 16.5%, having an average of 15.3%. There were promising results concerning its physical, mechanical and durability properties. This material has been the basis of this research as well.

2.2 Chemical Analysis and Fourier-Transform Infrared Spectroscopy Analysis

Cellulose content (Kürschner-Hoffer method), lignin content (Klason method), extractive content (Soxhlet extraction) and ash content (EN 15403:2011) were measured and compared in untreated and acetylated wood. The hemicellulose content was given by subtracting these chemical components from 100%. Formic acid, acetic acid, levulinic acid, and furfural were determined by HPLC-PDA-ESI-MS/MS (High-Performance Liquid Chromatography- Photodiode Array Detector- Electrospray Ionization- Tandem Mass

Spectrometry). The total soluble carbohydrate content (Dubois method), the total phenolic content (Folin-Ciocâlteu method), changes in pH (pH meter) and buffering capacity (titration) were also determined.

DRIFT (Diffuse Reflectance Infrared Fourier Transform) and ATR (Attenuated Total Reflectance) spectroscopy were also used during the research. The chemical changes in wood were evaluated observing the difference spectra, where the absorption increase was represented by positive band while absorption decrease was represented by negative band. The band assignments were made using the difference spectra.

2.3 Microscopic Studies

Untreated and acetylated hornbeam samples, sound and decayed specimens, and PVAc (Polyvinyl Acetate) and PUR (Polyurethane) -bonded lap joint samples were analyzed by bright-field microscopy and scanning electron microscopy (SEM). For bright-field microscopy, sections were cut with a sledge microtome after softening. They were stained with safranin and astra blue or potassium permanganate (1% solution) for two minutes and embedded in Euparal. For SEM, the sections were given a smooth surface by hand cutting with a razor blade after softening. Cell-wall thickness was also determined, 50 measurements for each sample.

2.4 Field Test

The test was conducted according to EN 252, with sample dimensions $20 \times 50 \times 300\ mm^3$. There were 12 stakes of each type: untreated and acetylated hornbeam, supplemented with beech and Scots pine sapwood according to standard.

The stakes were buried in the outdoor exposure testing field at the University of Sopron (47°40′41.4″ N 16°34′32.6″ E) in April 2016, and it is ongoing since then. The difference in the rate of degradation was determined visually and by calculating density and mass loss.

2.5 Wetting, Shear Strength and Bonding Strength Assessment

The Sessile Drop Technique was used with water (polar) and diiodomethane (DIM) (nonpolar) as test liquids. Before measuring the contact angle, the surfaces were sanded in the same way as if they were to be bonded with adhesive (EN 205: 2016). A PGX goniometer was used for the test. The volume of the probe liquids was 5 μL, and the contact angles were determined at 0, 0.1, 1, 2, 5, 10, 20, 30, 40, and 50 s. The surface energy was calculated by the Owens–Wendt equation. Tangential and radial surfaces of untreated and acetylated hornbeam samples were tested as well.

The shear strength was determined according to MSZ 6786-6:1977 in the conditioning sequences (CS) defined by EN 204:2016. The shear strength was tested with an Instron 4208 universal testing machine, and a Memmert WNB 7-45 water bath was used to heat the samples. The testing speed was 1 mm/min.

Bonding strength was determined according to EN 205:2016. The adhesives were a PVAc D3 adhesive (‘Ponal Super 3’), a PVAc D4 adhesive (‘Ponal Super 3’ + 5%

'Ponal D4 Hardener'), and a one-component PUR adhesive ('Ponal Pur-Leim'). They were applied on the tangential surface, where 0.5 N/mm^2 pressure was applied for at least 1 h at 20 °C and 65% relative humidity. The bonding strength was tested using a Tinius Olsen H10KT universal testing machine. A Memmert WNB 7-45 water bath was used to heat the lap joint samples. The durability classes were determined according to EN 204:2016. There were 20 samples of each species for each conditioning sequence (CS 1, 3, 4, and 5). The testing speed was 50 mm/min. The microscopic images were examined manually in AutoCAD 2019 software to determine the effective penetration and the maximum penetration.

2.6 Photostability Without Coating and With Coating

Untreated and acetylated samples were exposed to weather on a metal stand tilted at 45° in the Botanical Garden of Sopron (47°40′41.4"N 16°34′32.6″E) for 1.5 year. In another test, they were exposed on a wooden stand tilted at 45° on the roof of The Central Library of the University of Sopron (47°40′54.3″N 16°34′40.6″E) for two years. In every case, the weather parameters were taken into consideration.

The UV irradiation test was performed in an artificial ageing chamber. In one test, there were two mercury-vapor lamps used (800 Watt, 76 W/m^2), in the other test, two xenon lamps (482 W/m^2) were used. The lamps were 64 cm above the samples and the equipment temperature was set to a maximum of 50 °C.

The color was expressed in CIE (International Commission on Illumination) L*a*b* color space with a portable colorimeter. The colorimeter's sensor head was 8 mm. The color was measured and calculated based on the D65 illuminant and 10° standard observer. Color was measured after 0–5–10–20–30–60–120–200 h of irradiation, and also during weathering tests in the first weeks, then after months. The color coordinates were calculated according to ISO 11664:2019. Photos and scans were also taken of the samples in their current state.

There was a preliminary photostability test with boiled linseed oil. It was applied two times according to the manufacturer's instructions. Later, another photostability test was conducted with various oils and stains: a part of them were linseed oil-based coatings; the rest were mainly polymer made from natural oils, fatty acids and resins. The finishes were applied using a paintbrush according to the manufacturer's instructions. Weathering and irradiation tests were conducted as well.

2.7 Statistical Analysis

Statistical analysis was performed using the Dell Statistica software. Correlation tests, t-tests and factorial analyses of variances (ANOVA) combined with the Fisher's LSD (Least Significant Difference) test was conducted, and the differences were considered significant at $p < 0.05$.

3 Results and Discussion

3.1 Chemical Analysis and Fourier-Transform Infrared Spectroscopy Analysis

The proportional increase of the hemicellulose content (+11.10%) was compensated by the decrease of cellulose (−8.55%) and of lignin (−2.86%) contents. HPLC analysis revealed that during the acetylation of hornbeam wood it is the hemicellulose fraction that gets acetylated and which is primarily responsible for the WPG, while a slight degradation also occurs to it [7].The extractive content increased to 1.91% and the pH decreased to 4.73. The slightly darker, walnut-like color after acetylation was due to the alterations in the structure of lignin as well as the increase of the extractive content.

The increase of absorption of methine, methylene and methyl groups at 2970 cm^{-1} is due to the exchange of OH (hydroxyl) groups into acetyl groups, which corresponds to the decrease of moisture sorption in acetylated hornbeam. The hemicellulose fraction was acetylated by the higher absorption of the C=O (carbonyl) groups at 1769 cm^{-1} and of the C-O stretching at 1270 cm^{-1} in xylan. The absorption of the conjugated carbonyl groups at 1597 cm^{-1} decreased slightly in the acetylated wood, probably due to a minor degradation of xylans in acidic medium. As the total phenolic content decreased, the real decrease of the lignin content is only minor, and changes in the FTIR spectra only indicate internal structural changes of the lignin matrix. FTIR studies of lignin samples also confirmed this finding [7].

3.2 Microscopic Studies

The cutting of smooth, undamaged sections of acetylated hornbeam was only possible after swelling and softening in acetone, not water as in case of untreated wood. The wood microstructure was not damaged, and the cell-wall thickness increased by 7%. The pits of acetylated hornbeam were much narrower, due to permanent cell-wall swelling. New brown deposits in parenchyma cells of acetylated hornbeam. Each cell contained a small amount of an insoluble substance, which led to the hypothesis that a natural component of the living cells (e.g. lignans) became insoluble during acetylation [8].

3.3 Durability Tests

In a long-term field test, acetylated hornbeam exhibited high durability, while control samples of hornbeam, beech and Scots pine failed after 0.5–6 years with 23–39% of their mass lost. There was one acetylated stake, that showed signs of superficial brown rot decay after 1.5 years of exposure (Fig. 1).

FTIR analysis indicated that the affected area had a much lower acetyl content, which may have been a wet pocket in the raw material. The remaining acetylated hornbeam stakes are still in the soil for further testing, without any signs of decay or insect damage up to this point. Microscopic studies showed that fungi were able to colonize acetylated cell lumina in field test studies and in laboratory studies against *Coniophora puteana, Rhodonia (Poria) placenta,* and *Trametes (Coriolus) versicolor* [8].

Fig. 1. Stakes placed in soil in April 2016 in the outdoor exposure testing field (left). Acetylated hornbeam stake no. 7 which was taken from the test to examine its damage after 5.5 years (right).

3.4 Wetting, Shear Strength and Bonding Strength Assessment

Acetylated hornbeam is less wettable, its surface energy decreased from 64 to 55 mJ/m^2, due to the decrease of the polar component. The dispersive component did not change. The water contact angle increased significantly by 18° after acetylation. There was no significant difference between the wettability of radial and tangential surfaces.

The shear strength of conditioned samples decreased after acetylation, but due to its lower water uptake, acetylated hornbeam exhibited significantly enhanced shear strength after soaking in cold and boiling water, compared to untreated hornbeam.

It also retained its bonding strength after soaking in cold and boiling water, unlike untreated hornbeam. PUR adhesive had better results than PVAc adhesives, and it gave better results when used with acetylated hornbeam.

This was due to the lack of overpenetration, while it had lower PVAc bonding strength due to overpenetration (Fig. 2). With overpenetration, high maximum penetration values are obtained, while the effective penetration values (which affect bonding strength more) are lower [9].

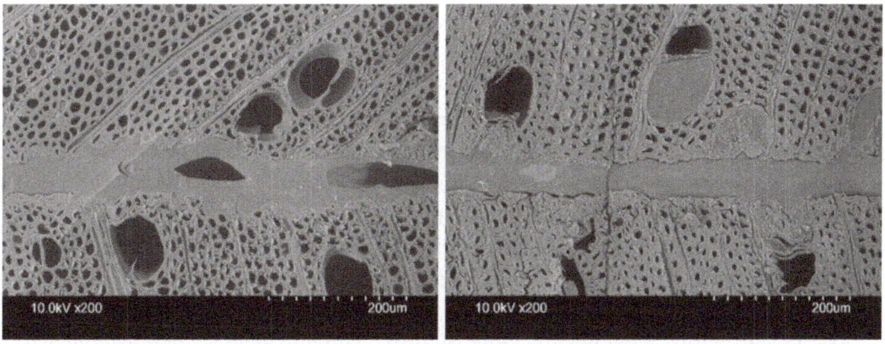

Fig. 2. Micrographs of PUR-bonded untreated (left) and acetylated (right) hornbeam wood. Hornbeam had overpenetration with starved bondlines and voids, unlike acetylated hornbeam.

3.5 Photostability Without Coating and With Coating

Acetylated hornbeam undergoes surface brightening during UV (ultraviolet) irradiation, and eventual graying outdoors. Here, the lightness (L*) increased, while red (a*) and yellow (b*) hue decreased. This was due to the transformation of extractives and degradation of lignin. These breakdown products were leached out during weathering. It resists dimensional changes caused by changing climate and precipitation, but it is susceptible to wasp stripping and can be stained by mold during weather exposure, but to a lower extent than untreated wood. The color resulted in 200 h of aging was achieved in one month or less when exposed to weather [10, 11].

The coating absorbance of hornbeam decreased after acetylation due to its lower water uptake. Its photostability can be largely improved by coating it with dark stains like Auro Wood stain No. 160 or Biopin Weather Protection Stain (for at least two years) (Fig. 3). On the other hand, colorless stains, and colorless or colored oils did not have a long-lasting effect on its photostability. Oli-Natura colorless oil and Auro colorless stain had no significant effect on the color stability during the test [10, 11].

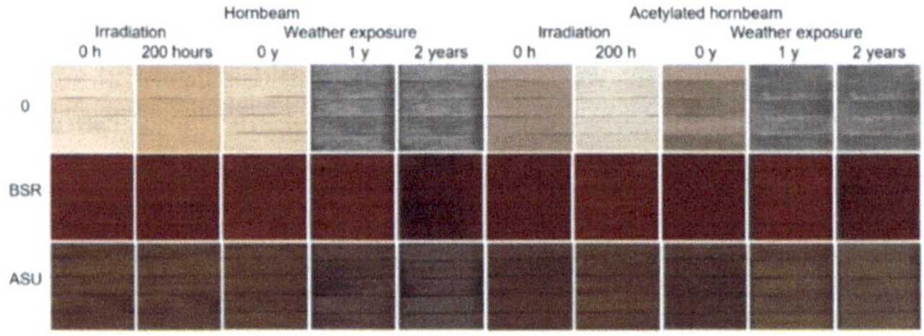

Fig. 3. Scanned pictures of hornbeam and acetylated hornbeam: uncoated (0), coated with Biopin Swedish Red Stain (BSR), and Auro Umbra Wood Stain Nr. 160 (ASU); before and after 200 hours of xenon lamp irradiation, and before and after one and two years of weather exposure.

3.6 Product Characterization and Conclusions

The most optimal countries (price, amount and distance from acetylation plant) to acquire 3 or 4-sides-clear hornbeam boards could be France, Germany, Poland and Romania [4]. The price (range) of acetylated hornbeam wood would be the same as Accoya®, because the raw material is similar or cheaper than the raw material used for Accoya® radiata pine, beech or alder production. Hornbeam can be optimized to similar acetylation settings to beech. The properties of acetylated hornbeam are similar to acetylated beech, and in some cases, even better [4]. Product groups may include: indoor furniture, floor and non-structural interior uses, exterior joinery, decking, fencing. It could possibly be used for outdoor furniture, and in-ground timber. Construction elements of acetylated hornbeam can be steps, stairs, flooring, and thresholds, etc. [4].

SWOT (Strengths, Weaknesses, Opportunities, Threats) analysis was established in [4]. Its main product competitors were identified, like thermo hardwood, furfurylated wood, tropical hardwood, WPC (Wood-Plastic Composite), etc. [12]. It has comparable properties to other commonly used materials, e.g. low thermal conductivity and high strength to density ratio.

Modified wood in general needs to find its niche market, where it is appreciated, and recognized as a superior product. It can even enter new non-wood markets as well. It is important to keep in mind that modified wood is not a replacement for preservative-treated wood; it is ideally a replacement for non-timber materials [13].

Acknowledgements. This article was made in frame of the project TKP2021-NKTA-43 which has been implemented with the support provided by the Ministry of Culture and Innovation of Hungary from the National Research, Development and Innovation Fund, financed under the TKP2021-NKTA funding scheme.

References

1. Rowell, R.M.: Acetylation of wood: journey from analytical technique to commercial reality. For. Prod. J. **56**(9), 4–12 (2006)
2. Mantanis, G.I.: Chemical modification of wood by acetylation or furfurylation: a review of the present scaled-up technologies. Bioresources. **12**(2), 4478–4489 (2017)
3. Varol, T., Cetin, M., Ozel, H.B., Sevik, H., Cetin, I.Z.: The effects of climate change scenarios on Carpinus betulus and Carpinus orientalis in Europe. Water Air Soil Pollut. **233**(2), 45 (2022)
4. Pozsgayné Fodor, F.: Acetylation of European Hornbeam (Carpinus betulus L.) Wood for Outdoor Applications. Doctoral thesis. University of Sopron, Sopron (2023). http://doktori.uni-sopron.hu/id/eprint/852/. Accessed 11 Feb 2024
5. Molnár, S., Bariska, M.: Magyarország ipari fái. Szaktudás Kiadó Ház Zrt, Budapest (2002)
6. Fodor, F., Lankveld, C., Németh, R.: Testing common hornbeam (Carpinus betulus L.) acetylated with the Accoya method under industrial conditions. iForest. **10**(6), 948–957 (2017)
7. Fodor, F., Németh, R., Lankveld, C., Hofmann, T.: Effect of acetylation on the chemical composition of hornbeam (Carpinus betulus L.) in relation with the physical and mechanical properties. Wood Mater. Sci. Eng. **13**(5), 271–278 (2018)
8. Rousek, R., Fodor, F., Németh, R.: Microscopic characterization of sound and decayed acetylated hornbeam (Carpinus betulus L.). Wood Mater. Sci. Eng. **18**(2), 640–649 (2023)
9. Fodor, F., Bak, M.: Studying the wettability and bonding properties of acetylated hornbeam wood using PVAc and PUR adhesives. Materials. **16**(5), 2046 (2023)
10. Fodor, F., Bak, M., Németh, R.: Photostability of oil-coated and stain-coated acetylated hornbeam wood against natural weather and artificial aging. Coatings. **12**(6), 817 (2022)
11. Fodor, F., Németh, R.: Testing the photostability of acetylated and boiled linseed oil-coated common hornbeam (Carpinus betulus L.) wood. Acta Sylv. Lignaria Hung. **13**(1), 81–94 (2017)
12. Larsson-Brelid, P.: Benchmarking and State-of-the-Art Report for Modified Wood. SP Report 2013:54. SP Technical Research Institute of Sweden, Borås (2013)
13. Jones, D., Carmo, J., Nunes, L., Kegel, E.: Products meeting needs—applying wood modification to its fullest potential. In: Proceedings of the 4th European Conference on Wood Modification, pp. 285–294. Technical Research Institute of Sweden, Stockholm (2009)

Effect of Lactic Acid Impregnation on Some Physical Properties of Wood

Miklós Bak[(✉)], Róbert Németh, and Mátyás Báder

University of Sopron, Sopron 9400, Hungary
bak.miklos@uni-sopron.hu

Abstract. This research deals with the combination of wood and polylactic acid (PLA), both of which are environmentally friendly materials issued from natural resources and quite easily recycled and degraded in good environmental conditions. Grafting of lactic acid oligomers and polymers onto wood and polymerization into the wood structure have been studied, considering that lactic acid contains carboxylic end groups, while solid wood cell walls contain hydroxyl groups. The main goal of this work was to increase dimensional stability. Weight percent gain (WPG) of beech and Scots pine samples showed 55–60% and 95–105% respectively. As a result, density increased 15–25% in beech and 35–45% in Scots pine. Equilibrium moisture content (EMC) decreased by a similar extent (45–55%) in both wood species, depending on the curing temperature. Higher curing temperature resulted in lower EMC. As a result of lower moisture uptake, anti-swelling-efficiency (ASE) decreased remarkably as well (60–80%). Surface hardness increased as a result of the treatments by 15–40% in cross section and 20–60% in tangential section, depending on wood species. Scanning electron microscopy (SEM) imaging showed that the impregnation was incomplete, as not all cell lumens were filled by the polylactic acid.

Keywords: Polylactic Acid · Impregnation · Wood-Water Relation · Anti-Swelling-Efficiency · Hardness

1 Introduction

There is a growing demand for the use of wood, especially in the use of valuable wood that is resistant to biological pests and has high dimensional stability. Impregnation with biopolymers to be used can significantly improve the targeted wood properties, especially dimensional stability. One of the most produced and processed biopolymers is polylactic acid (PLA), produced from lactic acid polycondensation [1].

Polylactic acid has already been used in combination with long natural fibers such as flax and hemp to make extruded and compressed composites, where the bond between them was stronger than conventional hydrophobic polymers such as polyethylene, polypropylene or polyvinyl chloride [2–4]. Treatment of wood with lactic acid was investigated as well, where one of the basic findings was that the treatment with LA oligomers leads to different results depending on the length of the heating cycle. A short

G. Goli et al. (Eds.): ECWM 2024, SPM 86, pp. 232–240, 2025.
https://doi.org/10.1007/978-3-031-99418-0_29

heating cycle results in a softened material, while a long heating cycle results in a hard material. In terms of chemical characterization, lactic acid oligomers appeared to polymerize in the walls of wood cells but did not appear to bind to the hydroxyl groups of the wood [5]. Overall, it can be seen that high treatment temperatures are required to induce further polymerization within the wood. In addition to the treatment temperature, the duration of the treatment is also an important factor in terms of polymerization, however, increasing these parameters increases the degradation of wood polymers [6, 7].

The treatment temperature significantly affects the hygroscopic behavior of the modified wood, which is caused by the polymerization within the material and the partial degradation of the wood polymers. The treatment time has less effect on the degree of polymerization than the treatment temperature [8]. In case of wood, the lactic acid oligomer treatment involves interesting chemical interactions and physical effects on the lignocellulose structure. The impregnation treatments were proven to further polymerize into the wood structure and were observed to set into the wood. The treatments give very high dimensional stability to the treated samples (approx. 95% ASE at high relative humidity or in water), but this was not maintained after prolonged water dissolution (25% ASE). The penetration of polylactic acid into the cell wall, as well as the reduced state of polymerization, causes a large modulus of elasticity (MOE) decrease during the bulking treatment [9]. Impregnation of wood at moderate temperatures reduces the viscosity of the oligomer, allowing mobility and penetration into the wood cell. This explains the swelling of the impregnated samples, which swell only during the heating steps. Continuous heating induces the polymerization of the oligomers inside the wood by maintaining the mass absorption of the oligomer [9].

Mass losses during treatment also occur, which increase as the temperature increases during the heating phase of the treatment. Three phenomena are responsible for the mass loss: evaporation of water during polycondensation, evaporation of some smaller lactic acid oligomers and wood degradation. The combination of these 3 phenomena together contributes to mass loss. In addition, lactic acid treatments provide significant biological resistance to wood, which was caused by the combined effect of the presence of lactic acid oligomers and heat treatment [7, 8].

This research deals with the combination of wood and PLA, both of which are environmentally friendly materials issued from natural resources and quite easily recycled and degraded in good environmental conditions. Grafting of lactic acid oligomers and polymers onto wood and polymerization into the wood structure have been studied, considering that lactic acid contains carboxylic end groups, while solid wood cell walls contain hydroxyl groups. The main goal of this work was to increase dimensional stability.

2 Materials and Methods

European beech (*Fagus sylvatica*) and Scots pine sapwood (*Pinus sylvestris*) was used for the tests. As a starting material of the modification, L(+) lactic acid monomer (90% purity, Acros Organics) was used. The initial step was a removal of water from the monomer, followed by two steps of oligomerization at elevated temperatures (75 °C, 100 °C and 130 °C respectively), under 150 mbar pressure and continuous stirring at

200 rpm. Samples were impregnated with the prepared oligomers at 100 mbar pressure at 90 °C for 60 min, followed by keeping the samples in the monomer for 60 min at atmospheric pressure. Subsequently, a curing step was applied at atmospheric pressure, using three different temperatures (110 °C, 120 °C and 140 °C) to test the effect of curing temperature on the properties of modified wood.

Weight percent gain (WPG) was determined using Eq. 1:

$$WPG = \frac{m_0 - m_{0c}}{m_0} \bullet 100\,[\%] \tag{1}$$

where m_0 is the oven-dry weight of samples before treatment and m_{0c} is the ovendry weight of samples after treatment, including curing step at elevated temperatures.

Oven-dry density (ρ_0) was determined using Eq. 2:

$$\rho_0 = \frac{m_{0c}}{V_{0c}} \left[\frac{g}{cm^3}\right] \tag{2}$$

where m_{0c} and V_{0c} is the oven-dry weight and volume of samples after treatment, including curing step at elevated temperatures.

Equilibrium moisture content (EMC) was determined using Eq. 3:

$$EMC = \frac{m_{Uc} - m_0}{m_0} \bullet 100\,[\%] \tag{3}$$

where m_0 is the oven-dry weight of samples and m_{Uc} is the weight of samples after conditioning (t = 20 °C, φ = 65%).

Anti-swelling efficiency (ASE) was determined using Eq. 4:

$$ASE = \frac{S_{cont} - S_{LA}}{S_{cont}} \bullet 100\,[\%] \tag{4}$$

where S_{cont} is the volumetric swelling of control samples between oven-dry and climatized state (t = 20 °C, φ = 65%) and S_{LA} is the volumetric swelling of treated samples between oven-dry and climatized state (t = 20 °C, φ = 65%).

Brinell hardness (HB) was determined using Eq. 5:

$$H_B = \frac{F}{D \bullet \pi \bullet h}\,[MPa] \tag{5}$$

where F is a constant load (500 N), D is the diameter of the steel sphere used (10 mm) and h is the indentation depth.

20 samples per treatment were used for each test. Untreated samples were used as control.

Scanning electron microscopy (SEM) (Hitachi S-3400 N) was used to investigate the location and distribution of the lactic acid in the cell wall structure. Applied acceleration voltage was 10 kV, sputter coating was not applied to the samples.

3 Results and Discussion

3.1 Weight Percent Gain and Oven-Dry Density

Beech and Scots pine samples showed weight percent gain (WPG) of 55–60% and 95–105% respectively (Fig. 1). There was no significant difference between the sample groups separated according to the curing temperature. This result supports that the impregnation was even among the samples and thus did not influence the effect of the treatment on the material properties. As a result, oven-dry density increased 15–25% in beech and 35–45% in Scots pine (Fig. 2). Other studies reported similar WPG values for beech (55–65%) and pine wood as well (80–90%). Besides, a similar 20% increase in beech density, increase in pine density was reported higher (50–60%) [5, 6, 9].

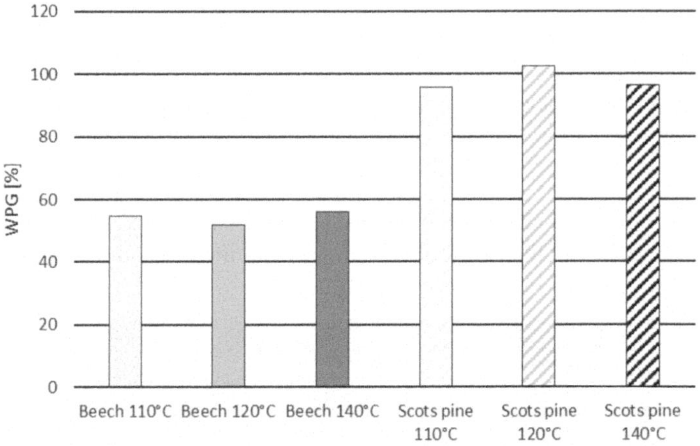

Fig. 1. Weight percent gain of beech and Scots pine sapwood as a result of LA impregnation before different curing temperatures.

In general, increasing curing temperature resulted in decreasing density of the modified samples. Previous studies found that degradation caused by the elevated temperature and the acidity of the treatment medium are behind this result [7, 8].

3.2 SEM Imaging

Beside testing the material properties, scanning electron microscopy was used to investigate the location and distribution of the lactic acid in the cell wall structure. It showed that the impregnation was incomplete, as beside some filled lumens, not all cell lumens were filled by the lactic acid (Fig. 3). In spite of that, empty cell lumens were reported in case of a similar LA treatment. WPG results were found similar, indicating different pore volumes or different penetration efficiencies between the two studies [5]. However, all cell types (tracheids, fibres, rays, vessels) were penetrated by the lactic acid. However, not all cell lumens were completely filled, lactic acid was present at least on the cell

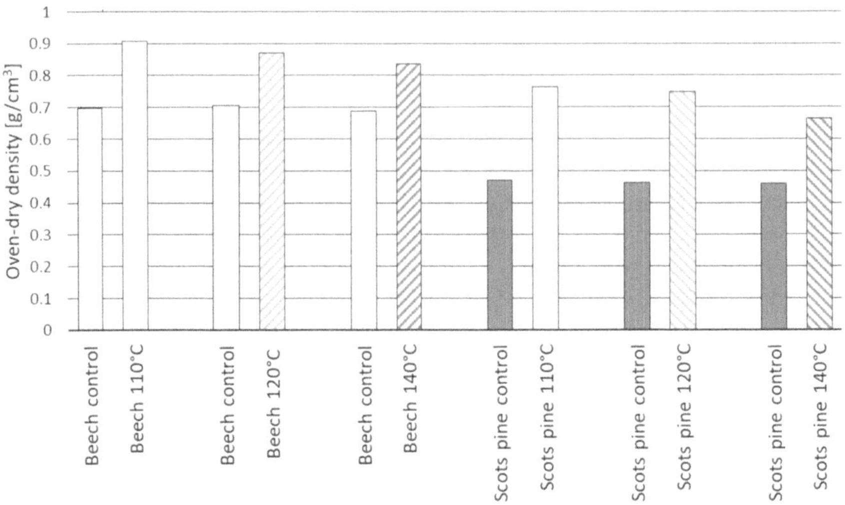

Fig. 2. Oven-dry density of beech and Scots pine sapwood as a result of LA modification after different curing temperatures.

wall surfaces as a thin coating, which might function as a protective layer against the penetration of water into the cell walls, resulting in an improved dimensional stability.

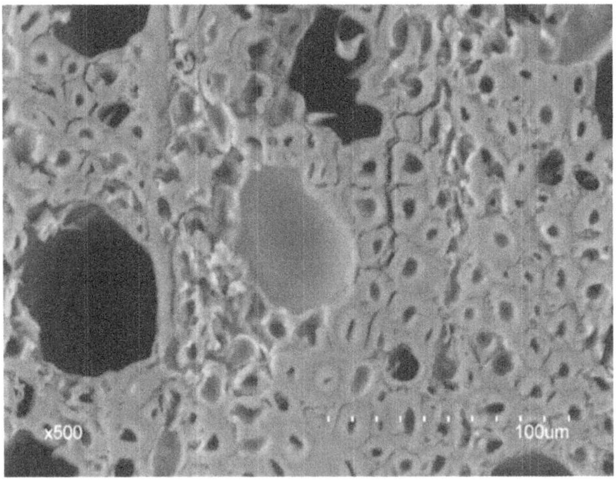

Fig. 3. Presence of LA in beech wood cross section, showing saturated, partly saturated and empty cells. Magnification: 500 × .

3.3 Equilibrium Moisture Content and Anti-swelling Efficiency

Equilibrium moisture content (EMC) decreased by 45–55% in both wood species, depending on the curing temperature (Fig. 4), which is a smaller reduction compared to other results. EMC is dependent on the cell-wall penetration of the LA oligomers, which, according to the SEM results, seems to be less effective [9]. Higher curing temperature resulted in slightly lower EMC. Additionally, the time necessary to reach EMC has tripled in treated samples, compared to the untreated ones, which is a clear proof of the change in moisture dynamics as well [10]. It shows that not only the moisture adsorption capacity, but the capability decreased remarkably as well. Since wood reacts slower to the moisture changes in its environment, it is a favorable property during utilization.

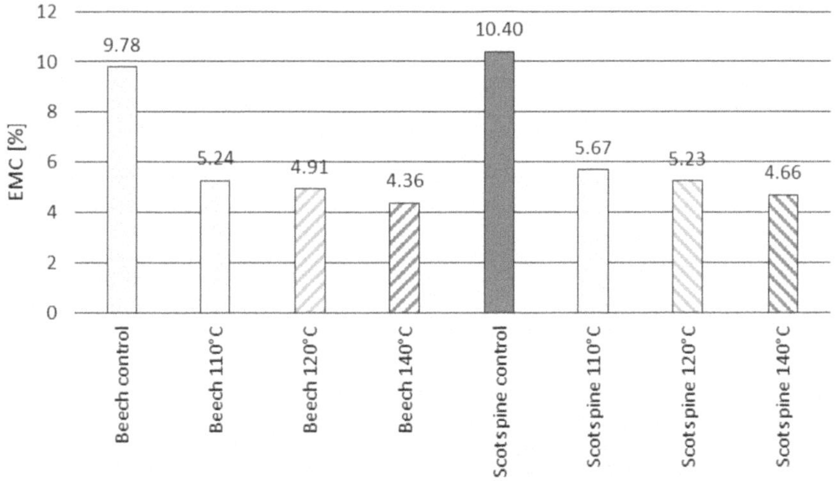

Fig. 4. Effect of LA impregnation and curing temperature on the EMC of beech and Scots pine sapwood (T = 20 °C; φ = 65%).

As a result of lower moisture uptake, anti-swelling-efficiency (ASE) decreased remarkably as well. ASE was influenced by the curing temperature, similarly to the EMC. Higher curing temperature resulted in higher ASE (Fig. 5). The PLA modification process resulted in 60-80% ASE, depending on wood species and curing temperature. SEM imaging (Fig. 3) showed that the LA penetrated into the cell lumens. However, it did not clog all cell lumens, which means that the impregnation was not uniform. But the polymer was present at least as a surface layer on the cell walls resulting in the exclusion of water. On the other hand, ASE values are remarkably higher compared to the EMC values, which indicates the bulking of the cell walls, creating empty volumes allowing the moisture to penetrate. This phenomenon was observed in case of LA treatment [9, 11] and carboxylic or cyclic anhydrides treatments as well [12–14].

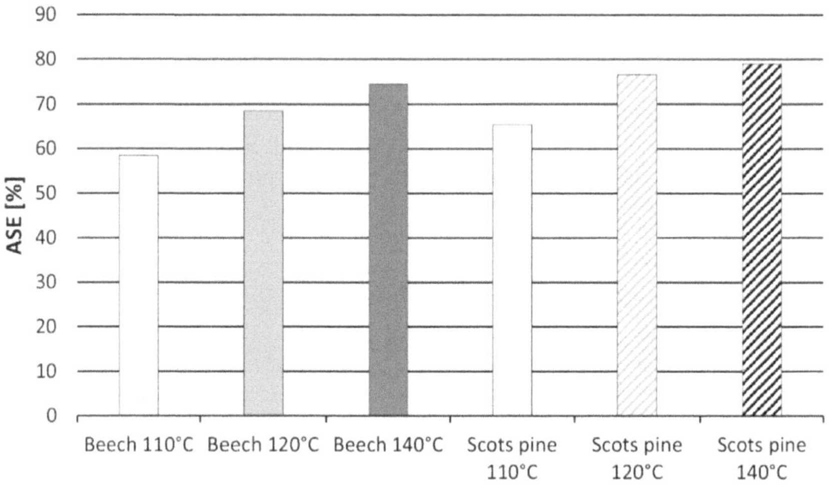

Fig. 5. Effect of LA impregnation and curing temperature on the ASE of beech and Scots pine sapwood.

3.4 Brinell Hardness

Surface hardness increased as a result of the treatments (Fig. 6). The improvement was 15–40% in cross section and 20–60% in tangential section, depending on wood species and curing temperature. However, there was a significant improvement in surface hardness, this effect was decreasing by increasing the curing temperature. It is assumed to be a sign of the cell wall degradation caused by the combined effect of LA impregnation followed by the high temperature curing [6–8]. One of the reasons for the improved surface hardness after PLA modification is the presence of the LA polymer in the wood tissue, filling the cell lumens, improving the lateral stability of the structure [15, 16]. On the other hand, the lower EMC of the modified samples contributed to the improved surface hardness as well. Mechanical properties of wood are inversely proportional to the moisture content [17]. Conditioning of the samples before hardness test was made under the same conditions ($T = 20$ °C; $\varphi = 65\%$), thus, the 45-55% lower EMC could improve the surface hardness considerably as well.

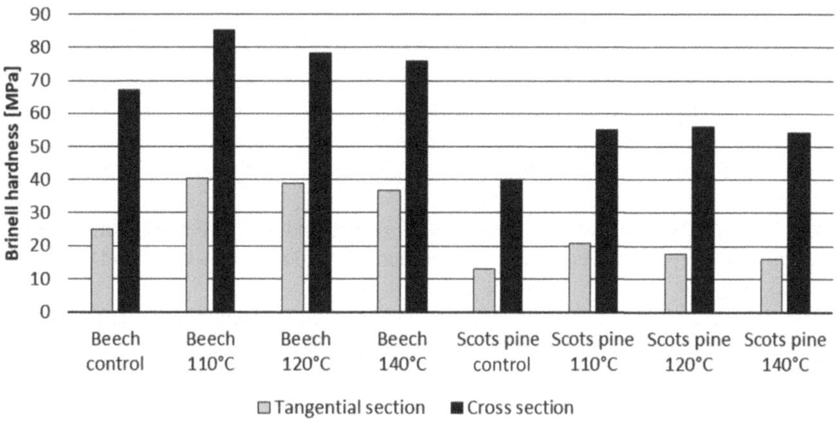

Fig. 6. Effect of LA impregnation and curing temperature on the Brinell hardness of beech and Scots pine sapwood.

4 Conclusions

The goal of this study was to increase dimensional stability by in-situ polymerization of lactic acid in the wood tissue. Results showed significant reduction in the equilibrium moisture content, depending on the curing temperature. Lower moisture uptake resulted in remarkably improved anti-swelling efficiency (60–80%). The treatment improved the surface hardness of the wood as well. However, this improvement was inversely proportional to the curing temperature.

Acknowledgement. This article was made in frame of the project TKP2021-NKTA-43 which has been implemented with the support provided by the Ministry of Innovation and Technology of Hungary (successor: Ministry of Culture and Innovation of Hungary) from the National Research, Development and Innovation Fund, financed under the TKP2021-NKTA funding scheme.

References

1. Horváth, T., Szabó, T.J., Marossy, K.: Synthesis of polylactic acid (PLA) by polycondensation method. Int. J. Eng. Manag. Sci. **5**(2), 286–293 (2020)
2. Ashori, A.: Wood–plastic composites as promising green-composites for automotive industries. Bioresour. Technol. **99**(466), 1–4667 (2008)
3. Fink, H.P., Ganster, J.: Novel thermoplastic composites from commodity polymers and man-made cellulose fibers. Macromol. Symp. **244**, 107–118 (2006)
4. Wong, S., Shanks, R.A., Hodzic, A.: Mechanical behavior and fracture toughness of poly(L-lactic acid)-natural fiber composites modified with hyperbranched polymers. Macromol. Mater. Eng. **289**, 447–456 (2004)
5. Noël, M., Fredon, E., Mougel, E., Masson, D., Masson, E., Delmotte, L.: LA-wood-based composite material. Part 1. Bioresour. Technol. **100**, 4711–4716 (2009a)
6. Noël, M., Grigsby, W., Volkmer, T.: Evaluating the extent of bio-polyester polymerization in solid wood. J. Wood Chem. Technol. **35**(5), 325–336 (2015a)

7. Grosse, C., Noël, M., Thévenon, M.-F., Rautkari, L., Gérardin, P.: Influence of water and humidity on wood modification with lactic acid. J. Renew. Mater. **6**(3), 259–269
8. Grosse, C., Grigsby, W.J., Noël, M., Treu, A., Thévenon, M.F., Gérardin, P.: Optimizing chemical wood modification with oligomeric lactic acid by screening of processing conditions. J. Wood Chem. Technol. **39**(6), 385–398 (2019)
9. Noël, M., Grigsby, W., Vitkeviciute, I., Volkmer, T.: Modifying wood with bio-polyesters. Int. Wood Prod. J. **6**(1), 14–20 (2015b)
10. Bak, M., Németh, R.: Changes in swelling properties and moisture uptake rate of oil-heat-treated poplar (*Populus* × *euramericana* cv. Pannónia) wood. Bioresources. **7**(4), 5128–5137 (2012)
11. Noël, M., Fredon, E., Mougel, E., Masson, D., Masson, E., Delmotte, L.: LA-wood-based composite material. Part 2. Bioresour. Technol. **100**, 4717–4722 (2009b)
12. Hill, C.A.S., Mallon, S.: The chemical modification of Scots pine with succinic anhydride or octenyl succinic anhydride. I. Dimensional stabilisation. Holzforschung. **52**, 427–433 (1998)
13. Hill, C.A.S., Jones, D.: Dimensional changes in Corsican pine sapwood due to chemical modification with linear chain anhydrides. Holzforschung. **53**, 267–271 (1999)
14. Papadopoulos, A.N., Hill, C.A.S., Gkaraveli, A.: Analysis of the swelling behaviour of chemically modified softwood: a novel approach. Holz. Roh. Werkst. **62**, 107–112 (2004)
15. Ghorbani, M., Poorzahed, N., Amininasabh, S.M.: Morphological, physical, and mechanical properties of silanized wood-polymer composite. J. Thermoplast. Compos. Mater. **54**(11), 1–10 (2019)
16. Islam, M.S., Hamdan, S., Hasan, M., Ahmed, A.S., Rahman, R.: Effect of coupling reactions on the mechanical and biological properties of tropical wood polymer composites (WPC). Int. Biodeter. Biodegr. **72**, 108–113 (2012)
17. Niemz, P.: Physik des Holzes und der Holzwerkstoffe, 1st edn. DRW-Verlag, Leinfelden-Echterdingen (1993)

Densification and Mineralisation

Effect of Thermal Post-Treatment on the Properties of Densified *Ceiba pentranda* Wood

Larissa do Vale and Cláudio Del Menezzi(✉)

Department of Forest Engineering, Faculty of Technology, University of Brasília, Brasília 70910-900, DF, Brazil
cmenezzi@unb.br

Abstract. Kapok wood (*Ceiba pentandra*) is a tropical hardwood species characterized by low density, ease of machinability, and low natural durability. It is primarily utilized in aeromodelling due to its exceptionally high strength-to-weight ratio. However, since this wood species is readily available in the market, it is important to promote its use for alternative applications. In this context, the aim of this study was to evaluate the effect of a thermal post-treatment on the properties of previously densified kapok wood. The densification treatment (DT) was performed using a hydraulic press at controlled temperature and pressure. A total of forty boards ($\rho = 0.34$ g/cm^3) measuring 400 mm x 200 mm x 35 mm (*l* x *w* x *t*) were densified at 180 °C according to four schedules: two duration times (5 and 10 min) and two pressures (1.07 MPa and 2.15 MPa). Subsequently, half of the boards underwent a thermal post-treatment (PT) for 25 min at 180 °C under a pressure sufficient only to ensure contact with the heated plates of the hydraulic press. It was found that all mechanical properties were significantly improved compared to the undensified material. The thermal post-treatment was effective in yielding further enhancements in mechanical properties, with the exception of the modulus of elasticity, which was negatively affected. It can be concluded that the densification processes significantly improved all mechanical properties; however. Dimensional stability did not show similar improvement. The proposed thermal post-treatment resulted in additional enhancement of the mechanical properties, while it was ineffective in reducing the dimensional stability of the densified boards.

Keywords: Wood Modification · Thermomechanical treatment · Mechanical Compression

1 Introduction

Kapok wood (*Ceiba pentandra*) is a tropical hardwood species known for its low density, easy machinability, and low natural durability. It is primarily utilized in aeromodelling due to its very high strength-to-weight ratio. Despite its availability in the market, it is crucial to promote its use for alternative applications. In this context, the modification of wood through thermal, chemical, or mechanical processes can play an important role in achieving this goal.

G. Goli et al. (Eds.): ECWM 2024, SPM 86, pp. 243–250, 2025.
https://doi.org/10.1007/978-3-031-99418-0_30

In recent years, research efforts have focused on thermal modification combined with mechanical compression, a process known as thermomechanical modification [1–3]. In these processes the wood is thermally treated while being compressed, leading to improved density. The main advantage is the minimization of the typical drawback of thermal modification: the reduction in mechanical properties. Densification generally maintains mechanical properties without adverse effects [4].

Conversely, wood treated in this manner can house a high level of compression stresses, which may be released upon contact with water, significantly increasing the thickness swelling. The phenomenon known as 'memory shape' is commonly observed in wood-based materials subjected to mechanical compression during manufacturing. Post-manufacturing thermal treatment has been researched to improve the dimensional stability of oriented strandboards [5, 6]. Despite of these findings, numerous studies have explored the utilization of the densification process to enhance properties of tropical hardwoods [1–3]. In this context, the aim of this study was to evaluate the effect of a thermal post-treatment on the properties of previously densified kapok wood.

2 Materials and Methods

2.1 Wood Material

Kapok wood boards were sourced from a local market at Brasília, Brazil. This material was compared with samples from Dr. Harry van der Slooten Xylotheque (FPBw) of the Forest Products Laboratory, Brazilian Forest Service. Forty boards ($\rho = 0.34$ g/cm^3) measuring 400 mm x 200 mm x 35 mm (*l* x *w* x *t*) were obtained and conditioned (20 °C; 65%) until a constant mass was achieved. Additionally, three samples measuring 150 mm x 50 mm x 35 mm (*l* x *w* x *t*) were taken to assess the perpendicular compression strength ($f_{c,90}$) according to [7].

2.2 Densification Process

The densification process was conducted using a simple-opening hot press (INDUMEC 1000kN) equipped with a temperature and pressure control system. The treatment was applied at 180 °C under two pressure levels: equal to 30% and to 60% of $f_{c,90}$. Initially the hot press was pre-heated to 180 °C and the boards were compressed at specific pressure of 1.05 MPa (30%) and 2.15 MPa (60%) for either 5 or 10 min, totaling four densification treatments (T1-T4) (Table 1). After this initial compression, the pressure was relieved and the boards were kept in contact with the press plate for additional 10 min.

Afterwards, the boards were kept under conditioning until reach constant mass then the compression ratio (CR) and densification ratio (DR) were calculated according to Eqs. (1) and (2), respectively.

$$CR(\%) = \left(\frac{Ti - Tf}{Ti}\right)x\ 100 \quad (1)$$

$$DR(\%) = \left(\frac{Df - Di}{Di}\right)x\ 100 \quad (2)$$

Table 1. Experimental design.

Treatment	Time under pressure (min.)	Pressure (MPa)	Number of Boards
Control	–	–	4
T1	5	1.07	8
T2	10	1.07	8
T3	5	2.15	8
T4	10	2.15	8

where: Ti: initial thickness of the boards, mm; Tf: final thickness of the boards, mm; Di: initial density, g/cm^3; Df: final density, g/cm^3.

2.3 Post-Treatment

Within each treatment (T1–T4), the densified boards were sorted in ascending order of density and subdivided into two groups of four boards each with similar mean density values determined using one-way ANOVA. One group was subjected to subsequent thermal post-treatment at 180 °C for 25 min using the same single opening hot-press. The pressure applied was just enough to ensure contact between the board and the press plate.

2.4 Material Characterization

The ASTM D143 [7] standard was employed to determine the following mechanical properties: perpendicular compression strength ($f_{c,90}$) and parallel compression strength ($f_{c,0}$), Janka hardness ($f_{H,0}$), static banding modulus of rupture (f_M) and static bending modulus of elasticity (E_M). The tests were conducted using a universal testing machine (EMIC, DL 30000). Non-destructive evaluation was performed using stress wave method (Metriguard, Stress Wave Timer 239) and dynamic modulus of elasticity (E_d) was calculated (Eq. 3). For each test at least 8 repetitions were done within each treatment.

$$Ed = v^2 \, x \, \rho \tag{3}$$

where: v: stress wave velocity, m/s; ρ: board density, g/cm^3.

Dimensional stability properties were assessed following the main guidelines of ASTM D1037 (2014) [8]. Wood samples were immersed in water for 2 and 24 h with mass and dimensions recorded before and after this immersion and then water absorption (WA) and thickness swelling (TS) were calculated. Equilibrium moisture content (EMC) was calculated after the samples reached constant mass at conditioning room (20 °C/65%).

2.5 Statistical Analysis

The results of the physical and mechanical properties were analyzed through two procedures. Firstly, a Dunnett test at a 5% significance level was run to compare the property values of densified against undensified boards (control) specifically evaluating the densification process effect. Subsequently, one-way ANOVA at a 5% significance level was run to compare the properties of densified boards before and after the proposed thermal post-treatment.

3 Results and Discussion

3.1 Densification Process

Table 2 presents the densities of the boards before and after the densification. It can be observed that for all treatments, the initial density was consistent (0.34 g/cm^3), indicating that the segregation procedure was effective. The thickness of the densified boards decreased by at least 20% (CR), and this value nearly doubled when higher pressure was applied.

Table 2. Values of the board density before (ρB), after densification (ρA), compression ratio (CR) and densification ratio (DR).

Treatment	ρB (g/cm^3)	ρA (g/cm^3)	CR (%)	DR (%)
T1 (5′/1.07 MPa)	0.34	0.41^a	19.9	19.9
T2 (10′/1.07 MPa)	0.34	0.41^a	20.8	20.4
T3 (5′/2.15 MPa)	0.34	0.56^a	42.5	64.8
T4 (10′/2.15 MPa)	0.34	0.57^a	44.6	71.7

^{a}ρB x ρA value significantly different at ANOVA 5% level

There was also a significant improvement in density (DR) for all treatment tested, with the wood transitioning from low density (<0.50 g/cm^3) to the medium density range (0.50 to 072 g/cm^3). This indicates that the densification process effectively reduced board thickness, resulting in an effective improvement of the density (Fig. 1).

Research about densification of tropical wood is limited. According to a previous study [9], four tropical wood species were thermomechanically treated for flooring applications. It has been observed that despite a high level of mechanical compression, two species could not be property densified, while the other two experienced density improvements of only 4.8% and 9.9%, which are significantly lower than values observed here. High density wood species have low porosity and voids, which means that the densification process cannot be performed properly.

3.2 Material Properties

Table 3 presents mechanical properties of the undensified boards (control), after densification and thermal post-treatment. The Dunnett test revelated that all mechanical properties were improved with densification, especially at higher pressure. The maximum improvement obtained were as follow: $f_{c,90}$ 64.5%; $f_{c,0}$ 73.3%, f_H 115.6%; f_M

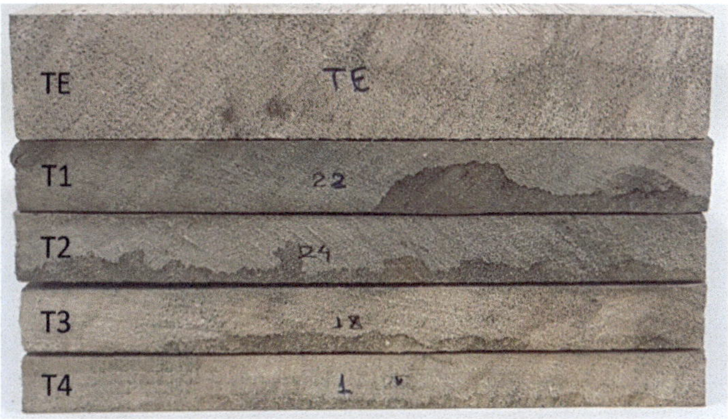

Fig. 1. Thickness profile of the boards before (TE, control) and after the densification (T1: 5′/1.07 MPa; T2: 10′/1.07 MPa; T3: 5′/2.15 MPa; T4: 10′/2.15 MPa).

76.4% and E_M 48.1%. This improvement indicates that kapok wood may have increased potential for additional applications, such as flooring, which requires specific levels of hardness and perpendicular compression. Additionally, the densified material became darker, which is valuable characteristic for this application.

Table 3. Values of the mechanical properties of undensified board, after densification (AD) and after post-treatment (APT).

Property	Situation	T1	T2	T3	T4	Control
$f_{c,90}$ (MPa)	AD[a]	6.03^{NS}a	$6.12\ ^{NS}$a	$7.12\ ^{NS}$a	8.21*a	4.99
		(1.01)	(0.98)	(2.17)	(3.18)	(0.69)
	APT[b]	7.94b	6.78a	9.28b	11.93b	
		(1.05)	(0.62)	(1.11)	(2.74)	
$f_{c,0}$ (MPa)	AD	28.62*a	29.19*a	36.81*a	36.19*a	21.25
		(2.24)	(2.00)	(4.05)	(4.95)	(2.28)
	APT	30.20b	32.25b	43.25*b	44.77*b	
		(2.85)	(3.36)	(3.76)	(3.84)	

(*continued*)

The dynamic modulus of elasticity was only affected when higher pressure was applied. In this case, the compression of wood cellular spaces reduced the voids, creating a more homogeneous pathway for stress wave propagation, which resulted in a higher E_d. Previous research by [10] indicated that both stress wave velocity and dynamic modulus of elasticity improved as a result of thermal treatment on the tropical wood *Simarouba amara*.

Table 3. (*continued*)

Property	Situation	T1	T2	T3	T4	Control
$f_{H,0}$ (MPa)	AD	18.73*a	19.81*a	25.79*a	29.88*a	13.86
		(2.01)	(1.92)	(3.43)	(2.28)	(2.93)
	APT	22.91*b	22.76*b	33.61*b	33.57*b	
		(2.50)	(2.32)	(1.53)	(3.13)	
f_m (MPa)	AD	53.13*a	49.71*a	67.41*a	65.76*a	38.12
		(6.03)	(5.58)	(5.87)	(7.18)	(3.95)
	APT	48.17b	56.41b	68.57a	75.44b	
		(4.48)	(8.33)	(5.08)	(7.71)	
E_M (MPa)	AD	6278.7*a	5803.1*a	7383.9*a	6494.8*a	4987.6
		(604.43)	(403.83)	(917.8)	(689.96)	(567.22)
	APT	4674.5b	5265.1a	5622.1*b	5923.7b	
		(298.17)	(1020.61)	(768.81)	(436.27)	
E_d (MPa)	AD	5880.6*a	5252.4 ^{NS}a	8377.3*a	8862.4*a	5154.5
		(796.37)	(493.22)	(648.03)	(645.30)	(507.99)
	APT	6543.7b	6732.2b	9468.8b	10429.2b	
		(468.24)	(382.09)	(563.61)	(197.81)	

*Control x treated means statistically different at 5% level according to Dunnet test; different letters compare AD x APT means for each property/treatment according to ANOVA; T1: 5′/1.07 MPa; T2: 10′/1.07 MPa; T3: 5′/2.15 MPa; T4: 10′/2.15 MPa; Standard variation values between parenthesis

The thermal post-treatment applied to the densified boards resulted in further improvements in mechanical properties, with the exception of the modulus of elasticity. The perpendicular compression strength experienced the most significant enhancement (up to 45.3%), followed by Janka hardness (30.2%), dynamic modulus of elasticity (25.2%) and parallel compression strength (23.7%). The modulus of rupture was only positively affected with longer treatment durations (14.7%). On the other hand, the post-treatment had a negative effect on the modulus of elasticity, with values decreasing by up to 25.6%.

Table 4 shows the results of the dimensional stability properties. The densification properties had a negative effect on thickness swelling; the values for the densified material were several times higher than those of untreated wood. After immersion in water for 2 h, the densified wood exhibited similar values to the untreated wood. However, after 24 h of immersion, the densified boards absorbed up to 52.3% more water than the undensified boards. The only property positively affected was the equilibrium moisture content, which was 33.7% lower in the densified samples. Thickness swelling decreased when higher densification pressures were applied (T3 and T4) and when the boards were immersed for 2 h (TS2h). Following post-treatment, the equilibrium moisture content further decreased, remaining below 10%.

Table 4. Values of the dimensional stability properties after densification (AD) and after post-treatment (APT).

Property	Situation	T1*	T2	T3	T4	Control
TS2h (%)	AD[a]	4.77^{NS}a	2.96^{NS}a	12.25*a	13.0*a	1.10
		(1.56)	(0.73)	(3.55)	(3.88)	(0.18)
	APT[b]	3.67b	4.10b	7.74b	5.76b	
		(1.61)	(1.00)	(3.06)	(1.57)	
WA2h (%)	AD	21.79^{NS}a	18.90^{NS}a	23.43^{NS}a	26.86^{NS}a	22.21
		(1.07)	(1.63)	(5.70)	(7.88)	(3.06)
	APT	19.04a	28.57b	21.58a	20.85a	
		(8.52)	(11.23)	(7.15)	(4.32)	
TS24h (%)	AD	17.31*a	17.94*a	45.37*a	39.40*a	2.27
		(7.34)	(8.49)	(11.14)	(4.58)	(0.38)
	APT	19.06a	18.68a	48.98a	42.73a	
		(7.91)	(7.98	(3.85)	(11.93)	
WA24h (%)	AD	79.92^{NS}a	88.17^{NS}a	114.45*a	113.48*a	75.42
		(12.41)	(24.77)	(27.04)	(13.15)	(9.85)
	APT	81.76a	94.76a	132.49a	108.65a	
		(24.05)	(27.11)	(42.25)	(23.30)	
EMC (%)	AD	11.54*a	11.36*a	10.41*a	10.20*a	15.37
		(0.17)	(0.49)	(0.56)	(0.35)	(0.46)
	APT	8.93b	8.90b	8.77b	8.14b	
		(0.67)	(0.67)	(0.75)	(0.38)	

*Control x treated means statistically different at 5% level according to Dunnet test; different letters compare AD x APT means for each property/treatment according to ANOVA; T1: 5′/1.07 MPa; T2: 10′/1.07 MPa; T3: 5′/2.15 MPa; T4: 10′/2.15 MPa; Standard variation values between parenthesis

4 Conclusion

The densification process significantly improved all mechanical properties of kapok wood. However, dimensional stability properties were not positively influenced, and the densified boards exhibited greater dimensional variation. The boards subjected to the proposed post-thermal treatment showed further enhancements in mechanical properties, except for the static bending modulus of elasticity. Furthermore, the equilibrium moisture content was significantly reduced following the post-treatment. While the post-treatment did not enhance dimensional stability properties overall of the densified boards, it was effective in reducing thickness swelling when higher temperatures were used during the densification process. Future work should explore the use of post-treatment under wet conditions to further improve the dimensional stability of the densified material.

References

1. Costa, M.A., Del Menezzi, C.H.S.: Effect of thermo-mechanical treatment on properties of parica plywoods (*Schizolobium amazonicum* Huber ex Ducke). Revista Árvore. **41**(1), 1–8 (2015)
2. Freitas, A.C., Gonçalez, J.C., Del Menezzi, C.H.S.: Tratamento termomecânico e seus efeitos nas propriedades de *Simarouba amara* (Aubl.). Floresta e Ambiente. **23**(4), 565–572 (2016)
3. Arruda, L.M., Del Menezzi, C.H.S.: Properties of a laminated wood composite produced with thermomechanically treated veneers. Adv. Mater. Sci. Eng. **2016**, 1–9 (2016)
4. Santos, P.C., Del Menezzi, C.H.: Effect of the densification process on properties of commercial oriented strand boards. Eur. J. Wood Wood Prod. **76**(6), 1707–1713 (2018)
5. Del Menezzi, C.H.S., Tomaselli, I.: Contact thermal post-treatment of oriented strandboard: a preliminary study. Eur. J. Wood Wood Prod. **64**(3), 212–217 (2006)
6. Del Menezzi, C.H.S., Tomaselli, I., Okino, E.Y.A., Teixeira, D.E., Santana, M.A.E.: Thermal modification of consolidated oriented strandboards: effects on dimensional stability, mechanical properties, chemical composition and surface color. Holz. Roh. Werkst. **67**(3), 383–396 (2009)
7. ASTM D143–22: Standard Test Methods for Small Clear Specimens of Timber (2022)
8. ASTM D1037-12: Standard Test Methods for Evaluating Properties of Wood-Base Fiber and Particle Panel Materials (2020)
9. Arruda, L.M., Del Menezzi, C.H.S., Andrade, A.: Utilization of a thermomechanical process to enhance properties of hardwood used for flooring. Ciência da Madeira. **6**(3), 223–231 (2015)
10. Del Menezzi, C.H.S., Amorim, M.R.S., Costa, M.A., Garcez, L.R.O.: Evaluation of thermally modified wood by means of stress wave and ultrasound nondestructive methods. Mater. Sci. **20**(1), 61–66 (2014)

X-Ray CT Scanning as a Method for Quantifying Mineralization in Spruce and Beech

Marcy Durimel[1,2(✉)], Liselotte De Ligne[1,2], Bogdan Parakhonskiy[1,3], Jan Van den Bulcke[1,2], Andre Skirtach[3], and Joris Van Acker[1,2]

[1] UGent-Woodlab, Department of Environment, Faculty of Bioscience Engineering, Ghent University, 9000 Ghent, Belgium
marcy.durimel@ugent.be
[2] Ghent University, Proeftuinstraat 86, 9000 Ghent, Belgium
[3] Nano-Biotechnology Laboratory, Department of Biotechnology, Faculty of Bioscience Engineering, Ghent University, 9000 Ghent, Belgium

Abstract. Wood and wood-based building materials are considered key elements of our future socio-economic environment. Wood is mainly used as a structural material, in building structure envelopes, and as insulation. Its biodegradability is an asset at the end of its service life, but also an important parameter to control to maintain a good product lifespan.

Wood mineralization has attracted a great deal of interest in recent years in the field of wood modification. Compared to traditional wood modification processes, mineralization is a simple, environmentally friendly process that improves wood properties. Minerals are incorporated into the biological matrix of wood which increases the fire resistance. X-ray CT scanning is a non-destructive technique and can be used to visualize and quantify mineralization in 3D, to quantify the difference in mineralization in early- and latewood in spruce and beech wood blocks, the average penetration depth of mineralization, and the mineralization in different regions of the woodblocks. Spruce samples have the highest weight gain which is represented by a higher normalized mean grey value in X-ray CT scanning. In the longitudinal direction, all specimens show a mineralization gradient with a higher normalized mean grey value a few millimeters from the surface and a minimal value in the middle of the block. For spruce specimens, the mineralization occurred to a greater extent in the earlywood region than in the latewood region.

Keywords: Earlywood · Latewood · beech · spruce · wood mineralization · X-ray CT scanning

1 Introduction

Wood and wood-based building materials are considered key elements of our future socio-economic environment. They are relevant nature-based solutions to climate change, as the continental usage of wood in construction could lead to the removal of about 21 million tons of carbon dioxide from the atmosphere [1]. Wood and wood materials are useful and readily available natural raw materials for the construction industry.

G. Goli et al. (Eds.): ECWM 2024, SPM 86, pp. 251–258, 2025.
https://doi.org/10.1007/978-3-031-99418-0_31

The susceptibility of wood to fire is one of the crucial challenges in the wood industry. Fire behavior of wood is a complex phenomenon due to the different constituents and their reaction to fire, and depends on many different parameters such as the composition of the wood, the density, the moisture content… In addition, certain wood species have a low resistance to bio-deterioration (fungi, insects, termites, …). This biodegradability is an asset at the end of its service life, but also an important parameter to control to maintain a good product lifespan [2]. Wood modification processes have been used to enhance and modify wood properties to use wood for different applications. Wood protection is undergoing major developments caused partly by environmental concerns regarding wood treated by certain preservatives, such as borates preservatives, copper systems, ... The main categories of wood modifications are chemical treatment, thermo-hydro treatment, thermo-hydro-mechanical treatment, treatment based on biological processes, and treatment with the use of electromagnetic irradiation or plasma [3]. The modification can be active meaning that the modification changes the chemical nature of the material or passive meaning that the modification results in a change in properties without altering the material's chemistry [4].

In situ mineralization is a simple, environmentally friendly process that improves wood properties. By incorporating minerals into the biological matrix of wood, increased fire resistance can be obtained [5, 6]. Weight gain of the material is often used to characterize the effectiveness of the treatment [7, 8]. The main techniques used to show the distribution of mineralization inside wood are visualization techniques such as scanning electron microscopy (SEM) and Raman spectroscopy. However, these techniques only visualize the distribution in 2D on slices of the treated material and do not allow to quantify it. In this paper, we propose the use of X-ray CT scanning, a non-destructive technique to visualize and quantify mineralization in 3D.

2 Materials and Methods

2.1 X-Ray Computed Tomography

X-ray computed tomography (X-ray CT) was used to visualize the internal structure of spruce and beech wood blocks before and after mineralization and to determine the mineralization gradient inside the wood blocks.

Specimens of spruce and beech (blocks of size 10 x 10 x 10 mm^3) were chosen for this study. Spruce and beech were chosen for their contrasting characteristics (density, pore size distribution, permeability). Three replicates of each species were studied. The specimens were oven-dried for 24 h at 103 °C and scanned before and after mineralization, with a Microfocus Hamamatsu X-ray tube (Nanowood scanner, UGCT), operated for 20 min at 70 kV voltage, 100 μA target current, and 400 ms exposure time. Octopus Reconstruction version 8.9.4.13 was used for reconstruction, with the cone-beam mode. The resulting scans were analyzed using ImageJ-win64 and Python.

The mineralization gradient inside the wood block was based on a grey value analysis of the scans, with higher grey values indicating a higher level of mineralization.

The normalized grey value (NGV) can be calculated.

$$NGV = \frac{G_m - G_d}{G_d} \tag{1}$$

with G_m the grey value after mineralization and G_d the grey value of the dry wood.

In addition, the weight gain percentage, the theoretical maximal uptake of the blocks, the penetration and pore area percentage were calculated. To do so, the X-ray CT images were transformed to binary images using thresholding, with black pixels corresponding to the pore area. Three samples of each species were scanned and compared.

$$Weight\ gain\ (\%) = 100 \times \frac{w_{dm} - w_{d0}}{w_{d0}} \tag{2}$$

with w_{d0} the mass of the oven-dry specimen before mineralization and w_{dm} the weight of the oven-dry specimen after mineralization.

The theoretical maximum uptake (TMU) corresponds to the uptake if all the pores are filled with minerals.

$$\mathrm{TMU} = \mathrm{V} \times \varnothing \times \mathrm{d} \tag{3}$$

$$\varnothing = 1 - \left(\frac{D_{dw}}{D_{cw}}\right) \tag{4}$$

With V the volume of the sample, Ø the porosity, D_{dw}= the density of dry wood and D_{cw}= the average density of the cell wall (1.5 g/cm^3), and d the density of the impregnation solution.

The penetration volume is the percentage of pore volume filled by minerals

$$Penetration(\%) = 100 \times \frac{real\ uptake}{T.M.U} \tag{5}$$

The pore area percentage is the part of the total surface of the samples that is pores.

$$Pore\ area(\%) = 100 \times \frac{cells\ area}{Total\ area} \tag{6}$$

2.2 Mineralization

In situ chemical precipitation of $CaCO_3$ by impregnation of electrolytes solutions ($CaCl_2$ and Na_2CO_3).

Anhydrous calcium chloride (CAS: 10043-52-4) and anhydrous sodium carbonate (CAS: 497-19-8) of analytical grade and de-ionized water were used.

The wood blocks were oven-dried at 103 °C for 24 h and vacuum-impregnated first with calcium carbonate solution for 30 min at room temperature and under 1 bar and 48 h at atmospheric pressure. The specimens were dried and impregnated with the second solution, sodium carbonate under the same conditions. Then the blocks were oven-dried at 103 °C for 24 h and the weight gain was calculated.

3 Results and Discussion

3.1 Visualizing Mineralization

On reconstructed X-ray CT images, high-density zones appear whiter (higher grey values) whereas low-density zones appear darker (lower grey values). Latewood regions are denser than earlywood regions because the cells have a smaller diameter and a thicker cell wall. The density of cellulose (1.53) is higher than the density of air (1) making the cell wall composed of cellulose a higher density region than the lumen composed of air. Consequently, the earlywood region has a lower grey value because the region is less dense than the latewood region. So latewood regions are clearer than earlywood regions on the scans (Fig. 1). The calcium carbonate minerals appear as white nodules, located inside the cells (Fig. 1). Because the only difference between the pre-treated block and post-treated one is the mineralization, the normalized grey value (formula 1) can be defined.

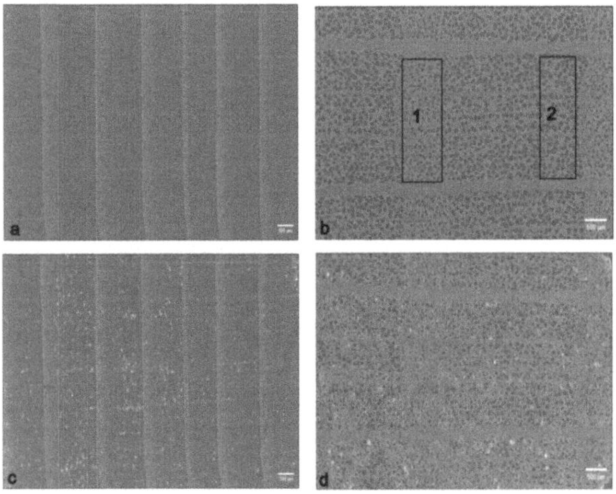

Fig. 1. X-ray CT scans spruce (a, c) and beech (b, d) before treatment (a, b) and after mineralization (c, d), 1 latewood region, 2 earlywood region.

3.2 Penetration

The theoretical maximal uptake (TMU) based on weight was higher than the TMU calculated based on the X-ray CT scans, mainly because the theoretical TMU is based on the density of the dry wood and the average cell wall density which overestimates the TMU values. The maximum penetration of beech samples (8%) was similar to spruce specimens (6–10%), while the weight percentage gain was two times lower (weight gain of 11% for beech, 18–22% for spruce) (Fig. 2). The difference of penetration for the spruce specimens is probably due to natural variation between the samples, different amounts of growth ring leading to different proportions of earlywood and latewood regions.

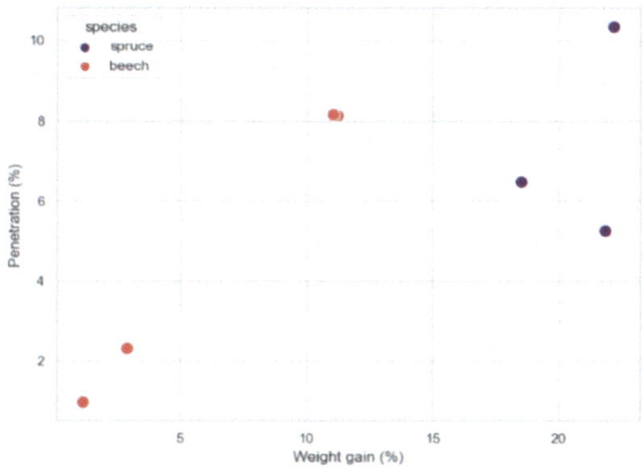

Fig. 2. Penetration compared to weight gain for spruce and beech specimens.

3.3 Mineralization Gradient

The normalized grey value represents the difference between the grey value post-treatment and the grey value before treatment and is thus an indicator of the degree of mineralization inside the wood blocks. The spruce samples are the samples with the highest weight gain percentage as well as the highest normalized grey value for the full block. Beech samples have a lower weight gain, translating into a lower normalized grey value.

The mineralization profile inside the wood blocks can be described by the normalized grey value profile. The normalized grey value profile in the longitudinal direction shows that the mineralization happened mostly on the sides of the block, the amount of mineralization in the middle of the block is lower. The longitudinal direction corresponds to the direction of mineralization. The sides contain more minerals than the middle. The mineralization profile inside the wood block presents two peaks on the two ends of the block, the first one is around 0.5 mm from the top and the second one is at 0.5 mm from the end of the block. The minimal value is reached at 5 mm from the surface of the block which corresponds to the middle. The normalized grey value decreased right after the peaks and reached a stable mean value of about 4% for beech 1 and 2% for beech 2. Beech 1 had a higher normalized grey value in the middle of the block which is related to the higher weight gain of beech 1 (2.9%), compared to beech 2 (1.2%). Moreover, the maximum normalized grey value is not on the surface of the wood block but a few millimeters from it, as indicated by the black line in Fig. 3.

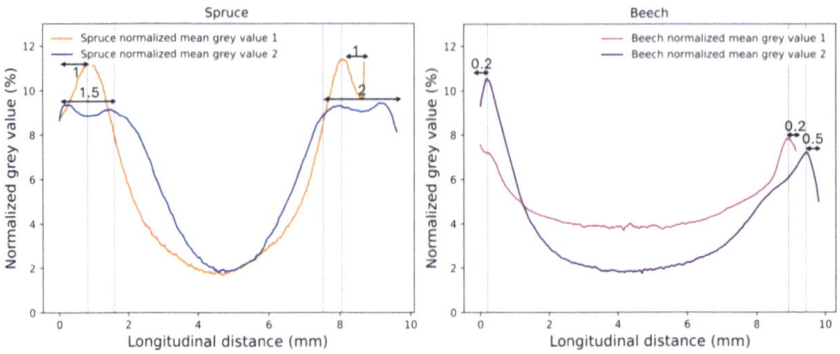

Fig. 3. Normalized mean grey value profile in beech and spruce wood blocks, 1 and 2 two replicates.

For spruce, similar observations were made. The mineralization occurred mostly in the first 2 mm, with the normalized mean grey value ranging from 8% to 12%. Similar to beech the normalized mean grey value reached a minimal value of 2% at 5 mm, the middle of the block. The main difference between the spruce and beech specimens is that the decrease in grey value (and thus mineralization) happened more gradually in beech than in spruce.

In the radial direction, the grey value profile gives information on the mineralization inside the early- and latewood region.

The grey values before treatment in the radial direction were higher in the latewood region compared to the earlywood region, as expected due to the higher density of latewood as compared to earlywood (Fig. 4). The grey value decreased in the transitional region between latewood and earlywood, where it reached its minimal value.

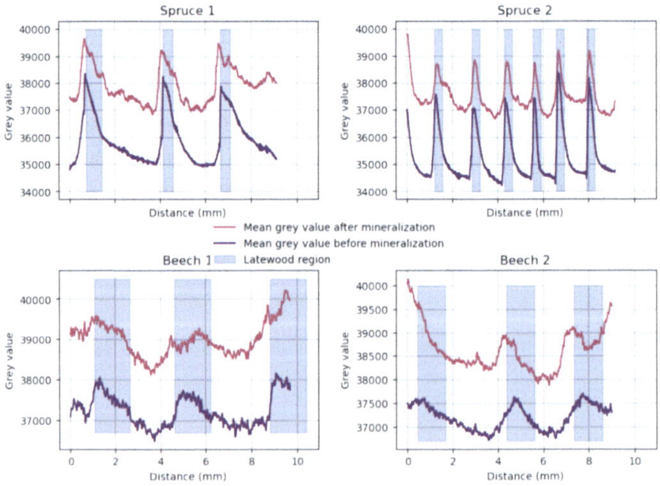

Fig. 4. Mean grey value before (pre) and after (post) mineralization for spruce and beech specimens, 1 and 2 two replicates.

The mean grey values of the mineralized specimen show a similar pattern, but were higher than the mean grey values of the wood before mineralization, as the density of the block after mineralization is higher than before mineralization. For spruce samples, in the latewood region, the difference in mean grey value is smaller than in the early wood region, indicating that the amount of mineralization was higher in spruce earlywood than in latewood. For the beech samples, there was no clear difference between the earlywood and latewood regions.

The mean normalized grey value presents the difference between the mean grey value post-treatment and the mean grey value before treatment. The normalized grey value shows a peak just before the latewood region with a sudden decrease in the latewood region followed by a progressive increase in the transitional region and earlywood. However, for beech samples, the difference between the maximum and minimum is less pronounced than for spruce samples, and in the transitional region the normalized mean grey value was stable. For all specimens, the normalized mean grey value pattern was the same in all the rings (Fig. 5).

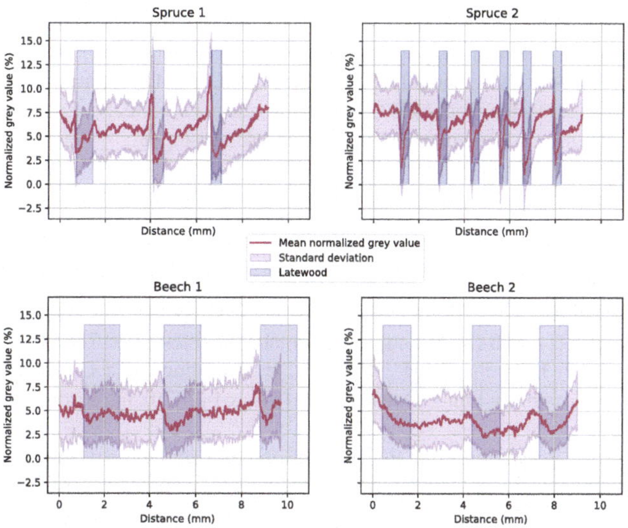

Fig. 5. Normalized grey value with standard deviation for spruce and beech specimens, 1 and 2, two replicates.

The normalized mean grey value for spruce samples was between 9 and 11% in the earlywood and between 1.5% and 2.5% in the latewood regions. The normalized mean grey value peak occurred at the border between the early- and latewood regions. In the latewood region, the normalized mean grey value decreased to reach a low value of 2%.

For beech, the normalized mean grey value was between 2 and 6%.

The normalized grey value was lower for beech than for spruce, which can be explained by the lower weight gain (and thus mineralization) of the beech samples. For spruce, there was a clear difference in mineralization between the earlywood and latewood, with the mineralization being higher in the earlywood region where the pores are larger, the difference in pore size between the early- and latewood regions was smaller in beech, and the difference in mineralization was likewise smaller.

4 Conclusion

The study showed that spruce samples have the highest weight gain and the highest normalized mean grey value for the full blocks. In the longitudinal direction, all specimens show a mineralization gradient with a higher normalized mean grey value a few millimeters from the surface and a minimal value in the middle of the block. In spruce specimens, the mineralization occurred to a greater extent in the earlywood region than in the latewood region. X-ray CT scanning in combination with other imaging techniques such as SEM could allow a detailed study of the minerals' distribution and characterization. X-ray CT scanning could be used to study the penetration and distribution of different minerals and wood treatments using vacuum impregnation, allowing a better understanding of the distribution of wood treatment inside of wood specimens.

Acknowledgments. Financial support of the UGCT Core Facility by the Special Research Fund (BOF.COR.2022.008). Financial support of the BOF Starting Grant JVdB (BOFSTG2018000701), FWO grant G009720N, FWO grant G019521N and FBW-CWO-UGent grant JVdB.

Research Foundation-Flanders (FWO) with reference G043322N, EnhanceWood: Nanochemical treatment of wood for enhanced fungal and fire resistance.

References

1. Mikhaylov, A., Moiseev, N., Aleshin, K., Burkhardt, T.: Global climate change and greenhouse effect. JESI. **7**, 2897–2913 (2020). https://doi.org/10.9770/jesi.2020.7.4(21)
2. Pomponi, F., Hart, J., Arehart, J.H., D'Amico, B.: Buildings as a Global Carbon Sink? A reality check on feasibility limits. One Earth. **3**, 157–161 (2020). https://doi.org/10.1016/j.oneear.2020.07.018
3. Khademibami, L., Bobadilha, G.S.: Recent developments studies on wood protection research in academia: a review. Front. For. Glob. Change. **5** (2022). https://doi.org/10.3389/ffgc.2022.793177
4. Sandberg, D., Kutnar, A., Mantanis, G.: Wood modification technologies—a review. iForest. **10**, 895–908 (2017). https://doi.org/10.3832/ifor2380-010
5. Huang, L., Yao, X., Huang, Y., Wang, Q.: The preparation of CaCO3/wood composites using a chemical precipitation method and its flame-retardant and mechanically beneficial properties. BioResources. **13**, 6694–6706 (2018). https://doi.org/10.15376/biores.13.3.6694-6706
6. Hernandez, V., Romero, R., Arias, S., Contreras, D.: A novel method for calcium carbonate deposition in wood that increases carbon dioxide concentration and fire resistance. Coatings. **12**, 72 (2022). https://doi.org/10.3390/coatings12010072
7. Merk, V., Chanana, M., Keplinger, T., Gaan, S., Burgert, I.: Hybrid wood materials with improved fire retardance by bio-inspired mineralisation on the nano- and submicron level. Green Chem. **17**, 1423–1428 (2015). https://doi.org/10.1039/C4GC01862A
8. Merk, V., Chanana, M., Gaan, S., Burgert, I.: Mineralization of wood by calcium carbonate insertion for improved flame retardancy. Holzforschung. **70**, 867–876 (2016). https://doi.org/10.1515/hf-2015-0228

Moisture Content Distribution of Densified Wood and the Impact of Various Heat Post-treatments on Brinell Hardness and Set Recovery

Elena Jäger(✉), Guillaume André, and Thomas Volkmer

Architecture, Wood and Civil Engineering, Bern University of Applied Sciences, Solothurnstrasse 102, Biel 2504, Switzerland
{elena.jaeger,guillaume.andre,thomas.volkmer}@bfh.ch

Abstract. This research focused on the moisture content distribution of densified poplar (*Populus Nigra* L.) prior to a post-treatment and the impact of various heat post-treatments (200 °C|3 h, 180 °C|10 h and 180 °C|7 h) on the Brinell Hardness and set recovery on densified wood. Therefore, lamellas with an average density of 383 kg/m^3 and length of 500 mm (longitudinal) x 100 mm (width) x 10 mm (thickness) without specific orientation were densified in the thickness direction with a targeted density of approximately 1040 kg/m^3.

The results of the moisture content distribution of the densified samples prior to the post-treatments showed a heterogeneous moisture content distribution (from 7% to 10.5%) right after the hot pressing. In order to reduce the local swelling during the post-treatment, additional pre-drying is needed. In addition, the post-treatment at 200 °C for 3 h allowed to reach the best compromise, significantly lowering the set recovery on the densified samples below 10% whilst not impacting statistically the Brinell hardness (50 N/mm^2) compared to the not post-treated samples (54.4 N/mm^2).

Keywords: wood densification · heat post-treatment · set recovery · Brinell hardness

1 Introduction

The ongoing climate disruption threatens many wood species historically used by the industry. Thus, local resilient wood could be used as an alternative, due to sustainable modification to improve the mechanical properties and mitigate raw material unavailability risk. One way to modify wood is through mechanical compression - densification. Due to the densification, the mechanical properties of the wood are improved (Niemz & Sonderegger, 2003). Nevertheless, ensuring the dimensional stability of densified wood remains one of the major challenges. Densified wood has the tendency to recover to its original state (so-called set recovery) when being exposed to high humidity and high temperature. Many researchers tried to reduce the set recovery (Kutnar & Kamke, 2012; Sandberg et al., 2013; Laine et al., 2016; Neyses et al., 2020; Neyses et al., 2020; Pelit & Yorulmaz, 2023) with more or less promising results.

G. Goli et al. (Eds.): ECWM 2024, SPM 86, pp. 259–266, 2025.
https://doi.org/10.1007/978-3-031-99418-0_32

According to Norimoto et al., 1993, there are three mechanisms essential in preventing shape deformation: a) formation of cross-linkages between molecules of the matrix in the compressed state, b) relaxation of the stresses stored in the microfibrils and the matrix, c) preventing softening of the wood by changing the hygroscopicity of the cell wall. Relaxation of the stresses stored in the microfibrils and the matrix can be done due to a heat treatment.

This study addressed first the moisture content (MC) distribution on densified poplar (*Populus Nigra* L.) samples prior to post-treatment, in order to reduce the risk of local swelling due to the high temperature. Then, it focused on comparing the effect of different heat post-treatments on Brinell hardness and set recovery of those densified samples.

2 Materials and Methods

2.1 Material, Densification and Post-treatment

Poplar lamellas with an average density of 383 kg/m^3 ($\sigma = \pm 54$ kg/m^3) and dimensions of 500 mm (longitudinal – L) x 100 mm (width) x 10 mm (thickness) without specific orientation were used. The specimens were stored at 20 °C and 65% relative humidity (RH) prior to the densification with a MC of 11–12%. The lamellas were densified in the thickness direction using a batch press (Höfer Presstechnik GmbH, Innkreis, Austria, type: HLOP 210) in a 2-stage process. In the first stage, the specimens were compressed between heated plates at 150 °C, until reaching a thickness of approximately 3.1 mm (compression ratio of 69%). Then, the second stage managed a cooling phase under a constant pressure of 1100 N/mm^2, until both metal plates reached a maximum temperature of 80 °C. Four batches of 9 lamellas were densified and subsequently stored at 20 °C and 30% RH. Afterwards, three sets of densified lamellas were heat post-treated in two phases. First, a pre-drying at 103 °C for 24 h allowed to reduce the MC, in order to minimize the risk of local swelling – densified cells to expand back. Second, three variants of high-temperature post-treatment (PT) were applied: first 200 °C for 3 h, second 180 °C for 10 h, and third 180 °C for 7 h.

Furthermore, changes in the microstructure of the densified wood after the post-treatments were observed with a light microscope (Leica Microsystems Wetzlar GmbH, Germany, type: DM LM/P).

2.2 Moisture Content Distribution

To investigate the MC distribution within poplar lamellas in respect to the densification process, three variants were used: a reference lamella which was not densified, a densified lamella measured immediately after pressing and a densified lamella after stabilization at 20 °C and 65% RH. The criterion for stabilization was 0.1% mass constancy within a minimum of two days difference, based on our own experience with densified wood. For each lamella, 8 sub-samples were cut out from various zones (sampling presented in Fig. 1) and their MC was calculated according to the Eq. (1):

$$MC\ [\%] = (m - m_0)/m_0 \tag{1}$$

where *m* - is the initial sample mass, and *m_0* - is the oven-dry mass.

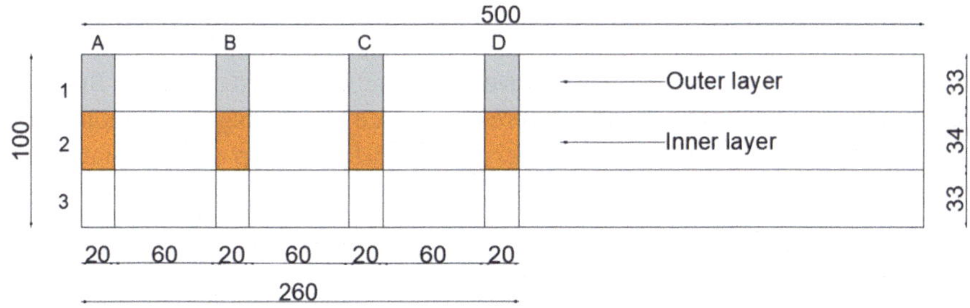

Fig. 1. Way of cutting the lamellas for tracking the moisture distribution in the samples (A1, B1, C1, D1 – outer layer; A2, B2, C2, D2 – inner layer).

2.3 Set Recovery Determination Method

The set recovery (SR) test was performed according to Inoue et al. (1993). Selected samples with dimensions of 10 mm (L) x 90 mm (width) were cut out from densified lamellas and then impregnated with water for 30 min with 8 bars pressure in a lab-scale autoclave (Büchi AG, Uster, Switzerland, type: Büchiglasuster pressure reactor) and subsequently left submerged for 210 min at 20 °C. Afterwards, samples were immersed in boiling deionized water for 30 min at 98 °C in a bath container (Grant Instruments Ltd., Shepreth, England, type: SBB Aqua 26) and finally oven-dried at 103 °C. The set recovery was calculated according to Eq. (2):

$$SR[\%] = 100 \times (T_R - T_c)/(T_o - T_c) \tag{2}$$

where *T_c* - is the oven-dry thickness after compression; *T_0* - is the oven-dry thickness before compression and *T_R* - is the oven-dry thickness after the SR cycle.

2.4 Brinell Hardness

Brinell hardness (BH) test was carried out according to the standard EN 1534, 2011. The measurements were conducted on a Zwick universal mechanical testing device (Zwick 30 kN RetroLine). The intender had a diameter of 10 mm. The maximum force of 1 kN was reached after 17 s and then held for 25 s. Subsequently, the diameter of the indentation was determined with a light microscope (Leica MZ16), by using the average along and across the grain. Finally, the Brinell hardness was calculated according to Eq. (3):

$$BH\left[N/mm^2\right] = \frac{2FF}{\pi\pi\pi\pi\left(\pi\pi - \left(\pi\pi^2 - dd^2\right)\right)} \tag{3}$$

where *F* - is the load corresponding to the maximum force applied, in N; *D* - is the diameter of the ball, in mm; *d* - is the diameter of the residual indentation, in mm, calculated as the average of the diameter along the grain and the one across the grain.

2.5 Statistical analysis

All measured values were represented as average value ± standard deviation. Statistical differences between the average values were determined by carrying out a variance analysis test (ANOVA, IBM SPSS Statistics, Version 29.0.0.0), where $p < 0.05$ is statistically significant.

3 Results and Discussion

3.1 Moisture Content Distribution

The reason for testing the MC distribution was due to the local swelling zones on the densified wood during the post-treatment (Fig. 2). By the local swelling zones, the vessels were less compressed in comparison to vessels of the non-swollen parts. The first assumption was, that the densified samples had different moisture gradients (higher MC in the middle, and lower at the edges), which resulted in swelling zones, especially in the middle of the specimen (inner layer) during the high temperature treatment.

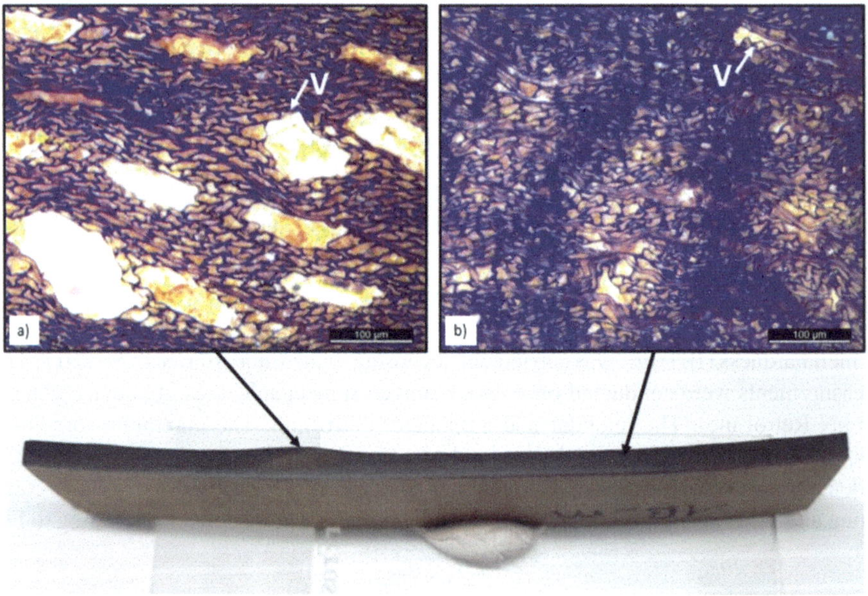

Fig. 2. Microscopy observation on densified poplar after the post-treatment at 180 °C|10 h: a) local swelling effect (less densified vessels – V) on the densified sample, b) densified part (high densified vessels – V) of the sample.

Therefore, the MC was determined at 8 various zones (outer layer and inner layer) of 9 lamellas using 3 different variants: reference poplar (normal wood), samples immediately after densification and densified samples after a conditioning period at 20 °C and 65% RH. The average MC of reference poplar was 11.1% ($\sigma = \pm 0.1\%$) (Fig. 3). The samples immediately after densification had a gradient of MC in a range from 7% (outer layer A – position beginning of the sample) to 10.5% (inner layer, D – middle part of the sample). The edges of the lamellas, especially zones A1 and A2, had the lowest MC between 7% and 8% compared to the middle zones (B, C and D) with MC above 9.5%. It is assumed that it is easier for the moisture to evaporate from the edges during hot pressing through the longitudinal pores, although they were petty closed due to the densification. Finally, even after being stabilized at 20 °C and 65% RH densified lamellas still had a MC gradient in a range from 9.8% to 10.9%.

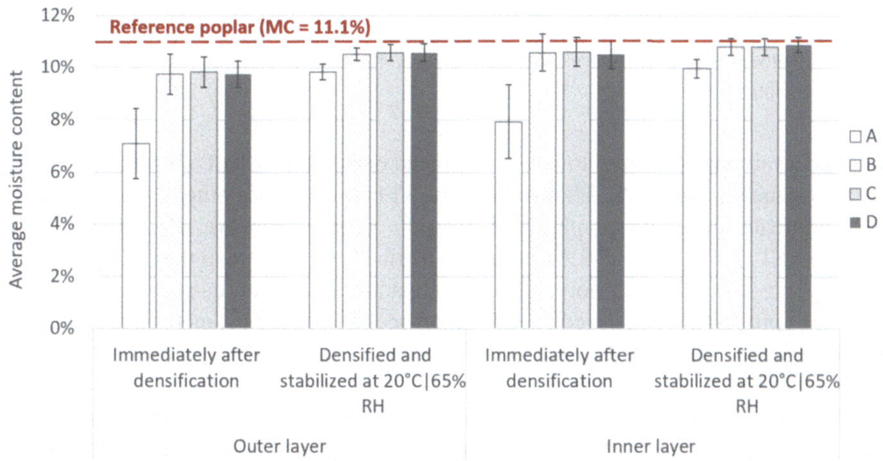

Fig. 3. Results of the average moisture content distribution in the outer layer and inner layer (A – position beginning of the sample, D – position middle of the sample) (n = 9 per position and variant, vertical bars ± standard deviation).

Consequently, those results showed that the densified wood might have a heterogeneous MC distribution even considering the state as stable according to our own criteria of mass variation less than 0.1% after a minimum of 48 h. Additionally, it could also mean that the densified wood needed a very long time to reach its moisture equilibrium, due to the slow movement of the moisture from the interior to the surface as a result of the high density (Bergman, 2010) and densified pores. The remaining trapped water might cause a local swelling effect in the densified wood when suddenly exposed to high temperatures. For that reason, a pre-drying step of the densified samples was applied to reduce the MC prior to the heat post-treatment.

3.2 Set Recovery Results

Selected densified samples with post-treatment (200 °C|3 h, 180 °C|10 h and 180 °C|7 h) and reference non-post-treated densified samples were compared on the set recovery test to determine the influence of the different heat post-treatments (Fig. 4).

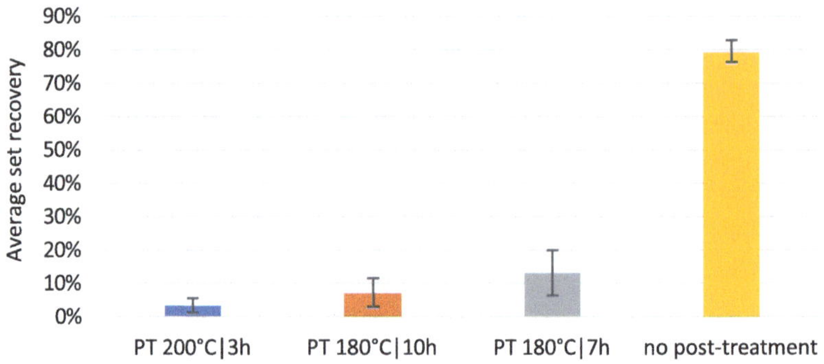

Fig. 4. Results of the set-recovery test (n = 18 per variant, vertical bars and horizontal bars ± standard deviation).

All PT variants compared to non-post-treated ones showed statically significant differences in the SR ($p < 0.05$). They reduced the ST below 15% in comparison to the ST of non-post-treated densified samples with 80% ($\sigma = \pm 3.3\%$). However, PT 200 °C|3 h had the lowest ST (3.5%, $\sigma = \pm 2.2\%$), but it was statically equivalent to the PT 180 °C|10 h (7.3%, $\sigma = \pm 4.2\%$). Furthermore, it is known that heat treatments can have an impact on the wood's properties such as reducing its hygroscopicity and thus, increasing its dimensional stability (Esteves & Pereira, 2009; Sandberg et al., 2017), in this case reducing the set recovery.

Moreover, the density of the post-treated densified samples was in the same range (956–1023 kg/m^3), but lower in comparison to the density of the non-post-treated samples (1100 kg/m^3).

3.3 Brinell Hardness Results

The impact of a heat post-treatment was also demonstrated on the BH (Fig. 5). There was no statistically significant difference on the BH results between non-post-treated and post-treated samples, except for PT 180 °C|10 h (44.2 N/mm^2, $\sigma = \pm 16.7$ N/mm^2) and the non-post-treated (54.4 N/mm^2, $\sigma = \pm 12.4$ N/mm^2).

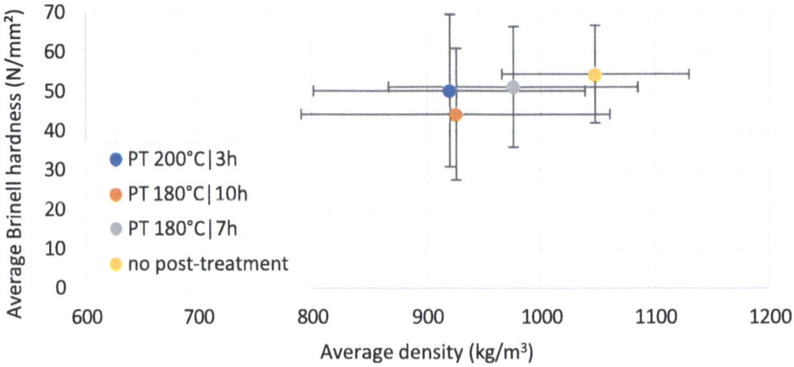

Fig. 5. Results of the Brinell hardness test (n = 36 per variant, vertical and horizontal bars ± standard deviation).

Niemz & Sonderegger, (2003) showed that in untreated wood the hardness correlates with the density. According to the literature (Esteves & Pereira, 2009, Sandberg et al., 2017) with increasing temperature and duration of the heat treatment, the hardness decreases. Consequently, the post-treatment of 200 °C|3 h, 180 °C|10 h and 180 °C|7 h, resulted in decreased density in a range of 920-975 kg/m^3 compared to the density of a reference densified samples (1040 kg/m^3). The different post-treatments did not show significant variations in their density. Differences of the BH for the PT 180 °C|10 h to the non-post-treated samples indicate a trend that long exposure to high temperatures might reduce the BH during a PT of densified specimens.

4 Conclusion

The result of the moisture content distribution of densified poplar samples (500 mm (L) x 100 mm (width) x 3-4 mm (thickness)) prior post-treatment showed a heterogeneous MC distribution (from 7% to 10.5%) right after the hot pressing. In order to reduce the local swelling before the post-treatment, additional pre-drying is needed.

The post-treatment at 200 °C|3 h reduced the set recovery below 10%, whilst statistically not affecting the BH value (50 N/mm^2, $\sigma = 19.3$ N/mm^2) compared to the not post-treated samples (54 N/mm^2, $\sigma = 12.4$ N/mm^2).

References

Bergman, R.: Chapter 13 - Drying and control of moisture content and dimensional changes. In: Wood Handbook - Wood as an Engineering Material, pp. 1–20 (2010)

EN 1534 Wood flooring – Determination of resistance to indentation – Test method (2011)

Esteves, B.M., Pereira, H.M.: Wood modification by heat treatment: a review. BioRes. **4**(1965), 370–404 (2009)

Inoue, M., Norimoto, M., Tanahashi, M., Rowell, R.M.: Steam or heat fixation of compressed wood. Wood Fiber Sci. **25**(3), 224–235 (1993)

Kutnar, A., Kamke, F.A.: Influence of temperature and steam environment on set recovery of compressive deformation of wood. Wood Sci. Technol. **46**(5), 953–964 (2012). https://doi.org/10.1007/s00226-011-0456-5

Laine, K., Segerholm, K., Rautkari, L., Wa, M., Hughes, M.: Wood Densification and Thermal Modification: Hardness, pp. 883–894 (2016). https://doi.org/10.1007/s00226-016-0835-z

Neyses, B., Karlsson, O., Sandberg, D.: The effect of ionic liquid and superbase pre- treatment on the spring-back, set-recovery and Brinell hardness of surface-densified Scots pine. Holzforschung. **74**(3), 303–312 (2020). https://doi.org/10.1515/hf-2019-0158

Niemz, P., Sonderegger, W.: Untersuchungen zur Korrelation ausgewählter Holzeigenschaften untereinander und mit der Rohdichte unter Verwendung von 103 Holzarten. Schweizerische Zeitschrift Fur Forstwesen. **12**(154), 489–493 (2003). https://doi.org/10.3188/szf.2018.s0001

Norimoto, M., Ota, C., Akitsu, H., Yamada, T.: Permanent fixation of bending deformation in wood by heat treatment. Wood Res. **79**, 23–33 (1993) http://hdl.handle.net/2433/53264

Pelit, H., Yorulmaz, R.: Influence of thermal pretreatments on dimensional change and humidity sensitivity of densified spruce and poplar wood. Maderas-Cienc Tecnol. **26** (2023). https://doi.org/10.4067/S0718-221X2024005XXXXXX

Sandberg, D., Haller, P., Navi, P.: Thermo-hydro and thermo-hydro-mechanical wood processing: an opportunity for future environmentally friendly wood products. Wood Mater. Sci. Eng. **8**(1), 64–88 (2013)
Sandberg, D., Kutnar, A., Mantanis, G.: Wood modification technologies—a review. iForest, 7823–7830 (2017)

Maximum Compressibility Along the Grain of Different Hardwood Species

Mátyás Báder(✉), Miklós Szauer, and Róbert Németh

University of Sopron, Bajcsy-Zsilinszky 4, Sopron 9400, Hungary
bader.matyas@uni-sopron.hu

Abstract. This study deals with beech and oak wood modified to become pliable. It is important to know the highest compressibility ratio because this makes it easier to adapt the modification ratio to a specific bending requirement. Hungarian logs were used from different locations. After sample preparation, they were stored frozen to preserve their moisture content. In each case, the specimens were steamed for 45 min and then compressed along the grain (aka. pleating) at a rate of 25%/min in a heated compression machine, followed by a fixation period for 1 min. The ratio of compression was carried out up to the preset value or until something unusual was observed on the graph plotted by the software of the material testing machine. All pleated samples were visually inspected and graded for any damage that may have occurred. The level of pleating was considered adequate if at least 90% of the specimens could withstand compression without damage. The highest compression ratio is 30% for beech, 27–28% for oak sapwood and 20–21% for oak heartwood. Thus, diffuse porous hardwood species have significantly higher compressibility than ring-porous hardwoods. Oak sapwood is easier to compress due to its different density, and chemical structure, etc. Logs of the same species but from different locations can bear pleating differently.

Keywords: compression ratio · different regions · pliability

1 Introduction

When a process improves the properties of wood and creates a new material, it is called wood modification. There are two main thermo-hydromechanical (THM) modification treatments for wood, compression parallel to the grain and compression perpendicular to the grain, which have no negative impact on the environment. In the first step, the wood is traditionally steamed and the heat and moisture softens the tissue between the cells. If compression is done perpendicular to the grain, it is possible to change the properties either at the surface or in the whole cross-section, and obtain a material with a higher density and hardness than the original one (Báder and Németh 2023). The main purpose of compression parallel to the grain is to obtain a more pliable material that is easier to shape. In general, only high quality and higher density hardwood species can be compressed along the grain if their moisture content is above 20%. High quality means defect-free raw material (no knots, insect holes, etc). The most important is that

G. Goli et al. (Eds.): ECWM 2024, SPM 86, pp. 267–274, 2025.
https://doi.org/10.1007/978-3-031-99418-0_33

the slope of the grain must be less then 7°, otherwise the wood will be damaged during compression and the device may be damaged as well. During the process, the wood cells are shortened due to compression, their walls are buckled as an accordion, and they can move relative to each other. It is important to keep the wood at least 80 °C temperature during the treatment. The properties of the final product for bending purposes are greatly influenced by the species, wood quality, moisture content, temperature, and compression ratio, etc. (Báder 2015).

For a wood specimen compressed along the grain by 20% compared to its original length, the modulus of elasticity is reduced by a third, together with the stiffness, compared to untreated wood. The deflection increases several times during the bending test, reaching a bending ratio of ¼. The modulus of rupture is halved, but the amount of energy absorbed is increased several fold. Thus, wood becomes less brittle and the specific impact work, the impact-bending strength and the dynamic fracture work can increase by one and a half to two times (Báder and Németh 2017).

Fixation is the process of holding the wood in a compressed state. The stresses occurring during the compression phase are gradually reduced during the fixation period. This step improves the effectiveness of the compression because the modulus of elasticity decreases significantly as the time of compression increases. After compression, it is decreased by 59%, after 1 min of fixation by 63% and after 18 h of fixation by 81% (Báder and Németh 2018). The longer the fixation time, the higher the remaining shortening of the specimens, so that the spring-back is less significant. Since fixation is a time-consuming process, longer fixation is only necessary for wood requiring extreme bending properties. Taking into account several applications and cost, Báder and Németh (2018) identified 1 min fixation time as the most advantageous, although this may vary depending on the property requirements of the product.

Wood that has been fixated for a short time retains its pliability when removed from the machine as long as its moisture content is above 20%. The remaining shortening in this case is between 3% and 5%. After fixation for at least 3 h, the specimens retain a significant degree of pliability even after conditioning under normal conditions (Báder et al. 2015). The remaining shrinkage is then between 10–20%.

2 Materials and Methods

2.1 Wood Species

In this study investigated the maximum compressibility along the grain of the mature wood and the sapwood of sessile oak (*Quercus petraea* (Matt.) Liebl.) and beech (*Fagus sylvatica* L.). Since both species are hardwoods and have similar densities, the most striking difference between them is that oak belongs to the ring-porous species and beech to the diffuse-porous species. The three oak logs examined were from Hungary: T1 from the Sopron area was a high-quality log (no wood failures except for a few small knots) with a diameter of 32 cm, so many specimens could be taken of both sapwood and mature wood. T2 came from Zala county, about 150 km away, and was also a high-quality and high diameter log. T3 also came from the forests of Sopron (near the town of Görbehalom). False heartwood was observed (the heartwood does not follow the annual ring boundary and is darker than usual), which may be due to some previous root damage or trunk damage of the tree. If this phenomenon, which is rare in oaks, has an impact on the test results, it will be revealed later by the comparison with other samples. In all cases, we prepared specimens from a section of the logs close to the ground with green moisture content, so most conditions were identical.

The situation was similar for beech, which also came from the Sopron region. Originally the experiment was designed for the testing of oak wood samples and their changes resulted by the THM modification, but including a beech sample seemed interesting for comparison. The false heartwood of beech not only causes a change in colour, but also greatly affects the moisture absorption of the wood and its mechanical properties (Divós and Göncz 2015), so we did not use it, only the healthy tissue.

We considered a compression of a given ratio to be successful if at least 90% of the specimens were suitable after the procedure. Surface failures are allowed, but failures that occur on the whole cross-section (visible from several sides) are not acceptable. Typical failures during the test are: the slope of grain, where the grain of the wood does not run parallel to the edges of the specimen and may cause splits at an angle during the modification procedure (Fig. 1), and small knots, which can sometimes significantly affect the structure of the wood, causing deep splits.

Fig. 1. Splitting of a specimen during the treatment due to slope of grain.

2.2 Compression Treatment

The dimensions of the specimens were 20 × 30 × 200 mm (T × R × L) according to the capacity of the compression device. They were stored frozen before the modification to minimize moisture loss. The ice was removed from the specimens, and the weight was measured to the nearest 0.01 g (Precisa XT 1220 M-FR, Precisa Gravimetrics AG, Switzerland) and the length to the nearest 0.1 mm.

The test specimens were then steamed for at least 45 min, after which they were placed in a preheated, custom-built laboratory compression equipment. The specimen was clamped to prevent buckling and minimize cross-sectional expansion. The compressive force was measured and monitored using an Instron 4208 (Instron Corporation, USA) universal material testing machine. After setting the compression ratio and a fixation time of 1 min, the machine operated at a rate of 25%/min in all cases and did the modification called pleating (Báder and Németh 2023). If larger jumps were detected on the graph of load against extension, the operation was stopped.

After pleating, the specimen was removed from the equipment and straightened if necessary, and its mass and length were measured again within 1 min. After recording any defects or deformations, the specimens were placed in a conditioning room at 20 °C temperature and 65% relative humidity (normal conditions) to set their moisture content.

3 Results and Discussion

Among the failures occuring during pleating, a high proportion of fibre separation was observed in oak (Fig. 2a), which is due to the fact that the wood cannot withstand too much compression, because its fibres do not have sufficient cohesive force and separate laterally, creating a gap or gaps in the wood. Particularly for beech, the most typical problems observed are the deep splits along the rays (Fig. 2b). Definite deformation occurs on the sides of the specimen, and the fibres, running at an angle of about 60 degrees, locally separate up to the entire cross-section, typically along the rays. In addition, there were oak specimens with deep radial splits on the end-grain. In this case, the fibres separate laterally due to non-uniform compression, and a small split appears at the end-grain, which expands as the compression increases (Fig. 2c).

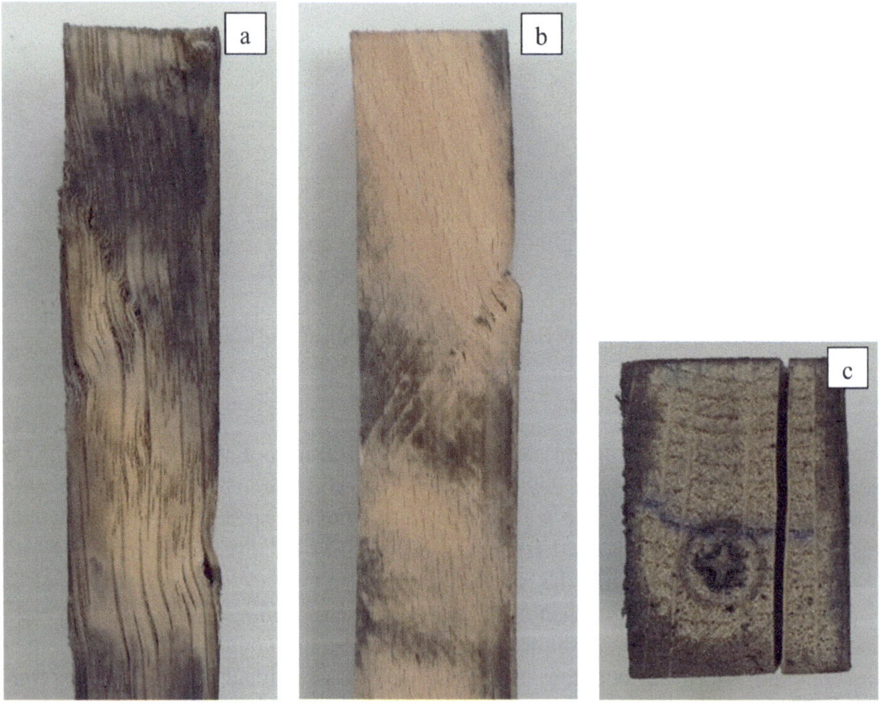

Fig. 2. Fibre separation in an oak specimen, resulted by pleating (a); deep splits along the rays due to pleating in a beech specimen (b); radial split on the end-grain (c).

Surface failures were more frequent in the form of short fibre separations and less frequent in the form of surface splits along the rays. The latter represents a failure visible on only one side of the specimen. These are tolerated failures, since the structural and strength properties of the wood are not affected by a surface failure, which is removed during the machining before final use.

3.1 Pleating of Oak Mature Wood

For the pleating of the T1 mature wood sample, we first tried a compression ratio of 28%. Since all the specimens were damaged, we used 23% to see if it could achieve a 90% yield, but only 53% were successful. Looking at the results of the 20% compression ratio, where only a small percentage of surface defects occurred, we concluded that the highest compression ratio for the T1 mature wood is 21–22%. Following the same methodology, the highest compression ratio for the T2 mature wood was found to be 17%. For T3 mature wood, the maximum compressibility was found to be 23% with 100% success, occurring only surface failures (Fig. 3).

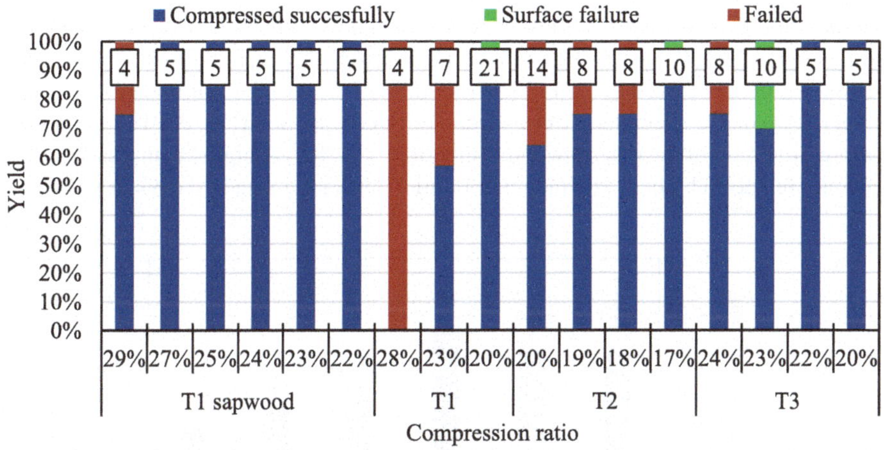

Fig. 3. Compression success of mature wood oak samples T1, T2 and T3 and sapwood oak sample T1, if losses due to specimen failures are not taken into account.

The remaining shortening after compression, 1 min fixation and spring-back due to different compression ratios is an indication of the extent of mechanical changes in the wood (Báder and Németh 2023). Remaining shortenings and their standard deviations increase with increasing compression ratio, and the remaining shortening increases at a much higher ratio than the standard deviation (Fig. 4). This is explained by the fact that as the compression ratio increases, the proportion of damage in the wood tissue also increases. It can be concluded that the false heartwood of oak mature wood has not caused problems in terms of compressibility, but the growing site and growing conditions influence the pleating properties. However, not to a large extent, with a difference of about 5 percentage points.

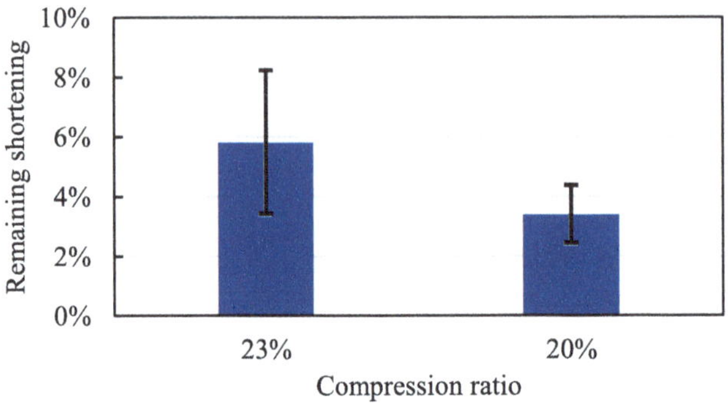

Fig. 4. Remaining shortening of oak mature wood T1.

3.2 Pleating of Oak Sapwood

The maximum compression ratio of sapwood of sample group T1 was initially underestimated at 22%. By increasing the compression ratio by one and then two percent, the first failures occurred at 29%, but these did not fit into the minimum 90% yield, as only 75% of the specimens were successfully compressed (Fig. 3). The maximum compression ratio of the oak sapwood is therefore 27–28%. A higher compression ratio generally results in higher remaining deformation and larger standard deviation (Fig. 5).

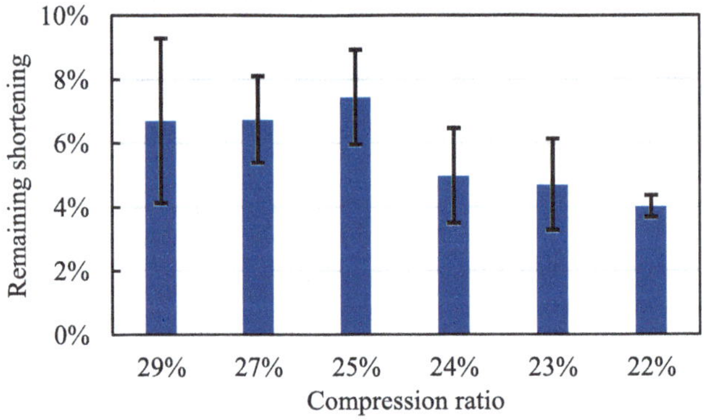

Fig. 5. Remaining shortening of oak sapwood T1 at different compression ratios.

From the results, it is clear that the sapwood is compresses significantly better than the mature wood. The explanation lies in its structure, as the density of the sapwood is lower than that of the mature wood, which is deposited with minerals and tyloses. Because sapwood is less dense and has more free volume in the cavities, it can be compressed more easily and in greater ratio, with more free space for cell deformations. The greater elasticity of the sapwood cells also explains this phenomenon: the outer part of the wood tissue, which is more recent and fresher, is less brittle than the inner part. This results in more malleability, i.e. easier compression. Furthermore, the phenomenon may be explained by the fact that the sapwood is easier to plasticise.

3.3 Compression of Beech

For beech, no distinction was made between sapwood and mature wood because of their negligible differences in structure and chemical properties. The initial compression ratio from the preliminary experiments was 32%, which resulted in too many failures and was therefore unsuccessful. Although at 31% compression ratio, 85% of the specimens were undamaged which was very close to the target of 90%. The final compression ratio of 30% was used to produce 91% undamaged specimens, and which was then set as highest compressibility limit. The diagram of the remaining shortening of beech (Fig. 6) is similar to that of the oak mature wood (Fig. 4), and the conclusions are the same: the averages and standard deviations increase with increasing compression ratio, and the magnitude of the remaining deformation increases at a higher ratio than the compression percentage.

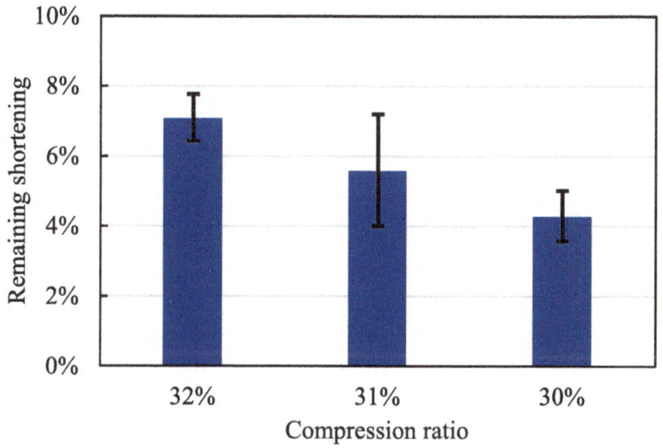

Fig. 6. Remaining shortening of beech.

Comparing beech with oak, it is clear that the former can be compressed to a much greater extent along the grain, with a maximum ratio of 30%, while oak can be compressed only by 21–22%. This can also be explained by their structure, since oak belongs to the group of ring-porous woods and splits easier, whereas beech is a diffuse porous wood and has a more homogeneous structure. The distributed arrangement of the vessels results in a more homogeneous structure and a higher compression ratio without material failures.

4 Conclusion

In this study, the maximum compressibility of oak and beech species were investigated to produce pliable wood, also called as pleating. Three sessile oak groups were studied from different locations and one beech group.

Oak sapwood can be compressed along the grain by 27–28%, while oak mature wood is compressible only by 17–22% depending on its place of origin. This means that the wood tissue significantly influences the properties of wood, which influences its pleating opportunities. Beech could be compressed 30% along the grain. This can be explained by its different structure, since oak is a ring-porous, while beech is a diffuse porous wood species and has a more homogeneous structure.

Acknowledgements. We dedicate special thanks to Imre Horváth for the preparation of the specimens.

References

Báder, M.: A faanyag rostirányú tömörítésével kapcsolatos elméleti és gyakorlati kérdések áttekintése. FAIPAR. **63**(1), 1–9 (2015). https://doi.org/10.14602/WOODSCI.2015.1.8

Báder, M., Németh, R.: Faanyagok rostirányú tömörítésének kísérleti körülményei – 1. rész. Gradus. **4**(2), 403–411 (2017)

Báder, M., Németh, R., Ábrahám, J.: Faanyag rostirányú tömörítésével kapcsolatos elméleti és gyakorlati kérdések áttekintése: II. rész: Történelem és szabadalmak. A tömörítési folyamat tulajdonságai és a tömörítést követő eljárások. FAIPAR. **63**(1), 10–20 (2015). https://doi.org/10.14602/WOODSCI.2015.1.13

Báder, M., Németh, R.: Further treatment option after longitudinal wood compression. In: Creemers, J., Houben, T., Tjeerdsma, B., Militz, H., Junge, B. (eds.) Proceedings of the 9th European Conference on Wood Modification, pp. 1–7. Practicum, Arnhem (2018)

Báder, M., Németh, R.: A review of wood compression along the grain—after the 100th anniversary of pleating. Forests. **14**(4), 763 (2023). https://doi.org/10.3390/f14040763

Divós, F., Göncz, B.: Bükk (Fagus sylvatica) álgeszt kimutatása elektromos feszültség- és ellenállásméréssel. FAIPAR. **63**(2), 29–35 (2015). https://doi.org/10.14602/WOODSCI.2015.2.40

Wood Modification via Geopolymer Impregnation: Effects on Decay, Mechanical Properties and Fire Retardancy

Aitor Barbero-López[1(✉)], Paivo Kinnunen[2], and Antti Haapala[1,3]

[1] Department of Chemistry, University of Eastern Finland, Joensuu, Finland
aitor.barberolopez@uef.fi
[2] Fibre and Particle Engineering Research Unit, Oulu, Finland
[3] FSCN Research Centre, Mid Sweden University, Sundsvall, Sweden

Abstract. This work aims to modify wood using dilute geopolymer suspensions for improving its decay resistance, leachability, mechanical properties and fire resistance.

Scots pine (*Pinus sylvestris*) sapwood specimens were impregnated with either 1, 3 and 5% geopolymer suspensions. The 3% and 5% geopolymer treated wood specimens presented lower mass loss caused by the decay fungus *Coniophora puteana* than control specimens, while 1% treatment increased the mass loss significantly. The geopolymer suspensions did not affect the mechanical properties of the sapwood specimens. The pine sapwood treated with geopolymer and subjected to leaching displayed the same mass loss to that of the control specimens exposed to leaching. This suggests that the dilute geopolymer suspension leached out from the wood. The geopolymer-suspension-impregnated wood presented an increased time to ignition compared to controls in the fire resistance test, but also a higher amount of smoke released. Results indicate that modification of wood using dilute geopolymer suspension is promising, but presents drawbacks that need to be addressed, such as leaching.

Keywords: biodegradation · durability · inorganic preservation · wood protection

1 Introduction

The susceptibility of wood to degradation due to different abiotic and biotic factors is a limiting factor when building outdoors, as it reduces its service lifetime. Out of all the factors causing wood degradation, the most relevant one in rural and urban areas is fungal decay [1]. A lot of research is currently aiming to find methods to avoid wood decay without using chemicals with toxic effects towards humans and the environment, as the use of traditional wood preservatives is being limited due to chemicals legislation [2]. Several potential bio-based antifungals that avoid wood decay, such as stilbenes [3] and hydrothermal liquids produced from spent mushroom substrate [4] have been identified. The market uptake of new chemical innovations is limited in part due to the expensive

G. Goli et al. (Eds.): ECWM 2024, SPM 86, pp. 275–282, 2025.
https://doi.org/10.1007/978-3-031-99418-0_34

authorization process for registering new preservative ingredients. Wood modification is seen as a potential alternative to improve the resistance of wood against decay fungi with successful results, such as thermal modification of *Pinus radiata* [5] and wood acetylation [6].

Enhancing the fire resistance of wood is also a common goal currently pursued by the wood industry and numerous researchers. The combustible nature of wood and the policies promoting its use in building sector makes the fire resistance of wood a very important topic. Wood modification can also reduce the fire resistance of the wood, such as thermal modification of *Tectona grandis* wood in the study performed by Čekovská et al. [7].

In addition to affecting the durability and the fire resistance, wood modification methods can also be used to improve the physical and mechanical properties. Wood furfurylation, developed several decades ago, causes several properties to improve, such as increment in hardness, modulus of elasticity and modulus of rupture [8]. In the study performed by Cao et al. [9], rubberwood showed improved mechanical properties, such as increased modulus of elasticity and compression strength after thermo-hygro-mechanical modification. While some results are positive, success has not been systematic. For instance, when various wood species from Quebec underwent the ThermoWood® heat treatment process, it led to an increase in the modulus of elasticity for species like fir, aspen, and birch, but a decrease in spruce and pine [10].

Geopolymers are aluminosilicate solids produced by reacting an aluminosilicate precursor with alkali activator which results in an amorphous solid [11]. The properties of geopolymers are like those of the Portland cement, and their use reduces carbon emissions from 20 to 50% [12]. In addition to their mechanical properties, they also present a high resistance to thermal stress and resistance to decay [13]. Research in geopolymer-wood composites has focused in adding wood as a filler to a geopolymer matrix [14] or in their including in sandwich panels as fire retardants [15]. However, there is not much research on the effects of wood modification using geopolymers.

The aim here is to study the effects of geopolymer dilute suspensions on the mechanical properties, resistance to fungal decay, and fire resistance of treated wood, and to assess the possibility of using this inorganic treatment by the wood industry.

2 Materials and Methods

2.1 Wood Samples

To check the performance of geopolymer-treated wood in decay, leaching, fire resistance and mechanical tests, Scots pine (*Pinus sylvestris*) sapwood was purchased in a local sawmill (Kerimäki, Finland). The sapwood specimens were cut in different sizes depending on the test: $5 \times 10 \times 40$ mm^3 size specimens were cut for the decay and leaching tests; $20 \times 20 \times 340$ mm^3 size specimens for the mechanical test; $100 \times 100 \times 20$ mm^3 size specimens for the fire retardancy test.

2.2 Geopolymer Preparation and Impregnation

The geopolymer suspension was prepared by adding Metakaolin (Metaver® M, Newchem GmbH, Austria) in ice bath to activate the solution of sodium hydroxide, sodium silicate and water stirred in a closed plastic container for 24 h with a ratio between Si:Al:Na:H_2O of 4:1:1:23 together with a surfactant (5% w/w) of Sodium dodecyl sulphate (SDS) purchased to ICN Biomedicals, OH, USA. The geopolymer suspension was then diluted to 1%, 3% and 5% using MilliQ water (Merck KGaA, Darmstadt, Germany) for using them as wood impregnants. The pH of the three dilutions was adjusted to 11 using NaOH 0.1 M. The wood specimens were divided into three groups, and each group was immersed in one of the three dilutions of geopolymer suspensions previously prepared. While submerged, the wood specimens were kept at 0.2 Bar for 20 min, and afterwards the pressure was increased to 10 bars for 1 h, following a modified Bethell process. The geopolymer-treated wood specimens were then taken out of the geopolymer solution and sealed into Ziplock bags and kept 24 h at 40 °C and afterwards 14 days at room temperature for curing. The dry mass of the wood specimens dried at 60 °C until constant mass was measured.

2.3 Mini-Block Test

The performance of geopolymer treated wood against decay fungi was tested following a mini-block procedure and a modified version of EN 113 [16], where flasks were substituted with Petri dishes. A second mini block test was performed to compare the leached and unleached specimens. For both tests Petri dishes containing 4% malt and 2% agar was inoculated with the wood-decaying fungus *Coniophora puteana.* The petri dishes were incubated at 20 ± 2 °C and 65 ± 5% RH until the mycelia reached the edge of the petri dish.

In the first mini-block test, 4 wood specimens with the geopolymer suspensions (1%, 3% and 5%) and controls were placed in each petri dish. A mesh with the same size of the wood specimens was used to avoid direct contact between the malt agar media and the wood. The petri dishes were sealed with parafilm™ and kept in a growth chamber at 20 ± 2 °C and 65 ± 5% relative humidity for 16 weeks. Afterwards, the wood specimens were carefully brushed to remove the mycelium and the dry mass at 60 °C was recorded to measure the mass loss caused by *C. puteana.* A total of 30 replicates were measured for each geopolymer concentration and control.

In the second mini-block test untreated and 5% geopolymer treated wood specimens were used. Half of the specimens (N = 8) were exposed to a leaching test according to the European norm EN 84 [17]. In brief, the wood specimens were submerged in 5:1 (v/v) of MilliQ water to wood. The water was changed first after 24 and 48 h and afterwards 7 times within the remaining 12 days. The specimens were taken out of the water after 14 days of leaching and their dry mass at 60 °C was recorded. Then, a mini-block test following the same procedure as in the first mini-block test was performed, putting in each petri dish one wood specimen from each treatment (control, leached control, 5% geopolymer and 5% geopolymer leached).

2.4 Mechanical Test

A Zwick Roell 050 materials testing setup (Ulm, Germany) was used for the mechanical testing of the wood specimens. T modulus of elasticity (E_{mod}), and modulus of rupture (MOR) were calculated by performing a three-point bending test. The mean values of the E_{mod} and MOR of the specimens treated with the different geopolymer suspension dilutions was calculated.

2.5 Cone Calorimetric Analysis

The fire resistance of the geopolymer-treated wood specimens was analyzed following the ISO 5660-2 Standard Test Method, with a cone calorimeter (ConeTool 1; SGS Govmark Ltd., Farmingdale, IL, USA) at a radiant flux of 50 kW/m^2. The measurement of the heat release rate was conducted by analyzing the oxygen consumption and flow rate within the combustion product stream.

3 Results and Discussion

The mass loss caused by *C. puteana* to control specimens was 9.9 ± 2.4% (Fig. 1), approximately half of what the European standard EN 113 [16] indicates, but we still consider the results good enough to allow us to compare the effectiveness of geopolymer treatments.

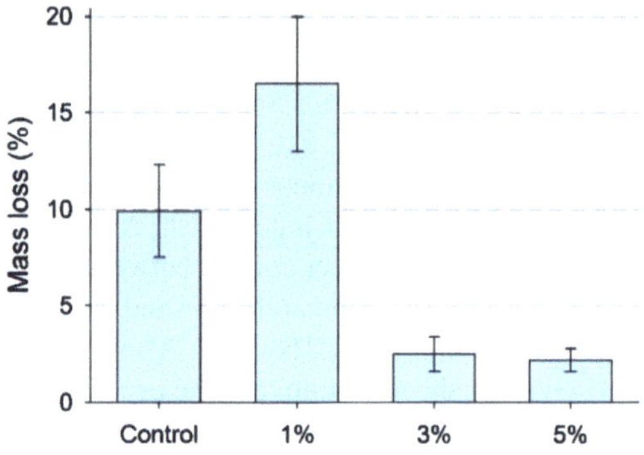

Fig. 1. Mass loss caused by *Coniophora puteana* to sapwood specimens with 1, 3 and 5% geopolymer and controls (N = 30).

The mass loss of the specimens treated with 1% geopolymer had a mass loss of 16.5 ± 3.5% indicating that the treatment at this low concentration was not effective enough to preserve wood, and even caused a slight increase in mass loss. This increase could be related to the low fixation of geopolymer in wood at low concentrations and the leaching of extractives, such as phenolics, during the geopolymer impregnation process. The 3% and 5% concentration of geopolymer reduced the mass loss caused by the decay fungus to 2.5 ± 0.9% and 2.2 ± 0.6% respectively, indicating that the treatment was already effective at 3% and did not improve its performance with a higher geopolymer concentration. The reason behind this effective inhibition was not studied in this work, but we hypothesize that the geopolymer suspension creates a physical barrier and fills the vessels with inorganic compounds such as aluminum and silica, like the silicon polymers acting as a barrier in commercialized OrganoWood® [18]. A study using silica–alumina sol added to fatty alcohol polyoxyethylene ether as a wood treatment was found to improve the wood durability, mechanical properties and fire resistance [19]. The silica-alumina sol has the same precursors despite being different products, which might be responsible of the fungal inhibition. However, further research is needed to understand he optimal concentrations to ensure encapsulation of wood in geopolymer and to understand if something else than the physical barrier is responsible for the improved wood durability.

The results from the second mini-block test prove that geopolymer leached out from wood, reducing their effectiveness as a potential wood modification method to avoid fungal decay on wood (Fig. 2). While the mass loss of wood caused by *C. puteana* got reduced when the wood was treated with 5% geopolymer, the mass loss of the wood after leaching was like that of the control specimens. Geopolymers are alumino-silicate polymers that should form a stratum around the wood, encapsulating the wood material, extending also to its pores. However, this did not happen in this experiment, as results show the leaching of the geopolymer. We hypothesize the reason might be the concentration of geopolymer suspension was not high enough to create the stratum necessary to fix the geopolymer suspension to wood, or the addition of SDS, intended to enhance the geopolymer adhesion to wood, negatively affected the formation of the stratum. Also, Na:Al -ratio used was higher than 1, which could explain some of the leaching. Higher concentrations of geopolymer should be tested to check which is the minimal concentration of geopolymer required to ensure the binding of geopolymers to wood, as well as the optimum Si:Al -ratio, which can be up to 20 in other composites. Additionally, the addition and different concentrations of SDS should be tested to improve the binding of geopolymer to wood.

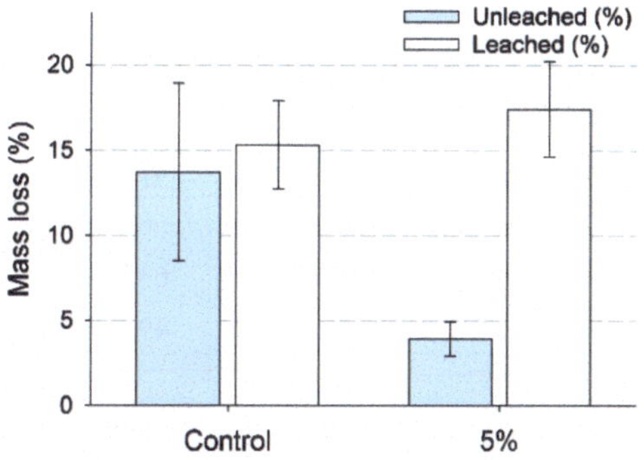

Fig. 2. Mass loss caused by *Coniophora puteana* to leached and unleached control and 5% geopolymer-treated sapwood specimens (N = 8).

The effect of the geopolymer suspension treatments on the wood specimens did not affect the mechanical properties, either E_{mod} and MOR, of the wood specimens (Table 1). These results were surprising as an inorganic filler can cause a change in the mechanical properties of the wooden matrix. As an example, the previously cited silico-alumina sol improved the mechanical properties of poplar wood [19]. This may be an indicator of the insufficient formation of a stratum of geopolymer in the wood or its low fixation.

Table 1. Elastic modulus (E_{mod}) and modulus of rupture (MOR) of sapwood specimens treated with 1, 3 and 5% geopolymer and controls (N = 30).

Treatment (%)	E_{mod} (GPa)	MOR (MPa)
1	12.8 ± 0.5	121.2 ± 6.0
3	13.1 ± 0.4	131.0 ± 5.7
5	12.4 ± 0.3	117.3 ± 4.8
Control	12.0 ± 0.4	113.0 ± 4.7

The time to ignition of 3% and 5% geopolymer suspension-treated wood specimens was higher than those of 1% geopolymer suspension-modified wood and control specimens (Table 2). However, the smoke produced by all the geopolymer-suspension modified wood was also higher than the smoke produced by the controls. While it is important to consider the limited number of replicates in this test, the results indicate that treated specimens with concentrations over 3% are promising treatments as fire retardants, but they also generate more smoke. This agrees with previous findings [20, 21], where inorganic wood treatments prolong the time to ignition of wood. Some of the precursors in geopolymer, such as aluminum, have also been successfully tested as part of fire retardancy formulations [22] and might be the reason for the fire retardancy of geopolymers.

Table 2. Fire test results of Scots pine sapwood specimens treated with geopolymer and untreated specimens (N = 2).

Sample	Time to ignition (s)	Mean heat release rate (kW/m^2)	Peak heat release (kW/m^2)	Mass loss (%)	Total smoke (m^2/m^2)
Control	15.5 ± 0.5	117.9 ± 2.6	184.0 ± 8.2	47.0 ± 0.2	184.7 ± 16.2
1%	16.5 ± 2.5	110.0 ± 11.4	176.9 ± 4.5	66.9 ± 15.2	344.5 ± 139.5
3%	21.0 ± 1.0	105.3 ± 3.8	213.0 ± 27.9	63.1 ± 16.0	356.0 ± 162.2
5%	20.5 ± 1.5	115.7 ± 5.1	183.5 ± 9.9	49.4 ± 1.9	254.7 ± 31.6

4 Conclusions

The wood modified with 3% or higher dilute geopolymer suspensions was more durable than untreated wood against the decay fungus *Coniophora puteana* but leaches out from wood when exposed to water. The geopolymer suspension treatments did not have any significant effect on the mechanical properties of wood, but concentrations of 3 and 5% were effective as fire retardants. We conclude that reported geopolymer modification of wood utilized here is an effective treatment against decay fungi and effective fire-retardant, which should be further studied to find better formulations which are still more effective and do not leach out from wood.

References

1. Broda, M.: Biological effectiveness of archaeological oak wood treated with methyltrimethoxysilane and PEG against Brown-rot fungi and moulds. Int. Biodeter. Biodegr. **134**, 110–116 (2018)
2. Barbero-López, A., Akkanen, J., Lappalainen, R., Peräniemi, S., Haapala, A.: Bio-based wood preservatives: their efficiency, leaching and ecotoxicity compared to a commercial wood preservative. Sci. Total Environ. **753**, 142013 (2021)
3. Lu, J., Venäläinen, M., Julkunen-Tiitto, R., Harju, A.M.: Stilbene impregnation retards brown-rot decay of Scots pine sapwood. Holzforschung. **70**, 261–266 (2016)
4. Barbero-López, A., López-Gómez, Y.M., Carrasco, J., Jokinen, N., Lappalainen, R., Akkanen, J., Mola-Yudego, M., Haapala, A.: Characterization and antifungal properties against wood decaying fungi of hydrothermal liquefaction liquids from spent mushroom substrate and tomato residues. Biomass Bioenergy. **181**, 107035 (2024)
5. Hermoso, E., Fernández-Golfín, J., Conde, M., Troya, M.T., Mateo, R., Cabrero, J., Conde, M.: Characterization of thermally modified *Pinus radiata* timber [Caracterización de la madera aserrada de *Pinus radiata* modificada térmicamente]. Maderas Cienc. Tecnol. **17**, 493–504 (2015)
6. Larsson Brelid, P., Simonson, R., Bergman, Ö., Nilsson, T.: Resistance of acetylated wood to biological degradation. Holz. Roh. Werkst. **58**, 331–337 (2000)
7. Čekovská, H., Gaff, M., Osvaldová, L.M., Kačík, F., Kaplan, L., Kubš, J.: *Tectona grandis* Linn. and its fire characteristics affected by its thermal modification of wood. Bioresources. **12**, 2805–2817 (2017)
8. Lande, S., Westin, M., Schneider, M.: Properties of furfurylated wood. Scand. J. For. Res. **19**, 22–30 (2004)
9. Cao, R., Marttila, J., Möttönen, V., Heräjärvi, H., Ritvanen, P., Verkasalo, E.: Mechanical properties and water resistance of Vietnamese acacia and rubberwood after thermo-hygro-mechanical modification. Eur. J. Wood Wood Prod. **78**, 841–848 (2020)
10. Shi, J.L., Kocaefe, D., Zhang, J.: Mechanical behaviour of Québec wood species heat-treated using ThermoWood process. Holz. Roh. Werkst. **65**, 255–259 (2007)
11. Sá Ribeiro, R.A., Sá Ribeiro, M.G., Sankar, K., Kriven, W.M.: Geopolymer-bamboo composite—a novel sustainable construction material. Construct. Build Mater. **123**, 501–507 (2016)
12. Assi, L., Carter, K., Deaver, E.E., Anay, R., Ziehl, P.: Sustainable concrete: Building a greener future. J. Clean. Prod. **198**, 1641–1651 (2018)
13. Morsy, M.S., Rashad, A.M., Shoukry, H., Mokhtar, M.M.: Potential use of limestone in metakaolin-based geopolymer activated with H_3PO_4 for thermal insulation. Construct. Build Mater. **229**, 117088 (2019)
14. Asante, B., Schmidt, G., Teixeira, R., Krause, A., Savastano Junior, H.: Influence of wood pretreatment and fly ash particle size on the performance of geopolymer wood composite. Eur. J. Wood Wood Prod. **79**, 597–609 (2021)
15. Giancaspro, J., Balaguru, P., Lyon, R.: Fire protection of flammable materials utilizing geopolymer. SAMPE J. **40**, 42–49 (2004)
16. EN 113: Wood Preservatives—Test Method for Determining the Protective Effectiveness Against Wood Destroying Basidiomycetes—Determination of the Toxic Values. European Committee for Standardization, Brussels (1996)
17. EN 84: Wood Preservatives—Accelerated Ageing of Treated Wood Prior to Biological Testing—Leaching Procedure. European Committee for Standardization, Brussels (1997)
18. Jones, D., Sandberg, D.: Wood Modification in Sweden. In: Goli, G., Todaro, L. (eds.) Procedings of COST Action FP1407 WG1 and WG4 meeting. Wood modification in Europe: processes, products, applications, Firenze, Italy (2018)

19. Yang, S., Liu, Z., Wang, Z., Wu, Y., Ji, N.: Organic–inorganic hybrid material containing silica–alumina sol and ether polymer for flame retardant wood with decay resistance and excellent mechanical properties. J. Build. Eng. **73**, 106676 (2023)
20. Son, D.W., Kang, M.R., Kim, J.I., Park, S.-B.: Fire performance of the wood treated with inorganic fire retardants. J. Korean Wood Sci. Technol. **40**, 335–342 (2012)
21. Uddin, M., Kiviranta, K., Suvanto, S., Alvila, L., Leskinen, J., Lappalainen, R., Haapala, A.: Casein-magnesium composite as an intumescent fire retardant coating for wood. Fire Saf. J. **112**, 102943 (2020)
22. Uddin, M., Alabbad, M., Li, L., Orell, O., Sarlin, E., Haapala, A.: Novel micronized mica modified casein–aluminum hydroxide as fire retardant coatings for wood products. Coatings. **12**, 673 (2022)

Removal of Non-cellulosic Wood Constituents and Subsequent Densification for Improved Mechanics

Matthias Jakob[1], Ulrich Müller[1], Felix Neudecker[1], and Wolfgang Gindl-Altmutter[1,2](✉)

[1] BOKU University, Tulln 3430, Austria
wolfgang.gindl@boku.ac.at
[2] Kompetenzzentrum Holz GmbH, Tulln 3430, Austria

Abstract. The delignification and subsequent densification of wood has been repeatedly demonstrated as being a useful and efficient route to improving the mechanics of wood and non-wood materials. Own work has shown that while delignification and densification almost always improve wood tensile mechanics in fiber direction, the side effects of the process with regard to bending strength, transverse strength and adhesive bonding are less studied. Also, most of literature applies rather harsh chemical treatment, focusing on almost complete removal of lignin. Own work with hardwood has shown that delignification is not always required and that significant improvements in strength can also be achieved by focusing on mild extraction of non-cellulosic cell wall polysaccharides. Thus, in the present paper, we briefly summarize our experience from five years of work on delignification and densification of wood and non-wood materials.

Keywords: Delignification · Densification · Strength

1 Introduction

Wood is the most abundant bio-resource, and wood utilisation has potential for substituting fossil-based materials at negative carbon emissions [1]. In order to benefit from this development, wood has to be used in the most resource-efficient manner [2], and wood products must be designed for optimum functionality and maximum service life.

Wood shows excellent mechanics at low density [3], which enables slender and wide-spanning constructions in building, and light fuel-saving structures in vehicles. Even though wood shows a high ratio of high strength at low density, which is very favourable in construction, its absolute strength is limited compared to many modern high-performance engineering materials. In applications where higher than natural wood strength is required, densification may be of advantage, because wood mechanical performance directly scales with density [4]. Densification procedures usually involve a softening step, followed by compression, setting and fixation [5]. Wood is known to show softening due to the influence of temperature and moisture [6]. At room temperature, softening of hemicelluloses occurs around 75% relative humidity [7], while softening

G. Goli et al. (Eds.): ECWM 2024, SPM 86, pp. 283–288, 2025.
https://doi.org/10.1007/978-3-031-99418-0_35

of lignin at full water saturation occurs at temperatures of 80–90 °C [8]. Therefore, most classic densification approaches make use of the softening of lignin at elevated temperature and moisture content [6, 9]. Alternatively, polymer infiltration, e.g. with phenol formaldehyde resin, also facilitates compression [10, 11]. More recently, high-performance applications of wood in engineering applications have attracted renewed attention to delignified and subsequently densified wood [12–15]. The combination of delignification and densification is of particular interest, because the partial or total removal of non-cellulosic cell wall polymers prior to densification results in a disproportionately high non-linear increase of tensile strength (up to 720 MPa) and stiffness (52 GPa) at densities between 0.8 and 1.3 g cm^{-3} (Fig. 1). Cellulose is the main structural polymer in wood cell walls, with an estimated tensile strength of several GPa [16, 17] and a modulus of elasticity between 140 GPa and 150 GPa [18, 19]. In comparison, the mechanical performance of the chemically heterogeneous "matrix" polymers hemicellulose and lignin is estimated to be at least by a factor 10 lower than that of cellulose [20, 21]. Hence it is obvious that removal of these components leads to improved overall strength as predicted by a rule of mixture approach [22]. However, the removal of wood cell wall components presents a severe disturbance of the native cell wall structure which may affect material properties beyond mere strength improvements.

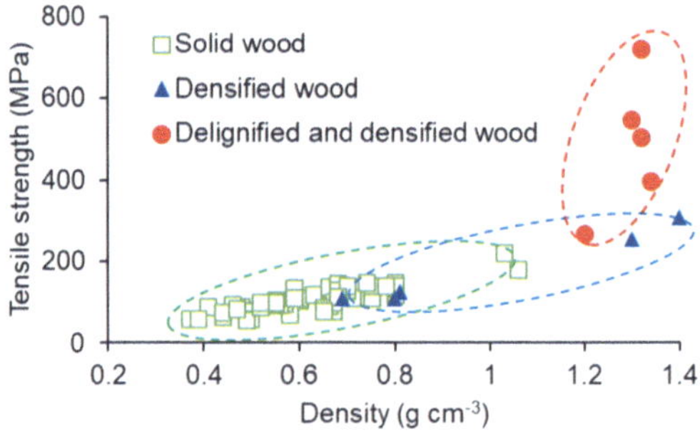

Fig. 1. The tensile strength of solid wood related to density for untreated solid wood, densified solid wood, and delignified and densified wood, respectively [15].

2 Own Work on Strong Delignified and Densified Wood

2.1 Densification and Microscopic Structure

In a study with one representative softwood (Norway spruce, *Picea abies*) and hardwood (silver poplar, *Populus alba*), respectively, the compression behavior in native and partially delignified state, wet and dry conditions, and three different orientations of wood grain were investigated [23]. It was demonstrated that the differences in density across annual growth rings of spruce hinder defect-free densification. Notably, densification in tangential direction and at inclined annual ring orientation lead to excessive shear deformations buckling of latewood bands, and intracellular microcracking. By contrast, the diffuse-porous hardwood poplar, which shows more homogeneous anatomy and hardly any variability of density across growth rings, behaved much more favorable in densification.

2.2 Transverse Tensile Strength of Delignified and Densified Softwood

Partial delignification and densification has been shown to convey exceptionally high axial tensile strength to wood [15]. However, transverse properties have been paid little attention to. In order to evaluate potential effects, partially delignified and densified spruce (*Picea abies*) wood veneers were characterized at varying degrees of fiber misalignment [24]. As expected, the tensile strength and the modulus of elasticity were clearly improved in parallel to the axis of wood fibers. However, this improvement quickly leveled off at misalignment angles $\geq 30°$ and for transverse tensile strength the performance of densified wood was even weaker than untreated wood, most probably due to abundant microscopic cracks in treated wood.

2.3 Tensile Strength of Delignified and Densified Woods Versus Bending Strength

Spruce (*Picea abies*), beech (*Fagus sylvatica*) and poplar (*Populus alba*) wood were delignified and densified according to well-known procedures [25]. As expected, impressive improvements in the axial tensile properties were achieved for all species, with final densities ranging from 1.0 g cm^{-3} to 1.2 g cm^{-3} and corresponding improvements by factors 3–5 for tensile modulus and 2.5–4.5 for tensile strength. Even when factoring out the pure effect of density-related improvements by calculating specific values, significant improvements persisted. Similarly, also in three-point bending mode improvements were seen after delignification and densification by factors 3.5–5.5 for bending modulus and 1.5–3 for bending strength. However, none of the variants examined showed an increase in bending properties in terms of specific strength or modulus. When compensating for differences in density, delignified and densified wood performed clearly inferior to natural unmodified wood both regarding specific modulus of elasticity and even more so regarding specific bending strength. It is presumed that the transverse loosening effect of matrix removal in the cell wall is largely uncritical with regard to tensile loads, whereas compression loads, which are sensitive to changing interfibrillar shear strength [26], are significantly affected. Hence specific bending performance, which is due to a combination of tensile and bending strength, clearly suffered in delignified and densified wood.

2.4 Wettability and Adhesive Bonding of Delignified and Densified Wood

In own unpublished work with delignified and densified spruce (*Picea abies*) wood, the surface roughness, wettability with water, and suitability for adhesive bonding were evaluated. As expected, densification led to a very significant reduction down to 1–2 μm of average roughness. As for wettability with water, no statistically significant effects of densification could be observed. Lap-shear specimens of spruce (*Picea abies*), beech (*Fagus sylvatica*) and poplar (*Populus alba*) wood were bonded with the representative construction resins phenol resorcinol formaldehyde (PRF), melamine urea formaldehyde (MUF), and one-component polyurethane (PUR). Regardless of wood species or adhesive used, no specimen group showed differences in bond strength compared to the untreated reference specimens. It can thus be said that delignified and densified wood bonds equally well as untreated wood.

2.5 Mild Extraction Prior to Densification of Hardwood

The majority of studies on delignified and densified wood aims for substantial or complete delignification [12, 13]. Extensive own work with alkaline delignification of hardwood showed that in terms of improvement of mechanics achieved there is a clear optimum of treatment severity at low mass loss. For birch (*Betula pendula*) wood treated with an alkaline delignification procedure, severe treatment resulted in 50% mass loss and twofold increase in tensile strength after densification. At an intermediate mass loss, strength improved by a factor 2.5, which further rose to 3.5 at low mass loss. Considering a densification factor of 1.8, this translates to a very impressive improvement also in terms of specific tensile strength.

2.6 Delignification and Densification of Non-wood Material

Non-wood materials such as straw or bamboo have been shown to be suitable substrates for delignification and densification [27–29]. Due to the specific lignin chemistry of grasses, only very mild alkaline treatment is sufficient to achieve excellent plasticization. In own unpublished work average tensile strength improvements from 70 MPa up to 480 MPa and modulus elasticity increase from 4 GPa to 27 GPa were observed. The well-known waxy layer covering the epidermis of straw is largely removed during the process, which enhances the strength of adhesive bond in composites made from delignified and densified straw.

3 Conclusions

Extensive studies with coniferous wood, coniferous wood, and non-wood material applying different delignification procedures and treatment severity have confirmed that improvements in tensile strength and modulus exceeding the increase expected due to the proportionality between density and mechanics may be regularly obtained. Furthermore, results show that delignified and densified biomass is suitable for adhesive bonding with commonly used wood adhesives. While tensile properties are improved very clearly, effects on the behavior in compression and in tension transverse to the fiber direction are not only positive. Finally, it was shown that complete delignification is not always required, because intermediate treatment severity also delivers excellent results, at least in alkaline delignification. Mild delignification procedures also carry the advantage of reduced chemicals consumption, which lowers economic cost and environmental footprint.

Acknowledgement. The authors gratefully acknowledge funding by the Austrian Research Promotion Agency FFG (grant FO999893356) as well as the contribution of the participating companies FACC Operations GmbH, LUXNER Engineering ZT GmbH, Volkswagen Aktiengesellschaft, and Weitzer Woodsolutions GmbH. Research on straw was funded by the Gesellschaft für Forschungsförderung Niederösterreich m.b.H. NÖ Forschungs-und Bildungsges.m.b.H. (GFF) grant FTI20-002.

References

1. Churkina, G., Organschi, A., Reyer, C.P.O., Ruff, A., Vinke, K., Liu, Z., Reck, B.K., Graedel, T.E., Schellnhuber, H.J.: Buildings as a global carbon sink. Nat. Sustain. **3**(4), 269–276 (2020). https://doi.org/10.1038/s41893-019-0462-4
2. Pramreiter, M., Nenning, T., Huber, C., Müller, U., Kromoser, B., Mayencourt, P., Konnerth, J.: A review of the resource efficiency and mechanical performance of commercial wood-based building materials. Sustain. Mater. Technol. **38**, e00728 (2023). https://doi.org/10.1016/j.susmat.2023.e00728
3. Ramage, M.H., Burridge, H., Busse-Wicher, M., Fereday, G., Reynolds, T., Shah, D.U., Wu, G., Yu, L., Fleming, P., Densley-Tingley, D., Allwood, J., Dupree, P., Linden, P.F., Scherman, O.: The wood from the trees: the use of timber in construction. Renew. Sustain. Energy Rev. **68**, 333–359 (2017). https://doi.org/10.1016/j.rser.2016.09.107
4. Niemz, P., Sonderegger, W.: Untersuchungen zur Korrelation ausgewählter Holzeigenschaften untereinander und mit der Rohdichte unter Verwendung von 103 Holzarten | Analysis of the correlation between selected wood properties among each other and the density of 103 wood species. Schweizerische Zeitschrift Fur Forstwesen. **154**, 489–493 (2003)
5. Morsing, N.: Densification of Wood: The Influence of Hygrothermal Treatment on Compression of Beech Perpendicular to Gain. Ph.D. thesis, Technical University of Denmark, Kgs. Lyngby, Denmark (1998)
6. Hillis, W.E., Rozsa, A.N.: The softening temperatures of wood. Holzforschung. **32** (1978). https://doi.org/10.1515/hfsg.1978.32.2.68
7. Olsson, A.-M., Salmén, L.: The association of water to cellulose and hemicellulose in paper examined by FTIR spectroscopy. Carbohydr. Res. **339**(4), 813–818 (2004). https://doi.org/10.1016/j.carres.2004.01.005
8. Horiyama, H., Miyoshi, Y., Kojiro, K., Furuta, Y.: Thermal softening properties of various wood species within an annual ring. J. Wood Sci. **69**(1), 30 (2023). https://doi.org/10.1186/s10086-023-02104-2
9. Uhmeier, A., Morooka, T., Norimoto, M.: Influence of thermal softening and degradation on the radial compression behavior of wet spruce. Holzforschung. **52** (1998). https://doi.org/10.1515/hfsg.1998.52.1.77
10. Stamm, A.J., Seborg, R.M.: Resin-treated, laminated, compressed wood. T Am. Inst. Chem. Eng. **37**, 0385–0398 (1941)
11. Shams, I., Yano, H., Endou, K.: Compressive deformation of wood impregnated with low molecular weight phenol formaldehyde (PF) resin I: effects of pressing pressure and pressure holding. J. Wood Sci. **50**(4), 337–342 (2004). https://doi.org/10.1007/s10086-003-0570-6
12. Song, J., et al.: Processing bulk natural wood into a high-performance structural material. Nature. **554**(7691), 224–228 (2018). https://doi.org/10.1038/nature25476
13. Frey, M., Widner, D., Segmehl, J.S., Casdorff, K., Keplinger, T., Burgert, I.: Delignified and densified cellulose bulk materials with excellent tensile properties for sustainable engineering. ACS Appl. Mater. Interfaces. **10**(5), 5030–5037 (2018). https://doi.org/10.1021/acsami.7b18646
14. Yano, H.: Potential strength for resin-impregnated compressed wood. J. Mater. Sci. Lett. **20**(12), 1127–1129 (2001). https://doi.org/10.1023/a:1010996424453
15. Jakob, M., Mahendran, A.R., Gindl-Altmutter, W., Bliem, P., Konnerth, J., Müller, U., Veigel, S.: The strength and stiffness of oriented wood and cellulose-fibre materials: a review. Prog. Mater. Sci. **125**, 100916 (2022). https://doi.org/10.1016/j.pmatsci.2021.100916
16. Wu, X., Moon, R.J., Martini, A.: Tensile strength of Iβ crystalline cellulose predicted by molecular dynamics simulation. Cellul. **21**(4), 2233–2245 (2014). https://doi.org/10.1007/s10570-014-0325-0

17. Saito, T., Kuramae, R., Wohlert, J., Berglund, L.A., Isogai, A.: An ultrastrong nanofibrillar biomaterial: the strength of single cellulose nanofibrils revealed via sonication-induced fragmentation. Biomacromolecules. **14**(1), 248–253 (2013). https://doi.org/10.1021/bm301674e
18. Nishino, T., Takano, K., Nakamae, K.: Elastic modulus of the crystalline regions of cellulose polymorphs. J Polym Sci B. **33**(11), 1647–1651 (1995). https://doi.org/10.1002/polb.1995.090331110
19. Sakurada, I., Nukushina, Y., Ito, T.: Experimental determination of the elastic modulus of crystalline regions in oriented polymers. J. Polym. Sci. **57**(165), 651–660 (1962). https://doi.org/10.1002/pol.1962.1205716551
20. Cousins, W.J.: Elastic modulus of lignin as related to moisture content. Wood Sci. Technol. **10**(1), 9–17 (1976). https://doi.org/10.1007/BF00376380
21. Cousins, W.J.: Young's modulus of hemicellulose as related to moisture content. Wood Sci. Technol. **12**(3), 161–167 (1978). https://doi.org/10.1007/BF00372862
22. Liu, G.R.: A step-by-step method of rule-of-mixture of fiber- and particle-reinforced composite materials. Compos. Struct. **40**(3), 313–322 (1997). https://doi.org/10.1016/S0263-8223(98)00033-6
23. Jakob, M., Gindl-Altmutter, W.: The effect of partial delignification on the stress–strain relationship in transverse compression. J. Mater. Sci. **58**(3), 1071–1085 (2023). https://doi.org/10.1007/s10853-022-08074-y
24. Jakob, M., Gaugeler, J., Gindl-Altmutter, W.: Effects of fiber angle on the tensile properties of partially delignified and densified wood. Materials. **13**(23), 1–9 (2020). https://doi.org/10.3390/ma13235405
25. Jakob, M., Czabany, I., Veigel, S., Müller, U., Gindl-Altmutter, W.: Comparing the suitability of domestic spruce, beech, and poplar wood for high-strength densified wood. Eur. J. Wood Wood Prod. **80**(4), 859–876 (2022). https://doi.org/10.1007/s00107-022-01828-0
26. Gindl, W., Teischinger, A.: Axial compression strength of Norway spruce related to structural variability and lignin content. Compos. A Appl. Sci. Manuf. **33**(12), 1623–1628 (2002). https://doi.org/10.1016/S1359-835X(02)00182-3
27. Neudecker, F., Jakob, M., Bodner, S.C., Keckes, J., Buerstmayr, H., Gindl-Altmutter, W.: Delignification and densification as a route to enable the use of wheat straw for structural materials. ACS Sustain. Chem. Eng. **11**(19), 7596–7604 (2023). https://doi.org/10.1021/acssuschemeng.3c01375
28. Li, Z., Chen, C., Mi, R., Gan, W., Dai, J., Jiao, M., Xie, H., Yao, Y., Xiao, S., Hu, L.: A strong, tough, and scalable structural material from fast-growing bamboo. Adv. Mater. **32**(10), 1906308 (2020). https://doi.org/10.1002/adma.201906308
29. Neudecker, F., Veigel, S., Puhr, C., Mihalyi, S., Guebitz, G.M., Buerstmayr, H., Gindl-Altmutter, W.: A biotechnological approach to upgrade wheat straw into high-performance binderless boards. Mater Today Sustain. **26**, 100744 (2024). https://doi.org/10.1016/j.mtsust.2024.100744

Analysis

Optical Properties of Spectrally Irradiated Wood

Hiroyuki Sugimoto(✉), Kai Maruyama, and Masatoshi Sugimori

Department of Forest Resources, Faculty of Agriculture, Ehime University, 3-5-7, Tarumi, Matsuyama, Ehime, Japan

sugimoto.hiroyuki.rw@ehime-u.ac.jp

Abstract. The appearance of wood is known to degrade when the wood is exposed to light. However, as the type and degree of degradation vary with the wavelength of the irradiated light, mechanisms such degradation are not fully understood. One of reason that it is difficult to elucidate the degradation mechanism caused by the irradiation light is that degradation due to factors other than the irradiation light occurs during sample preparation. To evaluate the color change caused by spectral light irradiation, samples were prepared to minimize the effects of other factors, such as other factors related to light and heat, the leakage of extractive components, and differences in cell structure. The samples were subjected to spectral light irradiation, and their optical properties were then measured. Both the transmittance and reflectance of wood changed according to the irradiation wavelength. A comparison of the differences before and after treatment showed a decrease in transmittance and reflectance treated with irradiated light in the 279–373 nm wavelength range and an increase in transmittance and reflectance treated with irradiated light in the 404–496 nm range. The position of the negative peak observed in the difference spectrum was not affected by the wavelength of the irradiation, but the position of the positive peak shifted with the irradiation wavelength. An understanding of these simple changes would shed light on the mechanisms of degradation in light-irradiated wood.

Keywords: appearance · degradation · optical properties · reflectance · transmittance

1 Introduction

Wood is a common material that is used for a variety of purposes in daily life, including wooden furniture, flooring, and decorative objects. However, one of the problems encountered with the use of wood is its susceptibility to degradation from light, which worsens its appearance. This degradation varies according to the wavelength of light [1–3]. In recent years, there has been a proliferation of indoor lighting sources such as fluorescent lamps, LEDs, and UV-cut window glass, each having unique wavelength characteristics. Given the need for wood to withstand exposure to various types of indoor lighting, it is essential to understand how its appearance changes according to the wavelength of irradiation. Some studies on the color change in wood due to irradiation with various wavelengths of light have revealed that the sapwoods of larch (*Larix decidua*),

G. Goli et al. (Eds.): ECWM 2024, SPM 86, pp. 291–296, 2025.
https://doi.org/10.1007/978-3-031-99418-0_36

hinoki (*Chamaecyparis obtusa*), and sugi (*Cryptomeria japonica*) darken when they are exposed to wavelengths of light in the ranges of 350–390 [1], 300–390 [2], and 246–403 nm [3]. Conversely, the sapwoods become lighter in response to wavelengths of light in the ranges of 390–580 [1], 440–560 [2], and 434–496 [3]. The variations in the lightness changes of these woods have been attributed to differences in their constituent molecules, such as lignin or extractives [4]. However, these previous studies did not consider the effects of light, heat, and the leakage of extracted components during sample preparation. To elucidate the mechanism of the discoloration caused by light irradiation, more rigorous sample preparation is required. Moreover, previous studies have directly linked the chemical constituents of wood to the mechanisms of color changes, with the aim to clarify these processes. However, the relationship between color change and chemical constituents tends to be complex, and the mechanisms have not yet been fully elucidated. Traditionally, color is analyzed using values such as CIE $L^*a^*b^*$, which represent optical properties that are relevant to human perception through a color-matching function. Therefore, CIE $L^*a^*b^*$ values may be inadequate for comprehensively evaluating material changes. However, optical properties directly "reflect" changes in the materials and offer a more direct measure of transformation. We showed that the transmission and reflection of incident light into a wood material is greatly influenced by the cell structure of the material, which suggests that light absorption may also be affected by the structure of the material [5].

Therefore, to evaluate only the color change caused by light irradiation, samples were prepared in this study to minimize the effects of other factors, such as other light history, heat, leakage of extractive components, and differences in cell structure. The resulting samples were then subjected to spectral light irradiation, and the changes in their optical properties, such as reflectance and transmittance, were measured and analyzed.

2 Experimental

2.1 Sampling

Within 24 h of felling a 78-year-old sugi (*Cryptomeria japonica*) tree from the University Forest at Ehime University, the log was debarked and made into flitch.

Thereafter, the flitches were processed into straight-grain veneer slices of predetermined dimensions (900 mm in the fiber (L) direction × 350 mm in the radial (R) direction × 0.3 mm in the tangential (T) direction). These processes were carried out carefully to minimize exposure to sunlight and lighting. The samples were then covered with plastic wrap to prevent the drying and leakage of extracted components during transportation between processing plants. The moisture content of the processed veneer slices was adjusted to approximately 9% by hanging them to dry for 5 days in a dark room in which air was circulating through a circulator. The samples were fabricated by continuously cutting 15 mm (L) × 25 mm (R) from the sapwood and of the heartwood parts of the veneer slices.

2.2 Irradiating Treatment

Using a multi-wavelength irradiation spectrometer (Bunko Keiki Co., Ltd., MM-3KA), we irradiated the samples for 8 h at eight different wavelengths between from 278 nm and 496 nm by using a diffraction grating (Table 1). Samples that had been continuous in the fiber (L) direction were used for the eight light irradiation conditions.

Table 1. Irradiating intensity at each wavelength.

Irradiating wavelength (nm)	279	310	342	373	404	435	466	496
Irradiating intensity (W/m^2)	118	178	233	275	268	252	282	187

2.3 Measurements of Optical Properties

Using an UV-visible near-infrared spectrophotometer (Shimadzu Corporation, SS3700) that was equipped with an integrating sphere, we measured the optical properties, reflectance, and light transmittance of the samples over the measurement wavelength range of 300–800 nm. A barium sulfate white plate was used as a reference. To confirm that there were no individual differences between the samples before irradiation, their optical properties were compared. As an example, Fig. 1 shows transmittance (τ_0) and reflectance (ρ_0) of sugi sapwood samples before irradiation depending on the wavelength (λ_m). Because we confirmed that there were almost no individual differences in the optical properties of the materials, we determined that the effect of light irradiation at each irradiation wavelength could be measured.

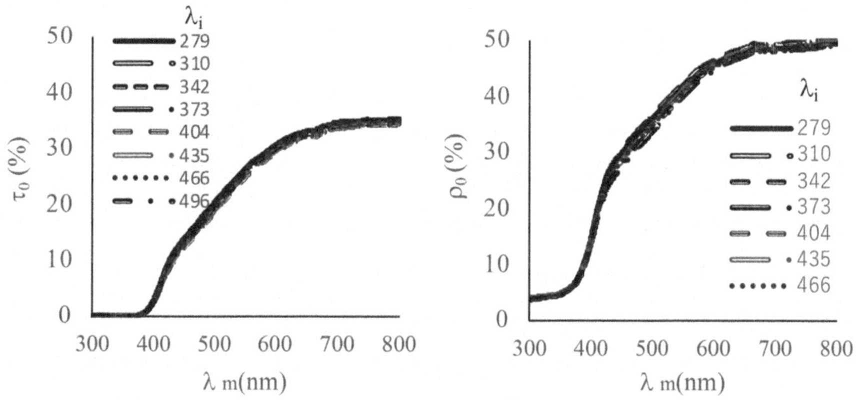

Fig. 1. Dependence on wavelength (λ_m) dependence of transmittance (τ_0) (left) and reflectance (ρ_0) (right) of sugi sapwood samples before irradiation. The numbers in the legend in the figure indicate that the sample will subsequently be irradiated by light of the corresponding irradiation wavelength (λ_i).

3 Results and Discussion

To investigate the effects of the irradiation wavelengths on the optical properties of the wood, the transmittance (τ) and reflectance (ρ) spectra of the samples irradiated at various irradiation wavelengths (λ_i) are shown in Fig. 2. The light irradiation treatment seems to change both τ and ρ of the wood at the measurement wavelength between 400 and 500 nm.

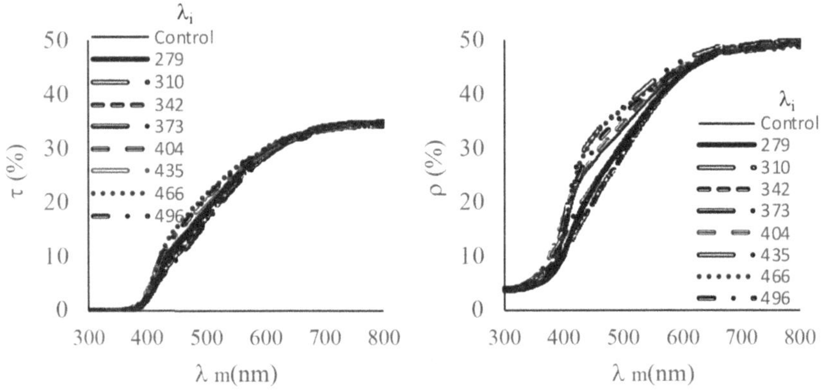

Fig. 2. Dependence on wavelength (λ_m) of transmittance (τ)(left) and reflectance (ρ)(right) of sugi sapwood samples irradiated with various irradiation wavelengths (λ_i).

To investigate this phenomenon in greater detail, the difference (Δτ, Δρ) between the changes in transmittance or reflectance of the samples after they were irradiated with various irradiation wavelengths (λ_i) (transmittance or reflectance of the treated sample minus that of the untreated sample) is shown in Fig. 3. For samples irradiated with λ_i of 279–373 nm, which gave a negative peak (α peak) in Δτ or Δρ spectra appeared at λ_m of approximately around 440 nm. The magnitude of the negative Δτ and the Δρ peak indicates that the irradiation treatment resulted in a decrease in the reflectance or transmittance, i.e., the color tended to become darker. This is presumed to be due to an increase in some constituent that absorbs light wavelengths of approximately 440 nm. According to the literature [1], the constituent in question is likely to be quinone(s) produced from the decomposition products of lignin.

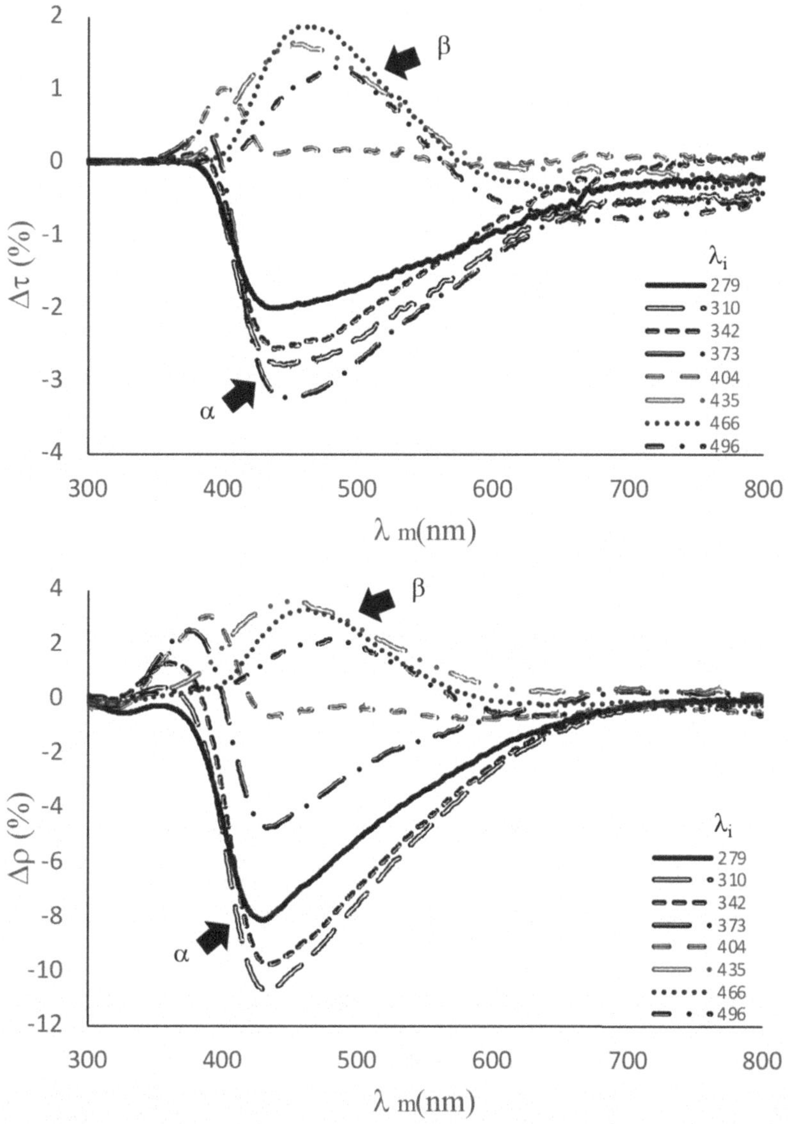

Fig. 3. Dependence on wavelength (λ_m) of the difference in transmittance ($\Delta\tau$: upper) and of the difference in reflectance ($\Delta\rho$: lower) of sugi sapwood samples irradiated with light of various irradiation wavelengths (λ_i). α: negative peak, β: positive peak.

Furthermore, for the samples that had been irradiated with a λ_i of 342–496 nm, a positive peak (β peak) appeared in $\Delta\tau$, $\Delta\rho$ spectra, and its position shifted to higher values of λ_m with an increase in λ_i. The position of the peak in $\Delta\tau$, $\Delta\rho$ shifted, which might have caused the values of a^* and b^* to change in a complex manner. From this, it is presumed that the structure, such as the constituent or the functional group, that corresponded to the energy state of λ_i had decrease or transformation. Moreover, as the wavelength range of the shift peak is wide, it seems that the structure that gives rise to the peak is not uniform but diverse.

4 Conclusion

To evaluate only the color change caused by spectral light irradiation, we prepared samples to minimize the effects of other factors such as other light and heat factors, the leakage of extractive components, and differences in cell structure. The samples were subjected to spectral light irradiation and their optical properties were then measured.

According to the obtained results, the following conclusions can be drawn:

1. Both transmittance and reflectance showed different changes depending on the irradiation wavelength.
2. A comparison of the differences before and after treatment showed a decrease in transmittance and reflectance due to irradiation with the light range in the 279–373 nm range and an increase in transmittance and reflectance due to irradiation with the light range in the 404–496 nm range.
3. The position of the negative peak in the difference spectra of transmittance or reflectance before and after irradiation treatment did not change with irradiation wavelength, but the position of the positive peak shifted with irradiation wavelength. These are thought to be caused by two different changes of constituent.

References

1. Minemura, S., Umehara, M.: Mokuzai no Choshoku. Hokkaido Shikenjo Hokoku. **68**, 92–145 (1979)
2. Mitsui, K.: Changes in the properties of light-irradiated wood with heat treatment. Part 2. Effect of light-irradiation time and wavelength. Holz. Roh. Werkst. **62**(1), 23–30 (2004)
3. Kataoka, Y., Kiguchi, M., Williams, R.S., Evans, P.D.: Violet light causes photodegradation of wood beyond the zone affected by ultraviolet radiation. Holzforschung. **61**(1), 23–27 (2007)
4. Nishihara, Y., Sugimoto, H., Sugimori, M.: Changes in visible light spectral properties of wood due to extraction treatments. In: Proceedings of the 73rd Annual Meeting of Japan Wood Research Society. Fukuoka, Japan, p. C000548 (2023)
5. Sugimoto, H., Kawabuchi, S., Sugimori, M., Gril, J.: Reflection and transmission of visible light by Sugi wood: effects of cellular structure and densification. J. Wood Sci. **64**(6), 738–744 (2018)

Study of the Machinability of Thermally and Chemically Modified Wood for Art Objects

Rostom Leila[1](✉), Damay Jérémie[2,3], Bender Tobias[4], Munk Christoph[4], Jousserand Michael[5], and Gérardin Philippe[1]

[1] Université de Lorraine, INRAE, LERMAB, 54000 Nancy, France
{leila.rostom,philippe.gerardin}@univ-lorraine.fr
[2] CIRAD, BioWooEB, 34000 Montpellier, France
jeremie.damay@cirad.fr
[3] EcoFoG, 97310 Kourou, Guyane, France
[4] Work Group Physics and Chemistry of Wood, University of Sustainable Development Eberswalde, Schicklerstraße 5, 16225 Eberswalde, Germany
[5] Buffet Crampon, 78711 Mantes la Ville, France
michael.jousserand@buffetcrampon.com

Abstract. Three hardwood species were chemically modified by furfurylation in order to assess their behavior to machining. Two treatment protocols were achieved: one being a combination of a thermal treatment prior to the chemical modification with furfuryl alcohol solution and the other one being only a chemical modification by furfuryl alcohol solution. Both treatments showed an overall improvement of the dimensional stability of all wood species and a decrease in the moisture content while density and Brinell hardness increased. Although, some differences have been observed between these two treatments with a better anti swelling efficiency and Brinell hardness of the samples subjected to a heat treatment prior to furfurylation. Nevertheless, the machining behavior of the latter samples was poor compared to samples modified only by furfurylation. Differences in machining behavior were also observed among the studied wood species and the furfurylated hornbeam appeared to be the most interesting modified wood.

Keywords: Furfurylation · Thermal Treatment · Chemical Treatment · Wood Modification · Wood machining

1 Introduction

Dimensional stability, decay resistance and great mechanical properties are the main characteristics of concern when using wood in many fields (buildings, boats, utensils, objects of art…). Research works have been engaged for centuries in the understanding of wood behavior and dimensional stabilization in order to develop the use of it and particularly of the local resources of wood. As part of many protocols, two main categories can be found: the thermal treatments and the chemical treatments. Thermal treatments are known for reducing the moisture content and increasing the dimensional stability

G. Goli et al. (Eds.): ECWM 2024, SPM 86, pp. 297–304, 2025.
https://doi.org/10.1007/978-3-031-99418-0_37

of wood by reducing the amount and availability of hydroxyl groups in the wood cell walls through degradation of the wood polymers [1–3]. Among chemical treatments, furfurylation has shown very good results in terms of dimensional stability and durability of wood, in addition to being a less harmful preservation method for the environment [4, 5]. To our knowledge, there have been many studies about furfurylation but none of them really shows the machining behavior of such modified wood, especially in terms of turning.

The purpose of this study is to:

- Better characterize the properties of furfurylated wood in combination or not with a heat pretreatment, as the harsher the thermal pretreatment is, the higher the solution uptake and weight percent gain could be for the studied wood species [5];
- Evaluate the machinability of the modified wood species to determine which treatment is best suited for machining purposes.

2 Experimental

2.1 Wood Materials

Three wood species were studied: alder (*Alnus glutinosa*), hornbeam (*Carpinus betulus*) and maple (*Acer pseudoplatanus*). Ten replicates of each species for each treatment and experiment were cut from mature heartwood free of any defect.

Two specimen dimensions were prepared: blocks of 195 x 40 x 40 mm^3 (long. x rad. x tang.) were used for wood characterization (Group 1), while cylinders of 280 x 40 x 40 mm^3 were used for the machining experiments (Group 2). Note that the latter samples have been turned and a hole was drilled in the center before the chemical treatment.

2.2 Treatments

All samples were dried at 103 °C for 48 h prior to any treatment. Their mass was weighed and their volume measured to determine the dry mass (m_0) and dry volume (V_0) of the specimens.

Two treatment protocols were investigated in this study. The first one is a chemical treatment by impregnation with a furfuryl alcohol (FA) solution containing tartaric acid as reactive catalyst (5%) and water (10%). The second treatment corresponds to the combination of a thermal modification preceding the impregnation of the furfuryl alcohol solution (TMT + FA).

2.2.1 Thermal Modification

The thermal modification process was performed through three stages: a progressive heating phase at the rate of 10 $K.h^{-1}$, a holding phase at the targeted temperature (200 °C) and finally a cooling phase until 20 °C. The holding phase duration was dependent on the species and on the targeted mass loss. Two target mass losses were applied for samples of Group 1: TMT-A which corresponds to an averaged mass loss of 8% and TMT-B that corresponds to an averaged mass loss of 12%. Two heat treatment processes were applied in order to evaluate the influence of the mass loss, since a higher mass loss would theoretically allow more solution to be impregnated in the wood. However, it is also known that a higher mass loss could decrease the mechanical properties of wood [4]. Therefore, for Group 2 (drilled cylinders for machining tests), only the TMT-A was applied in order to minimize the loss of cellulosic matter and to avoid extreme weakening of the mechanical properties.

2.2.2 Chemical Modification—Impregnation and Polymerization

The impregnation with FA solution was performed using a vacuum-pressure autoclave. Dried wood samples were introduced in the reactor and subjected to a vacuum of 10 mbars for 15 min. The FA solution was then injected into the reactor and the system is left under partial vacuum for 15 min. At the end of the vacuum step, a pressure of 12 bars was applied for 40 min. After the vacuum-pressure cycle, the system was brought back to atmospheric pressure to retrieve the samples, the latter were weighed (m_{imp}) to determine the impregnation rate (IR) according to Eq. (1), and their dimensions recorded to determine the swelling due to impregnation (S_{imp}) according to Eq. (2).

$$IR\ (\%) = \frac{m_{imp} - m_0}{m_0} \times 100 \tag{1}$$

$$S_{imp}\ (\%) = \frac{V_{imp} - V_0}{V_0} \times 100 \tag{2}$$

where m_0 and V_0 are the mass and volume at the native dry state (103 °C, 48 h) and V_{imp} is the volume of the samples after impregnation with FA solution.

The impregnated samples were air-dried at room temperature for 48 h to allow the residual water to evaporate before curing in the oven and to avoid cracks. The polymerization step was performed by curing the samples (wrapped in aluminum foil to avoid FA evaporation) in a ventilated oven. The heating process consist in heating at 40 °C for 12 h, then increasing progressively the temperature at the rate of 30 $°C.h^{-1}$ to the targeted temperature (120 °C), a holding phase of 18 h at 120 °C and finally a progressive cooling phase. Once the samples have cooled down to room temperature, they were placed in an oven at 103 °C for 48 h in order to fully evaporate the residual water. The mass of dry cured specimens was weighed (m_{pol}) and their dimensions measured (to obtain V_{pol}) in order to determine the weight percent gain (WPG) and the swelling after polymerization (S_{pol}) according to Eqs. (3) and (4), respectively.

$$WPG\ (\%) = \frac{m_{pol} - m_0}{m_0} \times 100 \tag{3}$$

$$S_{pol}\ (\%) = \frac{V_{pol} - V_0}{V_0} \times 100 \tag{4}$$

2.3 Characterization

Characterization in terms of mass loss, impregnation rate and weight percent gain was performed on samples from Group 1 and 2. Further characterization in terms of swelling in water (S_{sat}), anti-swelling efficiency (ASE) and hardness were performed essentially on samples from Group 1. To do so, the modified wood blocks were cut in small pieces as shown in Fig. 1. The samples of dimensions 10 x 20 x 20 mm^3 (long. x rad. x tang.) were used to evaluate the swelling and the ASE, while the samples of dimensions 100 x 20 x 20 mm^3 (long. x rad. x tang.) were used to measure the surface hardness.

Fig. 1. Cutting of the wood specimens from Group 1.

2.3.1 Swelling and Anti-swelling Efficiency

Dimensional stability of the treated specimens was determined by immersing the samples in water. In order to achieve a high degree of water saturation, the immersed samples were placed in a desiccator under vacuum for two steps of 30 min spaced by 1 h at atmospheric pressure. Then, the samples were left immersed in water for 24 h at atmospheric pressure. Once the vacuum-pressure cycle was finished, the samples were weighed and their dimensions were measured to determine the swelling due to water immersion (S_{sat}), according to Eq. (5).

$$S_{sat}\ (\%) = \frac{V_{sat} - V_{pol}}{V_{pol}} \times 100 \quad (5)$$

The anti-swelling efficiency (ASE) is another widely used parameter to determine the efficiency of a treatment on improving the dimensional stability by comparing the swelling of an untreated wood specimen after water immersion ($S_{untreated}$) to the swelling of a treated wood specimen after water immersion test ($S_{treated}$). The volumetric ASE is calculated according to the following Eq. (6).

$$ASE\ (\%) = \frac{S_{untreated} - S_{treated}}{S_{untreated}} \times 100 \quad (6)$$

2.3.2 Brinell Hardness

The surface hardness was measured following the DIN EN ISO 6506-1 (2014) using a TIRA test 28,025 E22 [6]. The details of the used parameters can be found in Damay *et. al* [5]. The studied specimens correspond to the wood samples of dimensions 100 x 20 x 20 mm^3 from Group 1. Ten replicates of each untreated and treated species were conditioned at 65% relative humidity at 20 °C until they reached constant mass before being tested.

2.3.3 Machining

The machining tests were performed on the samples from Group 2 using a numerically controlled machine, which is usually used for metal machining. Firstly, they were turned in order to obtain a thinner cylinder. Afterwards, small holes were drilled in the wood cylinders using small machining cutters. Finally, they were polished to smooth and improve the surface condition in order to study the esthetic of the processed wood.

To assess the overall condition and esthetic of the machined wood, each step of the machining was evaluated from the resistance to the turning, the aspect of the inside and outside drill surface and the aspect of the surface of the small holes. A score of 1 was attributed to a good surface condition with a smooth, neat and shiny surface, while 0 was for a rough, uneven surface. The total score for the machining steps was added to the total number of fully machined and unbroken wood samples in order to determine the ranking of the treated wood species and determine which could be the best wood-treatment association for wood machining. Note that these evaluations were based on visual observations.

3 Results and Discussion

Figure 2 displays the volume evolution of the treated wood samples (Group 1) from the native dry state (V_0) to the final dry polymerized state after curing and drying (V_{pol}). These results show that the volume of the samples increased by around 15% after impregnation for all three hardwood species, which indicates that the FA solution was able to penetrate in the wood cell walls and induce the swelling of the samples. Moreover, the volume of the samples after curing stays higher than the volume at native dry state for all three treated species. This shows that the FA solution was also able to polymerize into the cell walls as the blanks were still swollen after curing. These observations are consistent with the literature as Liu *et al.* showed deposition and polymerization of FA in the cell walls as well as in the lumen of pine wood by Scanning Electron Microscopy [7]. They are also consistent with the reduced swelling of the samples after FA treatment (Fig. 3), which is even more reduced when it is combined with a heat pretreatment, as there are less hydroxyl groups available and accessible to water.

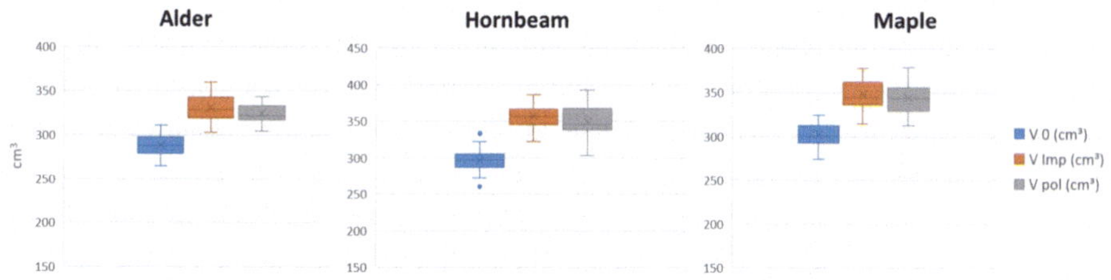

Fig. 2. Volume evolution of wood samples at the native dry state, after impregnation and dry polymerized state after curing and drying, for each treatment and wood species.

These results are also consistent with the WPG values in the Table 1 below, as the mass gain is up to 135% for alder. The WPG depends on the wood species, as alder is

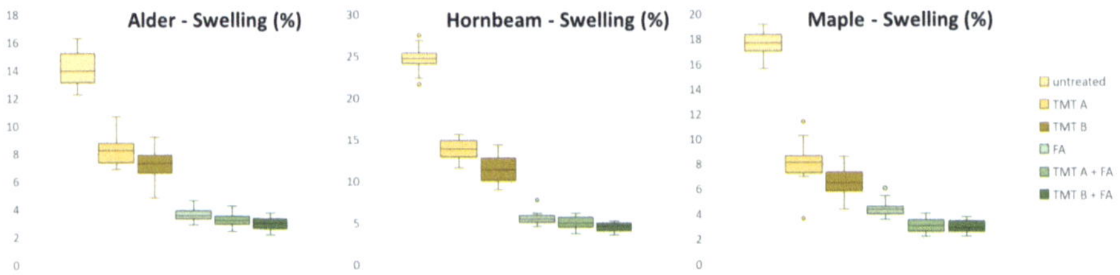

Fig. 3. Evolution of the swelling after water immersion of three hardwood species after a furfurylation treatment and a combined TMT + Furfurylation treatment.

more impregnable than maple or hornbeam, and hornbeam being the less impregnable species, the WPG values are lower. Those results are consistent with literature [5] for these hardwood species treated with FA solution and TMT + FA.

Table 1. WPG (%), ASE (%) and density values for three hardwood species treated by impregnation with FA solution and with a combined TMT + Furfurylation treatment.

		Group 1			
		Untreated	FA	TMT-A + FA	TMT-B + FA
Alder	WPG (%)	–	130.32 ± 9.31	135.57 ± 14.88	135.09 ± 6.92
	ρ (g/cm^3)	0.48 ± 0.03	1.01 ± 0.01	0.99 ± 0.01	0.97 ± 0.01
	ASE (%)	–	74.88 ± 2.33	77.75 ± 2.89	79.34 ± 2.52
Hornbeam	WPG (%)	–	37.15 ± 5.53	46.60 ± 9.44	55.60 ± 6.94
	ρ (g/cm^3)	0.76 ± 0.02	0.89 ± 0.04	0.95 ± 0.05	1.00 ± 0.05
	ASE (%)	–	77.60 ± 3.63	80.02 ± 2.66	82.20 ± 1.57
Maple	WPG (%)	–	77.69 ± 12.45	101.72 ± 9.06	104.33 ± 6.50
	ρ (g/cm^3)	0.55 ± 0.03	0.89 ± 0.05	0.98 ± 0.02	0.96 ± 0.01
	ASE (%)	–	75.02 ± 3.27	82.88 ± 1.93	83.38 ± 1.50

Density seems to be even greater with a heat pretreatment for hornbeam and maple, whereas for alder, this combination seems to be too harsh, degrading severely the lignocellulosic matter of wood. As the FA solution creates an acidic reactive environment, the cell walls are prone to be degraded in addition to the degradation resulting from the heat pretreatment.

Dimensional stability of treated wood species was greatly improved after these treatments, and even more so when the FA treatment is combined with a thermal pretreatment.

Surface hardness (Fig. 4) was notably improved after FA and TMT + FA treatments. However, there is no significant difference in surface hardness improvement between FA and TMT + FA in regards to the standard deviation.

The WPG values of samples from Group 2 were of the same order of magnitude as those of Group 1 for each treated species, which means that the characteristics of

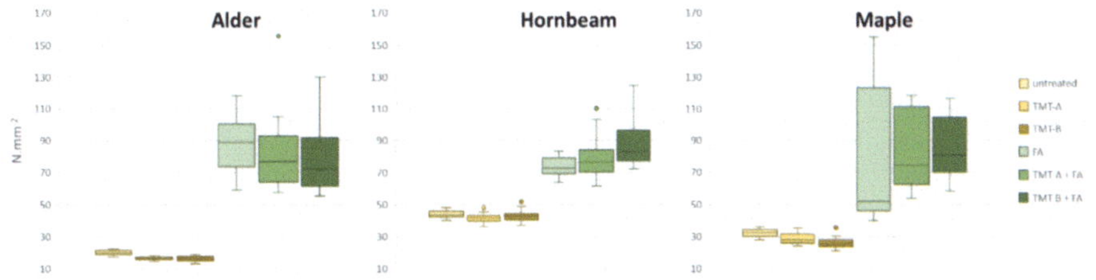

Fig. 4. Surface hardness according to Brinell for alder, hornbeam and maple.

the samples from Group 2 must be the same as those from Group 1 in terms density gain, dimensional stability and surface hardness. Therefore, as the combined TMT + FA treatment shows the best improvement in properties for all three wood species, we could suppose that these materials would be the best suited for machining.

The machining experiments will be decisive for determining the mechanical properties and resistance to cracks in real life application.

Contrary to our expectations due to the good results obtained previously, all the samples that were thermally pretreated experienced a very high percentage of splits during machining which made them unusable and unsuitable for the rest of the machining process (Fig. 5).

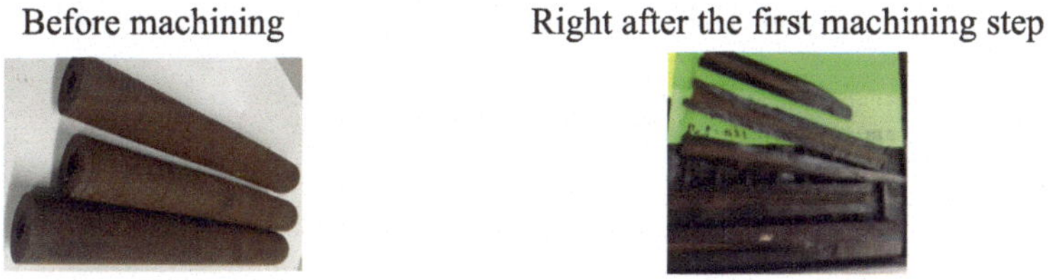

Fig. 5. Furfurylated samples before machining and after the first machining step.

On the contrary, the FA-treated samples behaved differently as it was possible to process them. The machining varied greatly with the treated wood species. Indeed, we observed a higher percentage of splitting from the treated alder as only 4 samples over 10 were able to be fully processed. Whereas, there was 6 fully processed samples for maple and 8 for hornbeam. These results show that the FA-treated hornbeam species seems to be the best candidate for machining and especially for turning.

The machined wood samples were graded by their esthetic as shown in Table 2 and it appears that maple has the best score in terms of surface condition with a total of 2.5 points against 2 points for alder and hornbeam. Unfortunately, maple is still more brittle than hornbeam for machining. Therefore, regarding the final score for each species, furfurylated hornbeam seems to be the best material for wood turning even though the WPG was lower than the other two hardwood species. This might be due to its intrinsic properties, such as its high natural density, and cellular organization.

Table 2. Scoring of the different steps of the machining and esthetics of the processed wood.

	Alder	Hornbeam	Maple
Drilling	0	1	1
Turning	0	1	0.5
Inner surface condition	0.5	0.5	0.5
Outer surface condition	1	1	1
Surface condition of the small holes	0.5	0.5	1
Number of fully machined blanks	4	8	6
Total Score	6	12	10

4 Conclusion

This study enabled to show that for these hardwood species, the thermal pre-treatment is not suited prior to furfurylation for machining purposes and especially for turning. However, furfurylation only enabled the machining of the treated wood, showing different behaviors according to each wood species. Furfurylated hornbeam showed the best results in terms of machining compared to alder and maple.

Acknowledgements. The authors thanks France Relance for the financial support allowed to this project.

References

1. Candelier, K., Thevenon, M.F., Petrissans, A., Dumarcay, S., Gerardin, P., Petrissans, M.: Control of wood thermal treatment and its effects on decay resistance: a review. Ann. For. Sci. **73**(3), 571–583 (2016)
2. Altgen, M., Willems, W., Hosseinpourpia, R., Rautkari, L.: Hydroxyl accessibility and dimensional changes of Scots pine sapwood affected by alterations in the cell wall ultrastructure during heat-treatment. Polym. Degrad. Stab. **152**, 244–252 (2018)
3. Rostom, L., Courtier-Murias, D., Rodts, S., Care, S.: Investigation of the effect of aging on wood hygroscopicity by 2D 1H NMR relaxometry. Holzforschung. **74**(4), 400–411 (2020)
4. Zelinka, S.L., Altgen, M., Emmerich, L., Guigo, N., Keplinger, T., Kymäläinen, M., et al.: Review of wood modification and wood functionalization technologies. Forests. **13**(7), 1004 (2022)
5. Damay, J., Bender, T., Munk, C., Jousserand, M., Creton, M., Fredon, E., et al.: Properties improvement of seven hardwood species by combination of thermal and chemical modifications. Eur. J. Wood Wood Prod., 1–14 (2023)
6. International Organization for Standardization (ISO): Metallic Materials—Brinell Hardness Test—Part 1: Test Method. DIN EN ISO 6506-1, pp. 1–22 (2014)
7. Liu, M., Lyu, S., Peng, L., Cai, L., Huang, Z., Lyu, J.: Improvement of toughness and mechanical properties of furfurylated wood by biosourced epoxidized soybean oil. ACS Sustain. Chem. Eng. **9**(24), 8142–8155 (2021)

Effects of Microwave Treatment on the Retention of a Preservative Product in Two Portuguese Wood Species

Fernando J. R. Mascarenhas[1,2](✉), Alfredo M. P. G. Dias[1,2], André L. Christoforo[3], and Rogério M. S. Simões[4]

[1] ISISE, ARISE, Department of Civil Engineering, Rua Luís Reis Santos—Pólo II, University of Coimbra, 3030-788 Coimbra, Portugal
fernando.mascarenhas@student.dec.uc.pt

[2] SerQ—Innovation and Competence Forest Centre, Rua J, Nº 9, Zona Industrial da Sertã, 6100-711 Sertã, Portugal

[3] Department of Civil Engineering, Federal University of São Carlos (UFSCar), Rodovia Washington Luís (SP-11 310), Km 235, São Carlos 3030-788, Brazil

[4] Department of Chemistry, Unit of Fiber Materials and Environmental Technologies (FibEnTech-UBI), Rua Marquês D'Ávila e Bolama, University Beira Interior (UBI), 6201-001 Covilhã, Portugal

Abstract. Wood is a crucial construction material. However, it usually requires protection against biological agents through preservative impregnation. Certain wood species have low permeability, which poses difficulties for impregnation. Eucalyptus and maritime pine are relevant wood species in Portugal due to their prevalence and economic importance. Among the different modification processes of wood, the microwave (MW) treatment is being researched as a method to improve wood's impregnability. Thus, this study aimed to improve the impregnability of both Portuguese wood species with MW treatment, using small clear wood specimens. Results showed significant improvement in preservative retention of MW-treated samples, allowing for application in Use Classes 1, 2, 3, and 4, with special emphasis on the remarkable increase in preservative retention of MW-treated eucalyptus compared to control ones. Finally, MW treatment has proven to be a valuable technique for enhancing impregnability in maritime pine and overcoming low permeability in eucalyptus wood. The improved impregnability resulting from MW treatment offers great potential for expanding the use of these wood species in construction applications.

Keywords: Eucalyptus · Maritime Pine · Microwave Treatment · Preservative Product · Impregnability

1 Introduction

Wood is an important material for construction, with structural or non-structural applications, due to its appreciated mechanical, acoustic, and thermal properties and sustainable characteristics. In recent years, standardization initiatives in numerous countries have

G. Goli et al. (Eds.): ECWM 2024, SPM 86, pp. 305–312, 2025.
https://doi.org/10.1007/978-3-031-99418-0_38

played a role in elevating wood to the status of a "modern" material that is suitable for meeting the most severe safety, industrial, and market standards [1].

Despite advancements in using solid and engineered wood in construction, depending on the environmental conditions where the wood elements are used, their susceptibility to biodegradation remains an issue that deserves attention. Then, chemical preservatives must be used to saturate the wood to protect it from attacks by various biological organisms like termites and beetles. These wood preservation techniques, which vary in intensity, facilitate the use of wood in a variety of settings, including interior, exterior, above-ground, ground contact, and marine environments [2]. Therefore, the treatment of wood to enhance its durability is a crucial step that must be undertaken before the timber is utilized in the construction industry [3]. Impregnation processes with biocides are still the most used method to protect wood elements [3, 4].

In Portugal, two wood species stand out due to their predominance in forests, different uses, and economic importance: eucalyptus (*Eucalyptus globulus* Labill.) and maritime pine (*Pinus pinaster* Ait.) [5]. Although eucalyptus wood has satisfactory mechanical properties, its heartwood is not durable against fungi and termites, and it is difficult to impregnate with preservative products [6]. It happens because eucalyptus has low permeability (treatability) [7], which impacts various commercial processes, such as wood drying, chemical treatment, and pulping [8].

In this sense, researchers have been conducting investigations in diverse fields of wood modification, looking for techniques to address the inherent limitations of wood. Microwave (MW) treatment (drying) of wood has gained projection in the field of wood modification as a tool to improve wood's drying quality and permeability [7, 9–12].

Jang and Kang [13] explain that when green wood enters into contact with the electromagnetic waves of a microwave, the water molecules inside the wood vibrate rapidly against one other, creating vapor pressure. Thus, tracheids and parenchyma cells are damaged, and pit membranes and tyloses in blood vessels are broken when these water and steam pressure leave the interior of the wood. The damaged tissue is responsible for increasing wood fiber permeability, which boosts drying quality [9, 10, 13–15].

MW power, frequency, applied energy, continuous exposure time, total drying time, and the initial moisture content (IMC) of the wood specimens are some of the parameters that impact the MW drying quality and kinetics [9, 16, 17]. In the case of the MW frequency, the thicker the wood specimens, the smaller the MW frequency. MW frequencies of 2.45 GHz are commonly used to dry small clear wood specimens [17].

Despite the knowledge that has already been produced about the MW treatment of different wood species, few studies focus on Portuguese eucalyptus and pine, particularly with regard to analyzing the effects of this treatment on increasing the impregnability of both species. Hence, assuming that MW treatment can make wood species more permeable to preservative products, this research work aimed to improve the impregnability of both Portuguese wood species after being MW-dried.

To do this, small clear wood specimens were used, and only one combination of MW power and continuous MW exposure time was employed. Although porosity and permeability were not analyzed in this work, the increase in impregnability of the specimens of both species was analyzed by the variation in the retention of the preservative product used, as has been done in other works in the literature [11, 12, 18].

2 Methodology

2.1 Materials

Small clear wood specimens of both wood species, containing only heartwood, measuring 20 mm x 20 mm x 320 mm, were used. The wood samples were divided into MW-treated and control groups (with a conventional drying method). Each wood group had 16 wood specimens. The control samples (conventional drying) from maritime pine and eucalyptus were named PP_Control and EG_Control, respectively. The MW-treated samples were named PP_400W_25min and EG_400W_25min, respectively, for maritime pine and eucalyptus.

2.2 MW Treatment

MW drying was performed using a conventional MW device with a 2.45 GHz frequency under a power of 400 W. Four wood samples were placed inside the MW oven, and successive MW cycles of 25 min were applied, with cooling intervals of 30 s to weigh the samples (Fig. 1).

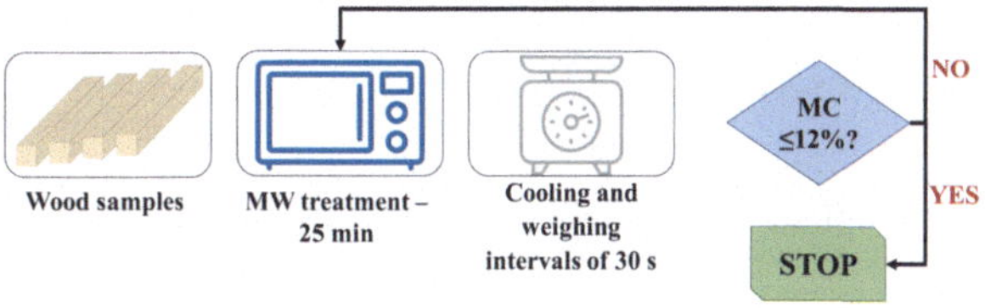

Fig. 1. MW treatment routine.

The IMC of the eucalyptus and pine samples were 60.8% (5.7%) and 95.4% (7.3%), respectively, and they were MW-dried until a final moisture content of 12% (Fig. 2).

Fig. 2. Wood samples and MW device used.

2.3 Impregnation Method

Once the MW drying was complete, the wood samples (control and MW-treated) were impregnated with a 5% aqueous formulation of wood preservative Tanalith E 8001, which is devoid of chromium and arsenic and incorporates copper alongside various other constituents Table 1), was employed (vacuum-pressure process) [18].

Table 1. Chemical composition of Tanalith E 8001.

Compound name according to IUPAC[a]	CAS Number[b]	Content (% w/w)
Copper(II) carbonate- copper(II) hydroxide (1:1)	12,069–69-1	14.57
1-(4-chlorophenyl)-4,4- dimethyl-3-(1,2,4-triazol- 1-ylmethyl)pentan-3-ol	107,534–96-3	0.16
1-[[2-(2,4-diclorofenil)-4-propil-1,3-dioxolan-2-il]metil]-1H-1,2,4-triazole (propiconazole)	60,207–90-1	0.16
Didecyldimethylammonium chloride (DDAC)	7173-51-5	0.50
Didecyl dimethyl ammonium carbonate (DDA Carbonate)	894,406–76-9	0.50
2-aminoetanol	205–483-3	26.91

[a]International Union of Pure and Applied Chemistry
[b]CAS: Chemical Abstracts Service

The retention and uptake were calculated using Eqs. 1 and 2, respectively:

$$R = \frac{(W_2 - W_1) \times C}{V} \times 10 \tag{1}$$

$$U = \frac{(W_2 - W_1)}{W_1} \times 100 \tag{2}$$

Where: R is the retention of the preservative product, in kg/m^3; W_2 and W_1 are the weight of the wood sample after and before the impregnation process, in g; V is the volume of the wood sample s, in cm^3; U is the uptake of the preservative product, in %.

In addition, the increase in the retention of the preservative product was calculated according to Eq. 3:

$$I_R = \frac{(R_{MW} - R_{Control})}{R_{Control}} \times 100 \tag{3}$$

Where: I_R is the increase in the retention of the preservative product, in %; R_{MW} and $R_{Control}$ are the retention of the MW-treated control samples, respectively, in kg/m^3.

2.4 Density

The densities of the wood samples before the MW treatment and the combined effect of MW treatment and Tanalith E 8001 impregnation were determined following the recommendations of ISO 13061-2 [19].

2.5 Statistical Analysis

The results of MW-treated and control samples were assessed by classical analysis of variance (ANOVA) at a 5% significance level with the Tukey Test.

3 Results and Discussion

The pine samples took 110 min to be dried, and the eucalyptus ones 100 min. Although their total MW drying time was approximately the same, due to the anatomical and chemical composition of pine samples, they dried faster than eucalyptus, as explained by Mascarenhas et al. [18].

The results demonstrated that the PP_Control retained 10.47 kg/m^3 of preservative product, whereas the corresponding samples treated with the MW (PP_400W_25min) retained 20.73 kg/m^3 (Fig. 3). This represents an increase in the retention of 98.0%.

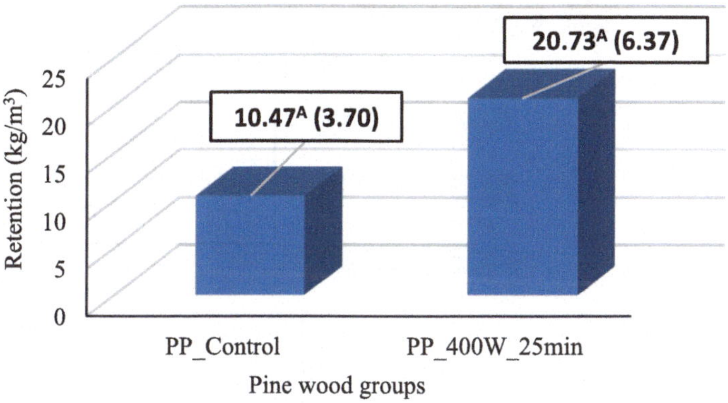

Fig. 3. Retention of the preservative product of maritime pine samples. Values associated with same letters in Tukey tests do not show significant differences with a p-value of 0.05. The values in parentheses are the standard deviation.

The results for the MW-treated eucalyptus samples were remarkable. The EG_Control samples only retained 4.94 kg of Tanalith E 8001 per cubic meter of wood, while the corresponding samples MW-treated (EG_400W_25min) retained 11.18 kg/m^3 (statistically different), showing that after MW treatment, the wood samples were satisfactorily impregnated with the preservative product (Fig. 4). This represents an increase in the retention of 126.3%. MW-treating Spanish eucalyptus, Hermoso and Vega [12] identified an increase in the retention of the same product by up to 147%.

The MW treatment led to an average increase of 71% and 132% in the uptake of preservative solution for maritime pine and eucalyptus wood samples, respectively. Although tests to assess the real durability of different degrading agents are necessary, based on the guidelines and recommendations for the use of the manufacturer of Tanalith E 8001, the MW-treated maritime pine samples could be applied in circumstances of Use Classes (UC) 1, 2, 3, and 4. The MW-treated eucalyptus wood samples could be

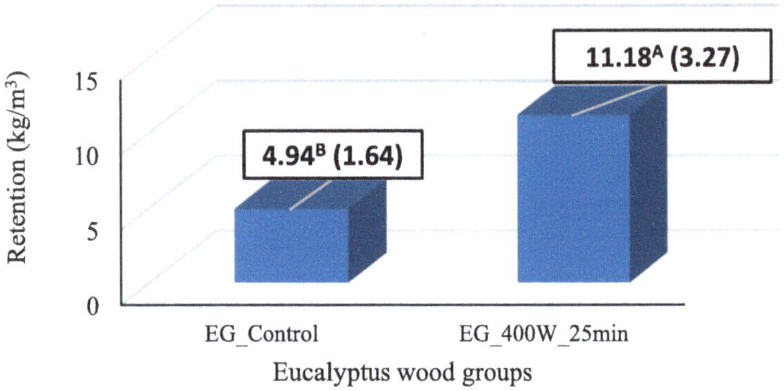

Fig. 4. Retention of the preservative product of eucalyptus samples. Values associated with distinct letters in Tukey tests show significant differences with a p-value of 0.05. The values in parentheses are the standard deviation.

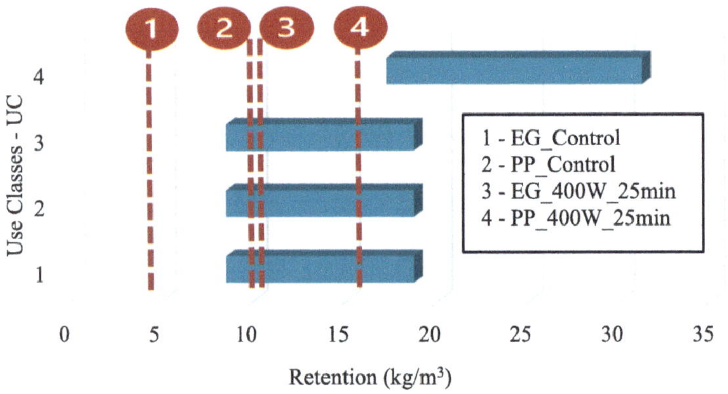

Fig. 5. The minimum and maximum quantity of the used preservative product for each UC according to the manufacturer [20] and the values retained by each wood group.

placed in conditions of UC 1, 2, and 3. At the same time, the EG_Control did not retain the minimum required quantity for any UC (Fig. 5).

The results found in this work agree with the well-known refractory nature of the eucalyptus wood regarding impregnation [7], and the present results have shown that the use of MW energy improved its impregnability. It is important to recall that Portuguese eucalyptus wood species lack natural durability and cannot be treated by regular impregnation processes [6]. The MW treatment generates heat almost uniformly in the sample, producing water vapor at appreciable pressure that moves from inside to outside, opening the wood structure, namely the tracheids, pit membranes, and parenchyma cells, increasing wood permeability [9, 15].

The conjugated effect of MW treatment and Tanalith E 8001 impregnation had a marginal effect on wood density. The density of PP_400W_25min ranged from 583 ± 9 kg/m^3 (before the MW drying) to 576 ± 6 kg/m^3 (after the MW treatment and impregnation), a decrease of 1.2%. EG_400W_25min had a density of 852 ± 30 kg/m^3

before the MW drying and 836 ± 42 kg/m^3 after the MW drying and impregnation, a decrease of 1.9%. The present results are in good agreement with the results presented by [7], which indicate that it is possible to increase permeability by 1.5 times with no statistically significant decreases in densities and mechanical properties.

Finally, the improvement in the impregnability attained through the use of MW technology can greatly enhance the applications of both wood species as building materials.

4 Conclusion

In conclusion, it has been demonstrated that microwave (MW) treatment has proven to be a highly effective technology within the field of wood modification to enhance the impregnability of Portuguese maritime pine and eucalyptus. This technology led to a substantial increase of 98% in the retention of maritime pine. Additionally, it resulted in a remarkable 126% enhancement in the retention of the preservative product in the eucalyptus samples, showcasing its potential to address the challenges posed by eucalyptus' low permeability. Ultimately, the considerable improvement in impregnability achieved through MW drying holds great promise in expanding the utilization of both maritime pine and eucalyptus as viable construction materials, thereby contributing to advancements in the field of wood science and technology.

Acknowledgments. This work is financed by national funds through Foundation for Science and Technology – FCT, from Portugal, under grant agreement 2021.07636.BD, the FCT, the Ministry of Science, Technology and Higher Education (Portuguese: Ministério da Ciência, Tecnologia e Ensino Superior or MCTES), the Programa Operacional da Região Centro (CENTRO 2020), and the European Social Fund (Portuguese: Fundo Social Europeu or FSE) attributed to the first author. This work was partly financed by FCT/MCTES through national funds (PIDDAC) under the R&D Unit Institute for Sustainability and Innovation in Structural Engineering (ISISE), under reference UIDB/04029/2020, and under the Associate Laboratory Advanced Production and Intelligent Systems ARISE under reference LA/P/0112/2020. The authors are thankful the ISISE Coimbra, the Department of Civil Engineering of UC, SerQ, the Department of Chemistry, FibEnTech-UBI unit of UBI, and Pinhal Nova Madeiras Tratadas for all support provided in this work.

References

1. Arriaga, F., Wang, X., Íñiguez-González, G., et al.: Mechanical properties of wood: a review. Forests. **14**, 1202 (2023). https://doi.org/10.3390/f14061202
2. Alade, A.A., Naghizadeh, Z., Wessels, C.B., Tyhoda, L.: A review of the effects of wood preservative impregnation on adhesive bonding and joint performance. J. Adhes. Sci. Technol. **36**, 1593–1617 (2022)
3. Ramage, M.H., Burridge, H., Busse-Wicher, M., et al.: The wood from the trees: the use of timber in construction. Renew. Sustain. Energy Rev. **68**, 333–359 (2017)
4. Damay, J., Fredon, E., Gérardin, P., Lemmens, P.: Evaluation of axial impregnation as an alternative to classical wood vacuum pressure impregnation method. Maderas Cienc. Tecnol. **17**, 883–892 (2016). https://doi.org/10.4067/S0718-221X2015005000077

5. Lopes, A.F., Batista, E., Hilário, S., et al.: Occurrence of Diaporthe species in Eucalyptus globulus, Pinus pinaster and Quercus suber in Portugal. For. Pathol. **51**, 1–13 (2021). https://doi.org/10.1111/efp.12674
6. CEN EN 350: Durability of Wood and Wood-Based Products - Testing and Classification of the Durability to Biological Agents of Wood and Wood-Based Materials Durabilité, Brussels (2016)
7. Torgovnikov, G., Vinden, P.: Microwave wood modification technology and its applications. For. Prod. J. **60**, 173–182 (2010). https://doi.org/10.13073/0015-7473-60.2.173
8. Ding, W.D., Koubaa, A., Chaala, A., et al.: Relationship between wood porosity, wood density and methyl methacrylate impregnation rate. Wood Mater. Sci. Eng. **3**, 62–70 (2008). https://doi.org/10.1080/17480270802607947
9. Mascarenhas, F.J.R., Dias, A.M.P.G., Christoforo, A.L.: State of the art of microwave treatment of wood: literature review. Forests. **12**, 1–31 (2021)
10. Kol, H.Ş., Çayır, B.: Increasing the impregnability of oriental spruce wood via microwave pretreatment. Bioresources. **16**, 2513–2523 (2021)
11. Kol, H.Ş., Çayır, B.: The effects of increasing preservative uptake by microwave pre-treatment on the microstructure and mechanical properties of oriental spruce wood. Wood Mater. Sci. Eng. **1–7** (2022). https://doi.org/10.1080/17480272.2022.2077656
12. Hermoso, E., Vega, A.: Effect of microwave treatment on the impregnability and mechanical properties of Eucalyptus globulus wood. Maderas Cienc. Tecnol. **18**, 55–64 (2016). https://doi.org/10.4067/S0718-221X2016005000006
13. Jang, E.S., Kang, C.W.: An experimental study on efficient physical wood modification for enhanced permeability–focusing on ultrasonic and microwave treatments. Wood Mater. Sci. Eng. **0**, 1–8 (2022). https://doi.org/10.1080/17480272.2022.2042852
14. Oloyede, A., Groombridge, P.: The Influence of microwave heating on the mechanical properties of wood. J. Mater. Process. Technol. **100**, 67–73 (2000). https://doi.org/10.1016/S0924-0136(99)00454-9
15. Weng, X., Zhou, Y., Fu, Z., et al.: Effects of microwave treatment on microstructure of Chinese fir. Forests. **11** (2020). https://doi.org/10.3390/F11070772
16. Mascarenhas, F.J.R., Dias, A.M.P.G., Christoforo, A.L.: Analysis of the influence of microwave treatment parameters of wood. Forests. **14**, 194 (2023). https://doi.org/10.3390/f14020194
17. Torgovnikov, G., Vinden, P.: High-intensity microwave wood modification for increasing permeability. For. Prod. J. **59**, 84–92 (2009)
18. Mascarenhas, F.J.R., Dias, A.M.P.G., Christoforo, A.L., et al.: Experimental investigation on the influence of microwave technology on the treatability and mechanical properties of Portuguese southern blue gum wood. Case Stud. Constr. Mater. **20**, e02698 (2024). https://doi.org/10.1016/j.cscm.2023.e02698
19. ISO 13061-2: Physical and mechanical properties of wood — Test methods for small clear wood specimens - Part 2: Determination of density for physical and mechanical tests Switzerland (2014)
20. YOU Solutions Germany GmbH: Resumo das características do produto biocida (SPC BP) (2023)

Physical, Mechanical and Biological Tests on Solid Wood and Bio-Composites with BioPCM and Thermal Characteristics of Small-Scale Models

Giovanni Aminti[1], Andrea Atena[1], Michele Brunetti[1], Paolo Burato[1], Fabio De Francesco[1], Gaye Köse Demirel[2], Özge Nur Erdeyer[2], Jakub Grzybek[3,4], Mohamed Jebrane[5], Meysam Nazari[5], Michela Nocetti[1,6], Güliz Öztürk[2], Sabrina Palanti[1](✉), Benedetto Pizzo[1], Thomas Schnabel[3,7], Federico Stefani[1], Ali Temiz[2], and Nasko Terziev[5]

[1] National Research Council Institute of Bioeconomy, Via Madonna del Piano10, Sesto Fiorentino, Italy
sabrina.palanti@cnr.it

[2] Department of Forest Industrial Engineering, Karadeniz Technical University, 61080 Trabzon, Turkey

[3] Department of Forest Biomaterial and Technology, Swedish University of Agricultural Sciences, Box 7008, 756 51 Uppsala, Sweden

[4] Department of Green Engineering and Circular Design, Salzburg University of Applied Sciences, 5431 Kuchl, Austria

[5] Department of Wood Science and Technology, Faculty of Forestry and Wood Technology, Mendel University in Brno, 613 00 Brno, Czech Republic

[6] Faculty for Design of Furniture and Wood Engineering, Transylvanian University of Brasov, Brasov, Romania

[7] Department of Forest and Wood Science, Stellenbosch University, Stellenbosch 7599, South Africa

Abstract. The research M-ERA Net project BIO-NRG-STORE is a collaboration of three universities and one research institute from Austria, Sweden, Turkey and Italy. The study includes physical-mechanical and biological tests on solid wood and wood composites, both impregnated with bioPCM (bio-based phase change material). BioPCM were used to impregnate solid wood or composite panel's raw material (wood particles). In addition, the biological resistance of the bioPCM-wood composites to decay fungi and termites was tested. The resistance to fungal attack, expressed in mass loss, was only high for the bio-composite containing ethyl palmitate ester. However, the used wood particles and bioPCM composite were resistant to termite attack. The results of the physical tests showed that the wood composites containing ethyl palmitate ester had a higher vapour permeability than similar products available on the market, while the other (the wood composite impregnated with capric acid) had a lower permeability. Wood composites with bioPCM can exhibit mechanical properties comparable to those of other commercial wood-based materials. Small wood building models were build using bioPCM inner envelope panels (flooring and wallboards). The results

G. Goli et al. (Eds.): ECWM 2024, SPM 86, pp. 313–320, 2025.
https://doi.org/10.1007/978-3-031-99418-0_39

of the ongoing field test show significant saving of thermal energy by the materials with incorporated bioPCM.

Keywords: bio-based phase change material · wood composite · decay fungi · termites · internal bonding · thermal energy storage

1 Introduction

In September 2023, before the opening of COP28, the United Nations reported the first two-year assessment of global progress in slowing down climate change, called the "global stocktake" where is emphasized that a phase-out of fossil fuels is needed [1]. Between the different measures to mitigate the climate change the ability of phase change materials (PCMs) to absorb and release heat during phase transitions makes them valuable in optimizing thermal performance in building construction. The use of PCMs aligns with sustainable building practices by optimizing energy usage and reducing the environmental impact associated with excessive energy consumption. In summary, integrating phase change materials in building construction is a practical and energy-efficient strategy. It not only contributes to improved thermal performance but also aligns with the broader goal of sustainable and environmentally conscious building design [2]. From a useful point of view, only the phase change solid–liquid is used in building envelope. From the literature review, the bioPCM material (pure substance or eutectic mixture), have phase change temperature in the range (20 °C, 60 °C) among those the one utilized for encapsulation on wood were mostly fatty acids, alcohols and esters [3].

The study goal is conducting laboratory, biological tests and prototype trials (housing models) to redact a comprehensive physical-mechanical characterization, to outline the wood degradation behaviour (biological tests) and finally to make field trial of the innovative treatments (bioPCM impregnated) compared to standard (non bioPCM impregnated) trough prototypes building and monitoring.

2 Materials and Methods

2.1 Physical-Mechanical Tests

Solid Wood

Specimens, Scots pine sapwood and poplar, with a cross-section of 20x20 mm^2 and a length of 300 mm, impregnated with two PCM, respectively ethyl palmitate ester (ES) and eutectic mixture capric acid and stearic acid (CA) were prepared for physical-mechanical characterization.

The first analysis conducted on each specimen was a tree point bending test, since it requires the entire length of the specimen (ISO 13061-3, ISO 13061-4) [4, 5]. Further tests were carried out on other 4 smaller specimens, cut from the original one, avoiding the damaged parts. Overall, it was performed 885 tests including physical and mechanical evaluations (ISO 3347, ISO 13061-17) [6, 7].

The samples were maintained in a conditioned environment (20 ° C temperature and 65% relative humidity) until constant weight was reached (equilibrium conditions). The moisture content and all other physical and mechanical properties are referred to these conditions (ISO 13061-1, ISO 13061-2) [8, 9].

The evaluation of dimensional stability was carried out adapting the procedure provided by standard EN 318 [10] for wood-based panels. The dimensions and weight of wood specimens were measured when specimens were in equilibrium to three different environmental conditions, 20 °C and respectively 65% as reference, 30% as "dry" condition and 85% as "wet" condition.

Biocomposites

The bio composites used for physical mechanical tests were respectively fibreboard based on *Pinus* spp. fibres impregnated with eutectic mixture stearic acid- capric acid (CA), of 8 mm thickness, density 300 kg/m^3, and chipboard based on *Pinus* spp. fibres and polymerized epoxidized linseed oil of 12 mm thickness, density 740 kg/m^3. In both cases 8 samples from two different panels were obtained and for comparison was used a chipboard 10 mm thick and a OSB 19 mm thick respectively.

Water vapor transmission test was carried out using glass squared containers (12x12 cm) filled with saturated solution of water and KCl in accordance with EN 12086 [11].

Bio composites were sealed on the edge and were glued on the container building a closed experimental system with constant relative humidity inside. All dimensions and weight of each specimen was recorded together with the weight of the experimental system. All specimens were positioned inside a climatic chamber obtaining a constant gradient of humidity between the board faces (35% outside and 85% inside). Each experimental system was weighed 6 times with interval of 24 h.

Square specimens of approximately 50 mm side and thickness equal to the thickness of the panel were prepared from the two panels of chipboards for testing internal bonding. The fibreboards panel, due to very poor cohesion of wood fibre, was considered not suitable for internal bonding tests.

These specimens were used to measure in turn:

- density, by measuring the mass and dimensions of the specimen using a centesimal calliper. In this case, the values were measured on a separate set of specimens and the average value was applied to them.
- Internal bonding (IB), according to standard EN 319 [12]:1993 "Particleboards and fibreboards - Determination of tensile strength perpendicular to the plane of the board.

2.2 Biological Tests

Samples of two biocomposites described in 2.1, 50 x 50 mm and thickness of each one were used for testing the resistance against fungal decay in accordance with EN 113-3 [13]. As reference samples were used 25 x 15 x 50 mm samples respectively of Scots pine and beech.

The results were based on the mass loss calculated on percentage of final dry mass with respect to initial theoretical one for bio-composites, instead effective for virulence controls. Due to the risk of leaching or chemical decomposition of PCM impregnated into bio-composites we didn't obtain directly the dry weight on the samples that then put in contact with fungi but on a group consisted by two different bio-composites in order to determine the theoretical average dry weight to apply in the calculation of final mass loss. The fungus utilized were *C. puteana, P. osteatrus, T. versicolor.*

The results shall be accepted as valid provided that the virulence control specimens have lost more than the minimum value given in EN 113-3.

The evaluation of resistance to subterranean termites, Italian species, *Reticulitermes lucifugus*, was carried out in accordance with European normative EN 117 [14]. Wood preservatives - Determination of toxic values against Reticulitermes species (European termites) (Laboratory method) 2023. The PMC impregnated wood samples were five replicates for treatments and was used three untreated reference controls.

2.3 Housing Models

The Housing Models (boxes) have been built to accomplish a specific delivery of BioNRGStore (ERA-NET) project among Italy Turkey, Austria and Sweeden. To evaluate similar experiments among different climate regions Each box, cubic shaped of 1 meter, shares the same specification of the other project partners in terms of dimension, shape, materials. The specification details are available in [15]. The Italian boxes laid in the middle part of the peninsula, in the CNR research area of Florence (coord.: 43.82052560704701, 11.19640449548179) (Fig. 1). Preparatory work for the experiment (box construction and sensor setting up), took approximately two months (from March to April 2023). Besides the same construction specification each partner's box had uses different BioPCM material: Italian boxes have cladding composed by Scots pine wood fiber impregnate by ethyl palmitate ester (EP). A "twin" box hosts an OSB panel cladding to have a "reference" test box. Meteorological data from LaMMA "climatic database" have joined to the database at 15 min interval by the "timestamp" field of each DB to associate to the inner box Temperature [°C] and RH [%] the nearest meteorological cabin data (T °C, RH, Wind Direction, Wind Speed, Diffuse Solar radiation Global Solar radiation, Total precipitation mm, etc.).

Fig. 1. House boxes at CNR IBE.

3 Results and Discussion

3.1 Physical Mechanical Tests

3.1.1 Solid Wood

Regarding the mechanical properties, no difference was noted between treated and untreated specimens, nor between ES and CA treated specimens. Only the ES specimens showed little higher hardness than the CA or untreated specimens.

On the other hand, the treatment with CA seemed to improve the mechanical performances of poplar wood. Only shear strength was like the control specimens, all the other properties were higher for the treated specimens.

As an example, Fig. 2 shows the results for compression strength.

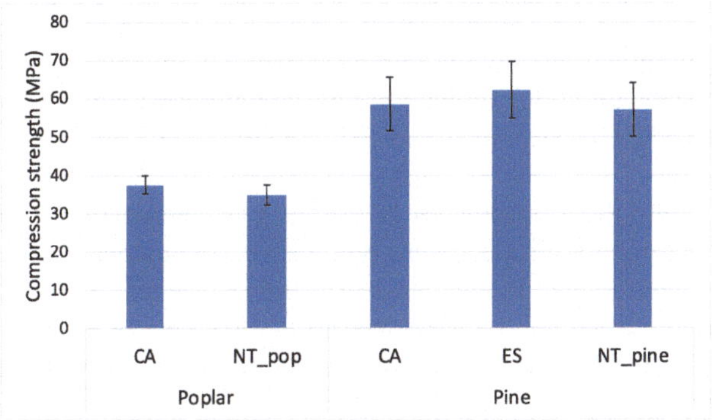

Fig. 2. Compression strength for treated and control specimens of poplar and pine.

These results suggested an involvement of the cell wall during the treatment. In fact, if the treatment had been limited to the cell lumens, no differences in the mechanical properties would have been found, which are mainly determined by the characteristics of the wall. Again, the density at equilibrium moisture content (i.e. specimens conditioned at 20 °C and 65% RH) was far higher for the treated specimens than the untreated. Looking at the variation in weight (Fig. 3) of the wood under dry and wet environment, the effect of PCM was very clear. In the dry environment (20 °C and 30% RH) the treated specimens lost about 2.5% of weight, contrary to the control which lost 3.5% (pine) and 3.8% (poplar).

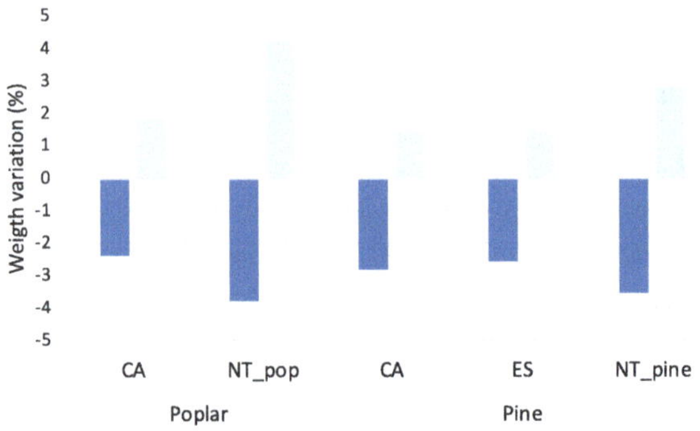

Fig. 3. Weight variation in dry (20 °C/30% RH, blue) and wet conditions (20 °C/85% RH, light blue).

In the wet environment (20 °C and 85% RH), both CA and ES pine specimens gained only 1.5% in weight comparing to the 2.8% of the untreated. This difference was even higher for poplar: the CA specimens showed an increase of weight of 1.9%, while the control gained 4.3%.

However, despite these differences in weight gain and decrease, the dimensional changes were the same for all specimens, both treated and untreated.

If in the dry environment the decrease of weight observed in PCM treated sample with respect the untreated can be addressed to the leaching of PCM, the same PCM can contribute to limit the variation of weight, the absorption of water, in moisturized environment.

Instead, we didn't observe any important difference in the limited dimensional changes, probably due to the encapsulation of PCM in cell wall and not in the lumen. The water molecules probably occupied the lumen without to contribute to the dimensional changes in a large measure, since the PCM is correctly as aim of the work only on cell walls.

Obviously, this was due to the presence of the PCM in the wood. This difference lowered when comparing the oven-dry density, but it was still significant, suggesting that part of the treatment was lost during drying but not all of it.

The same observation was done both for pine and poplar.

3.1.2 Biocomposites

All the properties tested on were significantly different between the treated boards and the correspondence commercial product. The vapor flow and the permeance were higher for the treated board, but these two parameters are linked to the thickness of the specimen tested (in our case the thickness was different between he untreated and treated board). So, to make a comparison of the material properties (regardless of the thickness of the specimen) the permeability and the resistance factor can be compared. As a results, the chipboard treated with ester had a higher permeability (lower resistance factor) compared to the similar commercial product, while the fiberboard treated with capric acid showed lower permeability (higher resistance factor) than the control.

The two biocomposites produced for the study, one (the chipboard treated with ester palmitate) showed greater vapor permeability than a similar product that can be found on the market, while the other (the fiberboard treated with capric acid) showed lower permeability.

The values of density and IB strength for the individual specimens prepared from the two panels were respectively 0.688 and 0.533. The average IB values were the same for the two panels, 0.06 while the average density values differed by only about 5%. This indicates that the procedures used to prepare panels were quite well reproducible.

The strength values obtained are obviously very low for a standard particleboard panel, but it should be borne in mind that these are insulation panels for which no specific cohesive strength requirements are normally specified.

3.2 Biological Test

The results of biological test against fungal decay expressed in mass loss showed that the chipboards were resistant, while the fiberboard showed a high mass loss percentage.

All specimens derived from bio-composite panels based on wood fibre and PCM were resistant to attack by termites and all the termites died after 3 weeks by beginning of the test. The validity of the tests was confirmed by the severe attacks suffered by the untreated pine control samples. The untreated specimens were maintained in the containers until the end of test, and they were attacked until 4 degrees.

3.3 Housing Models

BioPCM laid constantly at liquid state until early Autumn (October 2023), so EP did not prove to have a good transition interval for Mediterranean climate for his impossibility to reach a liquid-to-solid transition (even in the night). Since air temperature reached temperature of 39.4 °C, the activity aimed mostly, to reduce PCM leakage due to excessive heat. Air exchange rate was set at maximum and curtains have been placed to shield the boxes windows. In Autumn, the temperature of the box started to decrease allowing to the state-transition of the PCM, thus the thermal energy storage – release. The data showed that PCM improved the comfort maintaining the inner temperature of "BioPCM" box, lower than the "control" box. The house models, seems promising methodology to evaluate PCM cladding against real conditions environments. The caveat remains in the very simplified system, since a box could not reflect the complexity of a real building; from the other hand, the boxes represent the first applicative experiment able to give important feedback for further technology applications.

4 Conclusion

The most important outcomes are described in the following bullet points:

- Regarding the mechanical properties, no difference was noted between treated and untreated specimens, nor between ES and CA treated specimens.
- Bio-composite were resistant to attack by termites and fungi.

- The house models, seems promising methodology to evaluate PCM cladding against real conditions environments.

Acknowledgement. The study was carried out within the framework of the Smart Energy Systems Research and Innovation Program (ERA-Net E2B2) in the project "Bio-Based Phase Change Materials Integrated into Lignocellulose Matrix for Energy Store in Buildings (BIO-NRG-STORE)". The authors also acknowledge the financial support by FORMAS, project number 2017-00686, and by The Scientific and Technology Research Council of Turkey (TUBITAK) with the grant number 120N500.

References

1. Stanway, D., Alkousaa, R.: https://www.reuters.com/business/environment/un-says-more-needed-on-all-fronts-meet-climate-goals-2023-09-08/
2. Kuznik, F., David, D., Johannes, K., Roux, J.J.: A review on phase change materials integrated in building walls. Renew. Sustain. Energy Rev. **15**, 379–391 (2011). https://doi.org/10.1016/j.rser.2010.08.019
3. Baetens, R., Jelle, B., Gustavsen, A.: Phase change materials for building applications: a state-of-the-art review. Energ. Buildings. **42**, 1361–1368 (2010). https://doi.org/10.1016/j.enbuild.2010.03.026
4. ISO 13061-3: Physical and mechanical properties of wood—Test methods for small clear wood specimens—Part 3: Determination of ultimate strength in static bending (2014)
5. ISO 13061-4: Physical and mechanical properties of wood—Test methods for small clear wood specimens—Part 4: Determination of modulus of elasticity in static bending (2014)
6. ISO 3347: Wood—Determination of Ultimate Shearing Stress Parallel to Grain. International Organization for Standardization, Geneva, Switzerland (1983)
7. ISO 13061-17: Physical and mechanical properties of wood Test methods for small clear wood specimens. Part 17: Determination of ultimate stress in compression parallel to grain (2017)
8. ISO 13061-1: Physical and mechanical properties of wood. Test methods for small clear wood specimens. Part 1: Determination of moisture content for physical and mechanical tests (2014)
9. ISO 13061-2: Physical and mechanical properties of wood. Test methods for small clear wood specimens. Part 2: Determination of density for physical and mechanical tests (2014)
10. EN 318: Wood based panels. Determination of Dimensional Changes Associated with Changes in Relative Humidity. CEN, European Committee for Standardization, Bruxelles, Belgium (2003)
11. EN 12086: Thermal insulating products for building applications. Determination of Water Vapour Transmission Properties. CEN, European Committee for Standardization, Bruxelles, Belgium (2013)
12. EN 319: Particleboards and fibreboards. Determination of tensile strength perpendicular to the plane of the board (1993)
13. EN 113-3: Durability of wood and wood-based products. Test method against wood destroying basidiomycetes Assessment of durability of wood-based panels (2023)
14. EN 117: Wood preservatives. Determination of toxic values against Reticulitermes species (European termites) (Laboratory method) (2012)
15. Grzybek, J., Maffert, P., Petutschnigg, A., Schnabel, T.: Simulation-based performance analysis of impregnated wood with phase change material for energy efficient timber structures. FWIAFE. **16**, 99–116 (2023). https://doi.org/10.31926/but.fwiafe.2023.16.65.3.7

Resistance of Thermally and Chemically Modified Timber Against Soft Rot and Findings to Improve the Lab Test

Wolfram Scheiding[1(✉)], Kordula Jacobs[1], Christian Brischke[2], Susanne Bollmus[3], and Björn Weiß[1]

[1] Institute of Wood Technology Dresden (IHD), Zellescher Weg 24, 01217 Dresden, Germany
{wolfram.scheiding,kordula.jacobs}@ihd-dresden.de
[2] Thünen Institute of Wood Research, Leuschnerstraße 91, Hamburg-Bergedorf, Germany
christian.brischke@thuenen.de
[3] Wood Biology and Wood Products (UGOE), Georg-August-Universität Göttingen, Büsgenweg 4, 37077 Göttingen, Germany
sbollmu@gwdg.de

Abstract. With the 2016 version of EN 350 on determination and classification of biological durability, the scope of materials was extended from natural wood to various wood-based materials. However, there are some deficits to sampling, testing and data evaluation. In order to overcome these deficits, the research project DURATEST had been initiated. Various lab and field tests of the resistance of wood and wood-based materials against wood-destroying fungi were performed, applying the conditions of use classes 3 and 4. The results were statistically analyzed and critically evaluated. In addition, suggestions for improving the test procedure were deduced.

The tested wood species and materials represented typical decking materials, i.e. naturally durable, thermally and chemically modified, and impregnated. In all cases, both lab and field tests were carried out and included in the evaluation. This paper reports on the performed lab tests on resistance against soft-rot fungi according to CEN/TS 15083-2:2015. The resulting durability class of thermally modified Scots pine from MOE loss (4v) was one class worse than from mass loss (3v), but all DC of the CMT variants were equal in both parameters. In general, the soft rot tests confirmed the previous observations considering modified timber and pointed out the differences between thermally modified timber (TMT) and chemically modified timber (CMT). By evaluating the soft rot tests, different suggestions to improve the test method were deduced and provided to the responsible standardization committees of CEN/TC 38.

Keywords: thermally and chemically modified timber · lab test · soft rot

1 Introduction

The classification of biological durability is an important basis for the service life estimation of renewable materials and provides necessary information to select appropriate products for specific applications. The extended scope of EN 350:2016 allows the

G. Goli et al. (Eds.): ECWM 2024, SPM 86, pp. 321–328, 2025.
https://doi.org/10.1007/978-3-031-99418-0_40

classification of natural wood (as before) and also other materials, e.g. modified and preservative treated wood. However, there is a lack of detailed information e.g. about an accurate sampling and testing of these materials. In order to overcome these deficits and to improve the standard, the research project DURATEST had been initiated, with the full title "Development of sampling, testing and classification methods for determining the biological durability of wood and wood products". The project was carried out by the Institute of Wood Technology Dresden (IHD) and the Georg-August-Universität Göttingen, Wood Biology and Wood Products (UGOE) from 2020 to 2023 (Scheiding et al. 2020). The project was funded by the German agency for renewable resources (FNR) with means of the Federal Ministry of Food and Agriculture (BMEL). Within DURATEST, the durability of natural, preservative-treated and modified timber against wood-destroying basidiomycetes and soft rot fungi was investigated in lab and field tests. The samples were taken from different zones of trunks, boards and beams, and examined separately. The results were statistically analyzed and evaluated (see also Brischke et al. 2023a, 2023b). In addition, suggestions for improving the evaluation of test results were deduced.

The material's suitability for use in ground contact or under very unfavorable installation conditions (obstructed drying and heavy dust accumulation) can be determined in a lab test against soft rot and other soil-inhabiting microorganisms. The test with non-sterile soil (terrestrial microcosm, TMC) according to CEN/TS 15083-2 reproduces the outdoor exposure very well, but under defined environmental conditions. In addition to biological load by decay fungi as the main pathogen, infestation by basidiomycota (white and brown rot fungi) and bacteria is also relevant.

The DURATEST report (Scheiding et al. 2023) can be obtained via the FNR project database (see references). In this paper, the investigations of 4 types of modified timber on resistance against soft-rotting microfungi (performed at IHD) are reported.

2 Materials and Methods

2.1 Materials

According to the materials which are established on the market, these types of thermally and chemically modified timber were tested:

- thermally modified Black alder (*Alnus glutinosa*)
- thermally modified Scots pine (*Pinus sylvestris*)
- chemically modified Radiata pine, acetylated (*Pinus radiata*); Accoya®
- chemically modified Radiata pine, furfurylated (*Pinus radiata*); Kebony®

The materials were procured in form of decking boards; additional samples were taken from the outer and inner area of larger beams of acetylated wood (cross-sect. 78 mm × 160 mm), in order to test a gradient that would affect the durability. As refence material, specimens from Scots pine sapwood (*Pinus sylvestris*) and European beech (*Fagus sylvatica*) were used. From these materials, specimens were cut with a cross-section of (5 × 10) mm^2 and a length of 100 mm. For each material variant, 30 test specimens were used.

2.2 Methods

The soft rot tests were performed according to CEN/TS 15083-2:2015 "Determination of the natural durability of solid wood against wood-destroying fungi, test methods - Part 2: Soft rotting micro-fungi".

The specimens were kiln-dried at 103 °C for 48 h - in deviation from the standard - to determine the oven-dry mass before installation. All specimens were subjected to a leaching exposure according to EN 84.

The soft rot test was performed in non-sterile conditions, using boxes which were filled with a substrate, nominated as terrestrial microcosm (TMC). This is a mixture of soil, compost and sand in a mass ratio of 4:1:2 and of a water holding capacity of 55.2% (m/m). The specimens were placed 80 mm deep in the TMC and incubated at 27 °C and 70% rh over a period of 16 weeks (modified hardwood and beech reference) and 30 weeks (modified softwoods and pine reference). The modified hardwoods were removed after reaching a minimum mass loss of 20% of the reference beech specimens. Softwoods were removed after 32 weeks. The MOE after removal was determined on all softwood test specimens with sufficient remaining integrity after prior water-vacuum impregnation.

The test evaluation is usually based on the relative loss of dry mass (mass loss) for hardwoods and loss of bending modulus of elasticity (MOE loss) for softwoods. In this study, the mass loss was also determined for the modified softwoods. The durability classification is based on the x-value, which is the quotient of the median of the individual mass loss or MOE loss of the test specimens and the reference specimens.

The dry mass and the MOE of the specimens were determined before and after the test. The MOE test was performed by 3-point bending test according to EN 310[1] (test device TIRAtest 28,100). The specimens were tested in a water-saturated state with a span of 80 mm. The applied test load was 50% of the breaking load, which was determined for each material by previous bending tests with separate specimens. The resulting MOE was calculated from the recorded value pairs of the force-displacement diagram within the elastic range. The MOE was always calculated on the basis of the initial cross-section of the test specimens.

The allocation to durability classes was made by the calculated x-values according to CEN/TS 15083-2:2005 table E.1, which is equal with EN 350:2016 table 6. The classification was made on basis of the highest median mass loss or MOE loss determined for all the test specimens.

According to CEN/TS 15083-2 para. 7.6.3, the decay type is to be detected on the test material, to provide a proof that an infestation vector (rot attack) is present. For this purpose, specimens from all vessels (both reference and test materials) were selected randomly, and thin cross-section were cut from each and examined by means of incident light microscopy.

[1] EN 350:2016 mentions incorrectly EN 317 (swelling test for wood-based panels).

3 Results

3.1 Laboratory Tests

The resulting durability class (DC) of thermally modified Scots pine based on MOE loss (4v) was one class worse than that based on mass loss (3v), but all DC of the CMT variants were the same for both parameters. Generally, the soft rot tests confirmed the previous experiences with modified timber and pointed out the differences between TMT and CMT.

The results of the tests basing on dry mass loss are shown in Table 1 and Fig. 1, the results based on MOE loss are shown in Table 2 and Fig. 2. It should be noted that, according to the test standards, the duration of the incubation (exposure to the fungi) was different for hardwood variants (16 weeks) and softwood variants (32 weeks).

Table 1. Results of soft rot tests and durability classification based on mass loss (the x-value of alder refers to beech as reference wood species, the values of the modified softwoods refer to pine sapwood).

Material	x-val. (mass loss)	Frequency distribution of the number of single results [%] in the durability classes					DC by x-val.	DC by distrib.
		DC 1	DC 2	DC 3	DC 4	DC 5		
Therm. mod. Black alder (board)	0.11	47	43	10	0	0	2	1-2
Therm. mod. Scots pine (board)	0.19	20	33	47	0	0	2	2v
Furfuryl. Radiata pine (board)	0.01	100	0	0	0	0	1	1
Acet. Radiata pine (board)	0.00	100	0	0	0	0	1	1
Acet. Radiata pine (beam, outer)	0.00	100	0	0	0	0	1	1
Acet. Radiata pine (beam, inner)	0.00	100	0	0	0	0	1	1

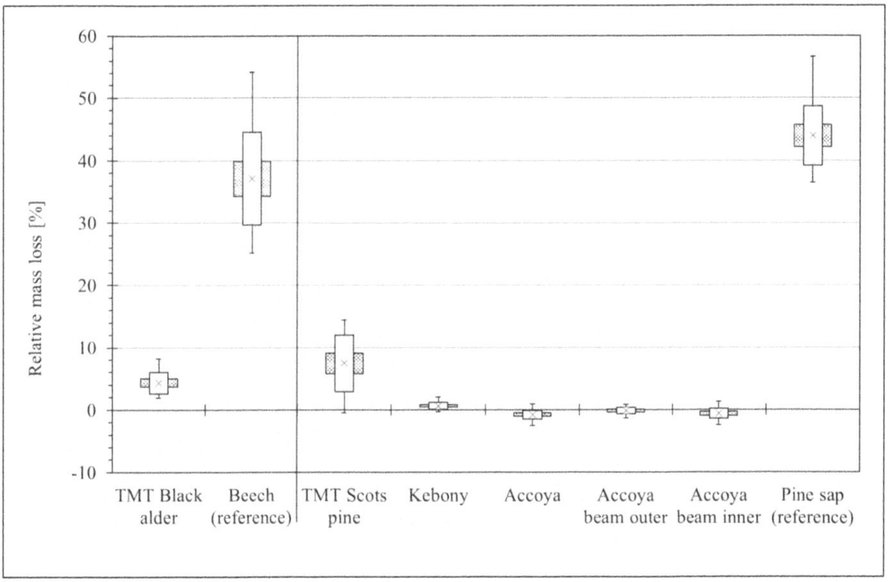

Fig. 1. Mass loss of modified wood after lab test acc. to CEN/TS 15083-2; accordingly, the hardwoods were incubated over 16 weeks, the softwoods over 32 weeks.

Table 2. Results of soft rot tests and durability classification based on MOE loss.

Material	x-val. (MOE loss)	Frequency distribution of the number of single results [%] in the durability classes					DC by x-val.	DC by distrib.
		DC 1	DC 2	DC 3	DC 4	DC 5		
Therm. mod. Scots pine (board)	0.27	10	23	67	0	0	3	3v
Furfuryl. Radiata pine (board)	0.03	97	3	0	0	0	1	1
Acet. Radiata pine (board)	0.00	97	3	0	0	0	1	1
Acet. Radiata pine (beam, outer)	0.00	100	0	0	0	0	1	1
Acet. Radiata pine (beam, inner)	0.00	100	0	0	0	0	1	1

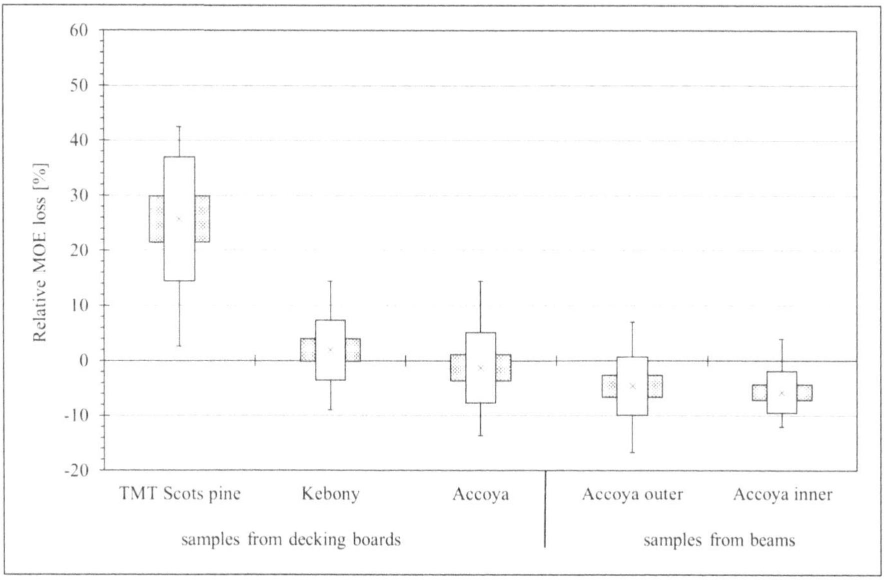

Fig. 2. MOE loss of modified softwoods after lab test acc. to CEN/TS 15083-2; MOE of pine sapwood (reference) was not measurable due to destruction and set as 100% (not displayed).

4 Shortcomings of the Test Methods and Improvement Options

By evaluating the performed soft rot tests, different suggestions were deduced to improve the test method and the standard documents.

As explained above, the criterion for the classification of durability (resistance) against soft rot fungi is different for softwood and hardwoods: MOE loss for the first and mass loss for the latter. The usually given explanation is that MOE would react more sensitively and thus a damage can be detected at an earlier point in time.

The MOE determination is difficult, time-consuming and fraught with higher potential for error. Hence, this test has long been viewed critically in testing laboratory circles. Specific adjustments required for use on the small-format test specimens are currently not defined, such as the definition of appropriate dimensions of the support rollers, the procedure for determining breaking load and span for a specific series of test specimens or ensuring the test specimen orientation in the test device when measuring before and after several months of exposure.

With regard to the bending test, the specimens are very small, which leads to very low bending forces and makes it very difficult to analyze the elastic range in the force-displacement diagram. In addition, many test specimens are not only greatly reduced in strength after the test, but also have a smaller cross-section, which varies over the length of the specimen and cannot be precisely determined. This would be necessary, since the cross-section is used as reference value for the MOE calculation. For this reason, the initial cross-section is usually used for the calculation, but this does not comply with the rule for MOE determination. Therefore, in case of a reduced cross-section of the specimens, this fact should be noted in the test report and this property could be labeled differently, e.g. "stiffness".

Due to the different evaluation parameters, test results for hardwoods and softwoods are not directly comparable. Rather, the project findings show that by using the MOE loss for the durability classification of softwoods, those may be rated worse than hardwoods and thus would be systematically disadvantaged.

The soft rot tests with softwoods, determining both MOE loss and mass loss, have shown that the mass loss is also a suitable parameter for softwoods. This is being further investigated in a study to compare it with the corresponding field test findings (Brischke et al. 2023b). Thus, considering the difficulties with MOE determination, the mass loss should be used – or preferred – as a criterion for both hardwoods and softwoods. The MOE loss could be kept as an optional parameter for testing of softwoods. The retention of the modulus of elasticity as a criterion in the mould rot test of softwoods is supported by the fact that it is more likely or more sensitive to indicate a fungal attack, which is particularly important in the case of a higher durability, e.g. of preservative treated timber, or in wood preservative testing.

Within the DURATEST project, as well further materials like native (untreated) hard- and softwood species and preservative treated timber were investigated, also by lab tests on durability against basidiomycetes. In result of all performed tests, some suggestions were deduced in order to improve the test standards:

- Both the durability against basidiomycetes and against soft-rotting microfungi should be determined by a comparative test with reference materials, classifying the durability on the basis of x-values. In addition, the class boundaries of the tests against basidiomycetes and soft rot should be unified. This would make the results of all durability lab tests comparable, which is currently only the case to a limited extent.
- The (optional) evaluation of durability tests on the basis of frequency distribution of individual values within the DC classes gives additional, useful information. For this purpose, EN 350:2016 proposes the use of fitted probability density functions. However, in order to apply those functions precisely, the procedure to do this is insufficiently defined. The application of fitted probability density functions is difficult to understand in practice and ultimately without added value and should be removed from the standard.
- The specification of the durability class (DC) as a range (DC x-y) is more meaningful than one class with the addition “variable”.
- From a practical perspective, for a product (commodity) the information “suitable for use class x” is much more useful than only the durability class. If a durability class is declared (specified), it should be stated which test standards were used and to which application (use class) the test related. As part of the planned revision of EN 350, a table is being discussed in which the applicable test standards are assigned to the (intended) use classes.

5 Conclusion

The high durability (resistance) of thermally and chemically modified timber against soft-rotting microfungi, which was determined in the reported investigations, met the expectations. The durability of the thermally modified materials was clearly less than of the chemically modified ones, what corresponds with the general experiences from durability tests against basidiomycetes.

Within the test of specimens received from the larger Accoya beams, no differences in MOE or mass loss were found for specimens which had been taken from the inner and the outer area. This reflects the complete and homogeneous modification of the investigated material throughout its cross section.

The conclusions and suggestions could contribute to improve the test methods and test standards. They are provided to the responsible working group WG 23 "Fungal testing (basidiomycetes-microfungi)" of the European standardization committee CEN/TC 38 "Durability of wood and wood-based products".

References

Brischke, C., Sievert, M., Schilling, M., Bollmus, S.: Laboratory durability testing of preservative-treated wood products. Forests. **14**, 1001 (2023a). https://doi.org/10.3390/f14051001

Brischke, C., Haase, F., Bollmus, S., Bächle, L.: Statistical analysis of wood durability data and its effect on a standardized classification scheme. Standards. **3**, 210–226 (2023b)

Scheiding, W., Jacobs, K., Bollmus, S., Brischke, C.: Durability Classification of Treated and Modified Wood – Approaching a Guideline for Sampling, Testing, and Statistical Analysis. The International Research Group on Wood Protection, IRG/WP/20-20676 (2020)

Scheiding, W., Jacobs, K., Brischke, C., Bollmus, S.: Entwicklung von Probennahme-, Prüf- und Klassifizierungsverfahren zur Bestimmung der biologischen Dauerhaftigkeit von Holz und Holzprodukten (DURATEST). Schlussbericht FNR-Verbundvorhaben (FKZ 22040718 und 2219NR372) (2023). www.fnr.de/projektfoerderung/projektdatenbank-der-fnr

EN 84: Durability of wood and wood-based products. Accelerated Ageing of Treated Wood Prior to Biological Testing—Leaching Procedure (2020)

EN 113-2: Durability of wood and wood-based product. Test Method Against Wood Destroying Basidiomycetes—Part 2: Assessment of Inherent or Enhanced Durability (2020)

EN ISO 178: Plastics - Determination of flexural properties (ISO 178:2019) (2019)

EN 310: Wood-based panels. Determination of Modulus of Elasticity in Bending and of Bending Strength (1993)

EN 317: Particleboards and Fiberboards; Determination of Swelling in Thickness After Immersion in Water (1993)

EN 350: Durability of wood and wood-based products. Testing and Classification of the Durability to Biological Agents of Wood and Wood-Based Materials (2016)

ISO 13061-4: Physical and mechanical properties of wood. Test Methods for Small Clear Wood Specimens—Part 4: Determination of Modulus of Elasticity in Static Bending (2014)

CEN/TS 15083-2: Durability of wood and wood-based products. Determination of the Natural Durability of Solid Wood Against Wood-Destroying Fungi, Test Methods—Part 2: Soft Rotting Micro-Fungi (2005)

The Chemical Interactions Between Phenolic Resin and Wood Studied by Liquid-State NMR spectroscopy

Carlo Kupfernagel[1](✉), Daniel Yelle[2], Morwenna J. Spear[3], Andrew Pitman[4], and Graham A. Ormondroyd[3]

[1] Institut für Holztechnologie Dresden, Dresden, Germany
carlo.kupfernagel@ihd-dresden.de, crk20scf@bangor.ac.uk
[2] Forest Products Laboratory, Madison, USA
[3] BioComposite Centre, Bangor, UK
[4] BM Trada, Buckinghamshire, UK

Abstract. Wood modification with thermosetting resins is an increasingly popular method aiming to expand the number of applications for readily available plantation timber [1]. Phenol urea formaldehyde (PUF) resin is impregnated into the timber, which is dried, and subsequently heat cured [2, 3]. This treatment renders the raw material dimensionally stable and imparts the ability to withstand wood destroying fungi and insects. The mode of action in doing so is typically considered a passive modification, but covalent bonds between the PUF resin and wood might form at low abundance. Since these covalent bonds are rare and structurally similar to linkages that occur in wood itself, a high-resolution method is required to accurately observe the formation of new bonds. In the current study we used liquid-state ^{1}H-^{13}C-heteronuclear single quantum coherence (HSQC) NMR experiments to identify chemical bonds in modified and unmodified wood with high accuracy. This method has been used in the past to detect even rare covalent bonds in modified wood [4, 5]. During the sample preparation, a ball-milled wood powder is dissolved in DMSO-d_6, which yields a viscous gel that can be handled like a liquid in solution-state NMR experiments [6, 7]. Lime and poplar samples, that were previously compared in anti-swelling efficiency tests, have been used for the NMR analysis [2]. While the same kind of covalent bonds are formed in different timbers, their abundance was shown to be significantly different, which could be due differences in the chemical composition of the two wood species.

Keywords: PUF resin · solution-state NMR · wood modification · cell wall solubilisation

1 Introduction

As a building material, wood offers several advantages in comparison to other materials, including a high strength-to-weight ratio [8], a carbon sequestration potential [9], and even health benefits to inhabitants [10]. However, as a naturally degrading material,

G. Goli et al. (Eds.): ECWM 2024, SPM 86, pp. 329–336, 2025.
https://doi.org/10.1007/978-3-031-99418-0_41

wood in certain applications (e.g., outdoor) is under the risk of fungal decay. Moreover, wood undergoes swelling and shrinking in a changing relative humidity environment, hence, it is dimensionally instable.

Recently developed and partly commercialized wood modification techniques are increasingly used to improve the natural durability and dimensional stability of wood species, that are often fast-growing plantation timbers (e.g. *Pinus radiata*) [11].

A promising wood modification technique is the impregnation modification with low molecular weight thermosetting resins, such as phenol urea formaldehyde (PUF) resin. PUF resin is impregnated into the timber, which is dried, and subsequently heat cured [2]. The resin cures within the cell wall, leading to a pronounced cell wall bulking effect, and as result to a reduction in the swelling coefficient, an enhanced biological durability, and a high surface hardness [12].

The modification with PUF resin is typically considered a passive modification, meaning that the resin is not required to react directly with the cell wall polymers in order to enhance the properties of the wood [1]. Vice versa, in an active modification, there are direct interactions between the treatment chemical and the cell wall polymers. An example for active modification is acetylation, where acetic anhydride forms covalent bonds with wood hydroxyl groups.

While passive modifications do not rely primarily on the formation of covalent bonds between the wood and the treatment chemical, they might still form at low abundance. In the wood modification with PUF resin, one likely reaction site for covalent bonds in the wood is the unsubstituted G_5-position in guaiacyl (G) lignin, where methylene bridges could be formed with the resin, as observed by Yelle & Ralph [5]. Another possibility for covalent bonding is an esterification reaction between wood carboxylic acids and resin hydroxymethyl group, which was observed by Yelle in the treatment of wood with hydroxymethylated resorcinol, which is chemically similar to PUF resin [4].

A previous study investigated the anti-swelling efficiency (ASE) of eleven wood species, that were treated with PUF resin under identical conditions. Notably, the timbers displayed relatively high variations, between 44% ASE in poplar 69% ASE in lime wood, that cold not fully be explained by differences in their density or in the liquid resin uptake. Thus, it was considered whether covalent bonding between the resin and the cell wall polymers plays a role, and this study aims at investigating the presence of such covalent bonds in poplar and lime wood.

Since these covalent bonds are rare and structurally similar to linkages that occur in native wood, a high-resolution method is required to accurately observe new covalent bonds. Two dimensional ^{1}H-^{13}C heteronuclear single quantum coherence (HSQC) solution-state NMR in combination with cell wall solubilization is a suitable method and has been used in the past for similar investigations. In this method, a ball-milled wood powder is dissolved in DMSO-d_6, which yields a viscous gel that can be handled like a liquid in solution-state NMR [6, 7].

2 Material and Methods

2.1 Wood Samples and PUF Resin

Small samples with the dimensions 20 (r) x 20 (t) x 5 (l) mm were used for this study. The investigated wood species were poplar (*Populus tremula*) and lime wood (*Tilia x europaea*). NMR scans were conducted both before and after the modification with PUF resin.

2.2 Resin Treatment

For the resin impregnation, pre-dried wood samples were placed in a 250 ml beaker in a desiccator attached to a vacuum line. A vacuum was drawn for 20 min prior to impregnation. Then, the resin was injected through a dropping funnel until all samples were fully submerged. A volume of 100 ml resin was used for each impregnation. Still under vacuum, the samples were immersed for 20 min, before the vacuum was released. Subsequently, the treatment solution was drained off and samples were removed from the beaker, excess resin was blotted with tissue and the mass and dimensions were noted. The samples were now carefully dried at 50 °C for 16 h to reduce the moisture content before cure. Heat curing took place at 150 °C for a duration of 8 h.

2.3 Cell Wall Solubilization

Each sample was weighed to the nearest of 0.1 mg and sliced in the radial direction to give approximately 1 mm thick sections using a knife. Six 50 ml milling jars, along with several 20 mm and 10 mm balls, made from ZrO2 were cleaned using hot water and mild detergent and finally rinsed with reverse-osmosis water. The jars and balls were dried at 50 °C for approximately 2 h prior to milling. Once dry, a clean and dry rubber gasket was placed into the groove of each jar cover. Each sliced sample was placed into each 50 ml jar followed by three 20 mm balls and secured in a Retsch PM-400 planetary ball-mill (Newtown, PA) and milled at 300 rpm for 24 h (20 min milling, 10 min pause). Afterwards, the three 20 mm balls were removed, shaken briefly in a copper sieve to recover material coating the balls, and replaced by ten 10 mm balls. Ball-milling was then continued at 300 rpm for a total of 24 h (20 min milling, 10 min pause). Following milling, the 10 mm balls were removed, shaken briefly in a copper sieve, and the milled sample scraped from the jar and weighed to the nearest 0.1 mg.

Approximately 30 mg of each ball milled sample was dissolved directly in 500 µl of a DMSO-d_6 and Pyridine-d_5 mixture with a ratio of 4:1 (v/v) within a 5 mm NMR tube. Each sample was sonicated at 35 °C for 1 h in an ultrasonic bath until a homogeneous gel evolved. The sample preparation is shown in Fig. 1.

2.4 Solution-State NMR Scans

Solution-state NMR spectroscopy was performed using a Bruker 400 MHz Ultrashield+ spectrometer. The spectra were collected in both ^{13}C DEPTQ and ^{1}H-^{13}C-HSQC mode.

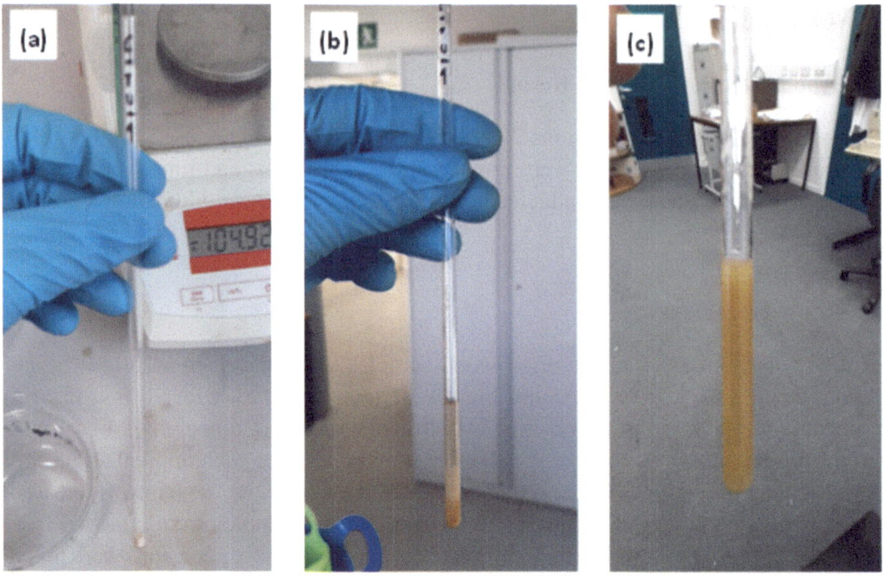

Fig. 1. Sample preparation for solution-state NMR. (a) Ball-milled wood powder (approximately 30 mg) is filled in 5 mm NMR tube. (b) Solvent is added to the tube resulting in an inhomogeneous gel. (c) Gel is homogeneous after mechanical mixing and sonication.

The pulse program used for ^{13}C DEPTQ NMR experiments was deptqgpsp.2. The number of scans was 20,000, the D1 delay was 0.5 s, and a scanning range between 0 and 250 ppm was used, requiring a total scan time of 5 h.

The pulse program for ^{1}H-^{13}C-HSQC NMR scans was hsqcedetgpsisp2.3. The number of scans was 150, D1 0.5 s, with a total scanning time of 17 h. Spectra were post-processed in MestreNova, where the gaussian broadening (GB) in F2 dimension was set to 0.001 and line broadening (LB) to −0.3. The chemical shift of DMSO-d_6 was set to 2.50/39.52 ppm. For a semi-quantitative analysis of the wood polymer structures present in the spectra, specific chemical shifts were integrated and referenced to the lignin methoxy signal at 56 ppm using MestreNova software.

3 Results and Discussion

If present, covalent bonds between wood and PUF resin are expected to form either ester linkages with the hemicellulose fraction or methylene bridges with the lignin fraction of the wood [4, 5]. Generally, ester linkages can be observed between 160–180 ppm in the ^{13}C NMR spectrum. In native wood, this region displays the carbonyl acetate group (among others). Methylene bridges would be expected to form between phenolic rings of the lignin and PUF resin. Depending on the substitution pattern in the phenolic ring, signals in the ^{13}C NMR spectrum may be observed between 100–150 ppm for both lignin in native wood and pure PUF resin. New covalent bonds between wood and resin would be displayed as a new signal in the corresponding region of the NMR scan, as was previously observed by [4, 5].

Figure 2 shows the region between 180–160 ppm in the ^{13}C DEPTQ NMR spectrum of poplar and lime, where ester bonds between wood hemicelluloses and hydroxymethylated resorcinol resin have previously been detected [4]. The ^{13}C DEPTQ NMR spectrum is used instead of the HSQC spectrum because quaternary carbons in an ester bond are not visible in the latter. However, the DEPTQ spectrum in Fig. 2 shows no novel peaks in the modified poplar and lime wood, despite a reasonable signal-to-noise ratio, so ester linkage formation could not be confirmed.

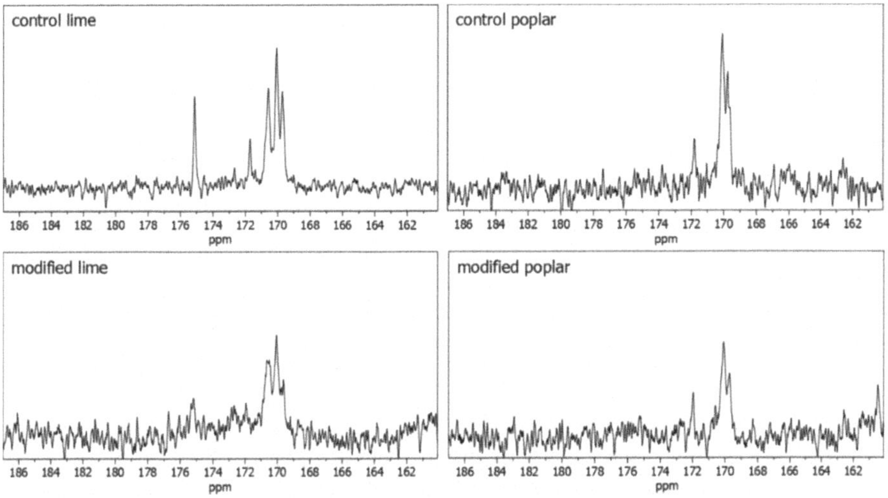

Fig. 2. Partial ^{13}C DEPTQ NMR spectrum of modified and unmodified lime and poplar wood in the ester bond region of the spectrum. No novel ester bonds could be detected.

Similarly, the aromatic region between 110–130 ppm where covalent bonds between resin and G-lignin have been observed previously, did not show any new signals in Fig. 3. Therefore, the formation of covalent bonds between resin and cell wall polymers is generally unlikely under the applied treatment conditions.

There are several possible explanations why previously observed covalent bonds have not been detected in this experiment. For one, previous studies investigated coniferous wood species, which are rich in G-lignin. Hardwoods, which were used in the current experiment contain an approximately equal share of G-lignin and S-lignin. The S-lignin generally shows a higher degree of substitution in the aromatic ring and is therefore less likely to react with the resin via methylene bridge formation.

Another possibility is that different process conditions enabled the formation of covalent bonds in the previous studies, whereas the conditions used in this experiment restricted covalent bonds. E.g., the ratio of resin solids to wood was 2:1 in Yelle & Ralph's study [5] compared to a ratio of approximately 1:3 in this experiment. The pH value, solids content, type of catalyst, and molecular weight of the resin were also different. Similar explanation can be found for the absence of ester bonds in Fig. 2.

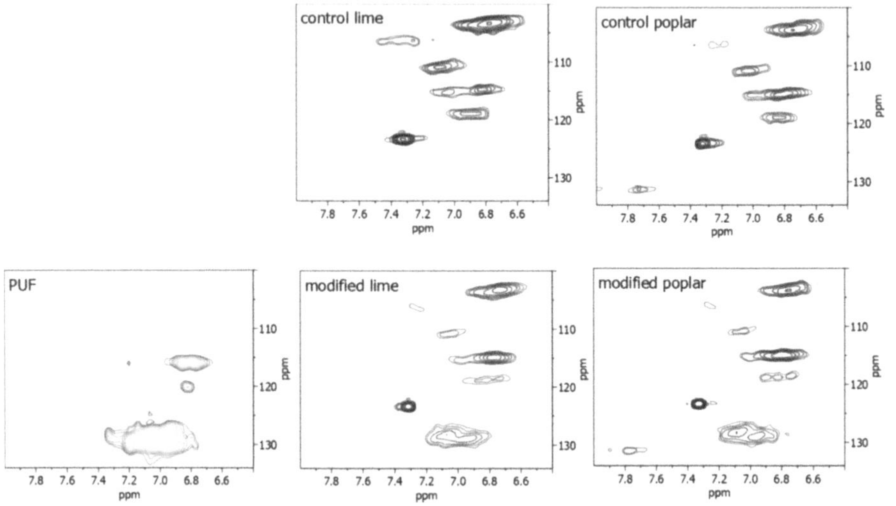

Fig. 3. Partial two dimensional ^{1}H-^{13}C HSQC NMR spectrum of PUF resin (left), modified and unmodified lime (centre), and modified and unmodified poplar (right). $PF_{2/4}$ and PF_6 signals overlap with $G_{2/6}$ signal of wood. Novel signals indicating the presence of methylene bridges between G-lignin and PUF resin were not detected.

Thus, the ester region in Fig. 2 displays signals that are exclusively assigned to the wood fraction, and the aromatic region in Fig. 3 shows signals of both wood and resin, but no novel signals that would indicate covalent bonding.

Figure 4 shows the alkyl region between 30–50 ppm, which is exclusively assigned to methylene bridges in the PUF resin fraction. PUF resin is a mixture of reactive hydroxyethyl phenols (HMP) and hydroxymethyl urea (HMU) molecules. The HMU fraction may react with itself via self-condensation, forming methylene bridges with *para-para* and *ortho-para* linkages. In Fig. 4, the corresponding NMR signals are indicated as $PF_{p\text{-}p}$ and $PF_{o\text{-}p}$, respectively. Moreover, HMP may also react with HMU via co-condensation, forming bis(hyrdoxyphenyl) urea, in various constellations. The NMR signals indicative of co-condensed structures are labelled PUF_A, PUF_B, PUF_C, and PUF_D,

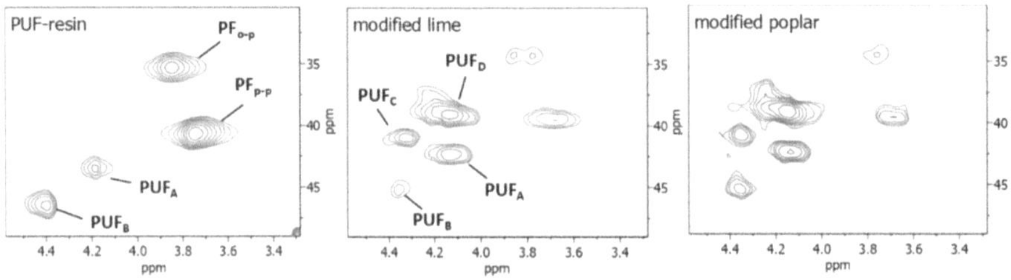

Fig. 4. Partial ^{1}H-^{13}C HSQC NMR spectrum of partly cured PUF resin, modified lime, and modified poplar in the methylene bride region. Structures PUF_A and PUF_B indicate novel methylene bridges that were not previously observed in the pure PUF resin. PUF_A and PUF_B might correspond to methylene bridges of resin interacting closely with the wood polymer surface.

Self-condensation and co-condensation occur in the neat PUF resin as well as in the modified timbers. But interestingly, the peaks corresponding to PUF_C at 4.34/41.1 ppm and PUF_D at 4.15/39.2 ppm were not detected in the neat resin, but only in the modified wood. Hence, while resin does not react directly with the cell wall, it reacts differently inside the cell wall, compared to pure PUF resin. This could be due to steric hinderance effects inside the cell wall, that favor different reaction pathways. It was observed that the relative intensity of the peaks PUF_A, PUF_B, PUF_C, and PUF_D differed between poplar and lime.

Additionally, it was observed that methylene bridge signals in modified wood displayed an upfield shift of approximately 1–1.5 ppm compared to the neat resin signals. It is possible that different conformations of the methylene bridges exist, meaning that sometimes the bonds will move and rotate, making neighbouring atoms either come closer or move farther away from each other. These conformational effects can lead to 1–1.5 ppm differences in chemical shifts of ^{13}C compared to the neat resin [13].

4 Conclusions

Following a previous investigation of the anti-swelling efficiency (ASE) of different timbers in the same treatment with low molecular weight PUF resin [2], the role of covalent bonding in the dimensional stabilization process was considered. Modified and unmodified specimens of lime and poplar wood were subjected to a whole cell wall dissolution technique to enable solutions-state NMR investigations of these specimens.

Solution-state NMR spectroscopy has shown that previously observed covalent bonds between resin and wood polymers (ester linkages and methylene bridges) did not form in this study. Hence, it is unlikely that covalent bonding between PUF resin and wood occurs, and consequently, covalent bonding bears no importance for the dimensional stabilization of the resin modified wood.

However, it was also shown that compared to the neat resin, different bonds are formed when the resin is cured inside the wood cell wall. Possibly, this is due to steric hinderance effects in the cell wall, that favor different reaction pathways.

Additionally, the signals from methylene bridges exhibit an upfield shift of 1–1.5 ppm in modified wood in comparison to pure resin. This shift may be attributed to conformational effects occurring within the cell wall.

References

1. Hill, C.A.S.: Wood Modification Chemical, Thermal and Other Processes. John Wiley & Sons, Ltd, Chichester (2006)
2. Kupfernagel, C., Spear, M.J., Pitman, A.J., Ormondroyd, G.A.: Wood modification with phenol urea formaldehyde (PUF) resin: the influence of wood species selection on the dimensional stability. Eur. J. Wood Wood Prod. **81**, 5–19 (2023)
3. Kupfernagel, C., Spear, M., Pitman, A., Ormondroyd, G.A.: Wood Modification with Phenol-Formaldehyde-Resin and its Influence on the Dimensional Stability of Homegrown and Imported Hardwoods. In: The 17th Annual Meeting of the Northern European Network for Wood Science and Engineering, Kaunas, Lithuania, pp. 12–15 (2021)

4. Yelle, D.J.: Solution-state NMR analysis of hydroxymethylated resorcinol cured in the presence of crude milled-wood lignin from Acer saccharum. J. Appl. Polym. Sci. **134**(41) (2017)
5. Yelle, D.J., Ralph, J.: Characterizing phenol-formaldehyde adhesive cure chemistry within the wood cell wall. Int. J. Adhes. Adhes. **70**(May), 26–36 (2016)
6. Yelle, D.J., Ralph, J., Frihart, C.R.: Characterization of nonderivatized plant cell walls using high-resolution solution-state NMR spectroscopy. Magn. Reson. Chem. **46**(6), 508–517 (2008)
7. Kim, H., Ralph, J.: Solution-state 2D NMR of ball-milled plant cell wall gels in DMSO-d 6/pyridine-d5. Org. Biomol. Chem. **8**(3), 576–591 (2010)
8. Wimmers, G.: Wood: a construction material for tall buildings. Nat. Rev. Mater. **2**(12), 17051 (2017)
9. State of Europe's Forests: State of Europe's Forests 2015 Published, Madrid, Spain (2015). [Online]. www.foresteurope.org
10. Kilpeläinen, P., Sutinen, V., Jokinen, E., Virtanen, V.: Health benefits of wooden buildings-long-term monitoring of physical and chemical properties of the building and stress of pupils in a wooden school and in two control schools. In: 18th Annual Meeting of the Northern European Network for Wood Science and Engineering (2022)
11. Sandberg, D., Kutnar, A., Mantanis, G.: Wood modification technologies—a review. iForest. **10**(6), 895–908 (2017)
12. Stefanowski, B.K., Spear, M.J., Curling, S.F., Pitman, A.J.: Properties of Lignia modified wood. In: Timber 2020 (2020)
13. Yelle, D.J.: Personal Communication (2023)

Author Index

G. Goli et al. (Eds.): ECWM 2024, SPM 86, pp. 337–339, 2025.
https://doi.org/10.1007/978-3-031-99418-0

The manufacturer's authorised representative in the EU is Springer Nature Customer Service Centre GmbH, Europaplatz 3, 69115 Heidelberg, Germany. If you have any concerns regarding our products, please contact ProductSafety@springernature.com

Printed and bound by CPI Group (UK) Ltd, Croydon, CR0 4YY
30/09/2026
02225590-0011